Electronic Logic Systems

A.E.A. Almaini

Department of Electrical and Electronic Engineering
Napier College of Commerce and Technology
Edinburgh, Scotland

Prentice/Hall International

ENGLEWOOD CLIFFS, NEW JERSEY LONDON MEXICO NEW DELHI
RIO DE JANEIRO SINGAPORE SYDNEY TOKYO TORONTO WELLINGTON

Library of Congress Cataloging in Publication Data

Almaini, A.E.A., 1945–
  Electronic logic systems.

  Bibliography: p.
  Includes index.
  1. Logic circuits. 2. Logic design. I. Title.
  TK7868.L6A38 1985   621.3815′3   85-12049
  ISBN 0-13-251752-3

British Library Cataloguing in Publication Data

Almaini, A.E.A.
  Electronic logic systems.
  1. Logic circuits
  I. Title
  621.3815′37   TK7868.L6

  ISBN 0-13-251752-3
  ISBN 0-13-251745-0 Pbk

Prentice-Hall, Inc., *Englewood Cliffs, New Jersey*
Prentice-Hall International, UK. Ltd., *London*
Prentice-Hall of Australia Pty., Ltd., *Sydney*
Prentice-Hall Canada, Inc., *Toronto*
Prentice-Hall of India Private Limited, *New Delhi*
Prentice-Hall of Japan, Inc., *Tokyo*
Prentice-Hall of Southeast Asia Pte., Ltd., *Singapore*
Editora Prentice-Hall Do Brasil Ltda., *Rio de Janeiro*
Prentice-Hall Hispanoamericana, S.A., *Mexico*
Whitehall Books Limited, *Wellington, New Zealand*

ISBN 0-13-251752-3
ISBN 0-13-251745-0 PBK

Printed in Great Britain at the
University Press, Cambridge

10 9 8 7 6 5 4 3 2 1

Contents

Preface

The development of electronic logic systems has been very rapid. Propositional and Boolean algebra techniques facilitated the analysis and design of sophisticated logic systems which advances in electronics (valves and relays in the forties, transistors and diodes in the fifties, large and very large scale integrated circuits since) have allowed to be implemented with constantly increasing power and flexibility.

In the sixties the engineer worked with small-scale integrated circuits with ten or so transistors per chip, and digital systems were a comparative by-way of electronics; today Very Large Scale Integration (VLSI) has created devices with over one million transistors per chip, and digital systems is the largest and fastest growing area of electronics, affecting almost every aspect of our lives. The concentrated effort to develop systematic design procedures and Computer Aided Design (CAD) tools is reducing design costs and simplifying the task of the design engineers, while high circuit density per chip, standardization and mass production continue to increase the availability and application of digital systems.

It has, consequently, become essential for every electronic and computer engineer to have a working familiarity with electronic logic systems. The greatly increased importance of the subject has led to the proliferation of introductory books designed for readers with a wide range of background knowledge and course requirements. A treatment of basic number systems, Boolean algebra and elementary logic gates can be found in most of these books, and these topics are taught to electronic and computer engineers increasingly early in the curriculum. For this reason *Electronic Logic Systems* does not attempt an in-depth treatment of basic material. Assuming that the reader has a grounding in electronics and mathematics, the fundamentals of logic gates and Boolean algebra are quickly revised to prepare the way for a comprehensive study of the principles and techniques of modern electronic logic systems. The hardware components, practical design methods and essential theory for combinational and sequential logic circuits are covered, providing the reader with sufficient information to undertake the design and analysis of sophisticated systems.

The book has been designed for use by undergraduates studying electronic or computer engineering in colleges and universities. Students in other engineering fields requiring a detailed knowledge of the subject will also find it valuable, and its practical orientation will make it of use to practicing engineers who find they need to know how logic systems operate. The advanced topics discussed towards the end of the book, which include practical optimization techniques and programmed algorithms, will be useful for research and graduate students. The material is as up-to-date as possible in this rapidly moving field, and the design examples interspersed throughout the text illustrate the different techniques required for use with small, medium and large-scale integration. To enhance the reader's grasp of the material, a large number of examples with worked solutions are provided. Almost all the circuits in the design exercises have been built and tested. Specific text references to entries in the list of references at the end of the book are shown by numerals in angle brackets, thus ⟨17⟩.

Finally, my thanks are due to my wife and children, my colleagues and students and all the many authors and researchers who have contributed directly or indirectly to this book. I should also like to thank the many manufacturers whose devices appear in this book for valuable data books and application notes which have helped to bridge the gap between theory and practice.

A.E.A. ALMAINI

Symbols and Abbreviations

ASCII	American Standard Code for Information Interchange
ASM	Algorithmic state machine
BCD	Binary coded decimal
CAD	Computer aided design
CDI	Collector diffusion isolation
CML	Current mode logic
cMOS	Complementary metal oxide semiconductor
chMOS	Complementary high-performance MOS
DIP	Dual in-line package
DTL	Diode transistor logic
EAROM	Electrically alterable read only memory
ECL	Emitter coupled logic
EEPROM	Electrically erasable and programmable read only memory
EPROM	Erasable and programmable read only memory
FET	Field effect transistor
FPLA	Field programmable logic array
FSR	Feedback shift register
glb	Greatest lower bound
HAL	Hard array logic
hMOS	High-performance metal oxide semiconductor
HNI	High noise immunity
IC	Integrated circuit
iff	If and only if
I^2L	Integrated injection logic
L	Latch
LNI	Low noise immunity
lsd(b)	Least significant digit (bit)
LSI	Large-scale integration − a chip containing 1000−100 000 transistors
lub	Least upper bound
MEG	Maximum exclusive group
MOS	Metal oxide semiconductor
MOSFET	MOS field effect transistor
msd(b)	Most significant digit (bit)

MSI	Medium-scale integration − a chip containing 100−1000 transistors
MTT	Multiple transition time
NCR	Non-critical race
nMOS	n-channel metal oxide semiconductor
NS	Next state
PAL	Programmable array logic
PALASM	PAL assembler and simulator
PLA	Programmable logic array
pMOS	p-channel metal oxide semiconductor
PROM	Programmable read only memory
PS	Present state
P/T	Programming testing
P.T.	Product term
ROM	Read only memory
RTL	Resistor transistor logic
SM	Sequential machine
SOC	Single output change
SOI	Silicon on insulator
SOS	Silicon on sapphire
s.p.	Substitution property
SSI	Small-scale integration − a chip containing up to about 100 transistors
STT	Single transition time
TTL	Transistor transistor logic
ULA	Uncommitted logic array
ULM	Universal logic module
USTT	Unicode single transition time
VLSI	Very large-scale integration − a chip containing more than 100 000 transistors
vMOS	V-groove metal oxide semiconductor
π	A partition having the substitution property (closed partition)
τ	A partition on a set of states
τ_o	Output consistent partition
τ_i	Input consistent partition
τ_s	State consistent partition
τ, τ'	A partition pair
$\lceil g \rceil$	The smallest integer greater than or equal to g
R_{ij}	Element i, j of a residue matrix
$\cup$	Set union
$\cap$	Set intersection
$A \subseteq B$	A is a subset of B
$A \subset B$	A is a proper subset of B
$a \in A$	a is an element of A
$b(\pi)$	Number of states in the largest block of π
$\#(\pi)$	Number of distinct blocks in a partition π

PART ONE

Principles of Logic Systems

1

Combinational Logic

1.1   INTRODUCTION

Consider the following statement:

All girls in the room have brown eyes.
Nadia is a girl in the room.
Therefore Nadia must have brown eyes.

This system of reasoning is an example of what we mean by logic.

In logic systems, variables, circuits, statements, etc., can be in one of two distinct states: true or false, conducting or non-conducting, on or off, energized or not energized, yes or no, present or absent, and so on.

Engineers know that it is easier to design two-state devices than multi-state devices. The valve, the transistor, the diode and the switch can be either on or off, which is very convenient for the implementation of logic circuits. The early pioneers of switching theory during the nineteen forties and fifties must have realized that their ideas contained the seeds of a modern industrial revolution. The design of logic systems refers to the processes of assembling switching or gating circuits in accordance with well-defined logical rules to achieve a given specification. This becomes an electronic logic system if electronic, rather than mechanical or fluidic, devices are used. In what follows the voltage level of the electronic device will be assigned a true value (logic 1) when the voltage is high and a false value (logic 0) when the voltage is low. This is known as *positive logic*.

1.2   LOGIC GATES

AND gate   The AND gate is a circuit which gives a high output (logic 1) if all
                its inputs are high. A dot ($\cdot$) is used to indicate the AND opera-
                tion. In practice, however, the dot is usually omitted.

OR gate   The OR gate is a circuit which gives a high output if one or more of
                its inputs are high. A plus sign ($+$) is used to indicate the OR
                operation.

EX−OR gate   The EXclusive-OR gate is a circuit which gives a high output if
                either of its two inputs is high, but not both. An encircled plus sign
                ($\oplus$) is used to indicate the EX−OR operation.

2

NOT gate   The NOT gate is a circuit which produces at its output the negated (inverted) version of its input logic. The circuit is also known as an *inverter*. If the input variable is A, the inverted output is written as $\bar{A}$.

NAND gate   The NAND gate is a NOT–AND circuit which is equivalent to an AND circuit followed by a NOT circuit. The output of the NAND gate is high if any of its inputs is low.

NOR gate   The NOR gate is a NOT – OR circuit which is equivalent to an OR circuit followed by a NOT circuit. The output of the NOR gate is low if any of its inputs is high.

The functions of these basic building blocks are summarized by means of a *truth table* as shown in Table 1.1. The table shows all possible input/output combinations for two inputs. A truth table with n inputs (logic variables) has 2^n rows.

The symbols for the above-mentioned logic gates, as used in this book, are summarized in Fig. 1.1. Though two inputs are shown, the AND, OR, NAND and NOR gates can have more than two inputs.

Inputs		Outputs				
A	B	AND	OR	NAND	NOR	EX-OR
0	0	0	0	1	1	0
0	1	0	1	1	0	1
1	0	0	1	1	0	1
1	1	1	1	0	0	0

Table 1.1   Truth table representation of two-input logic gates

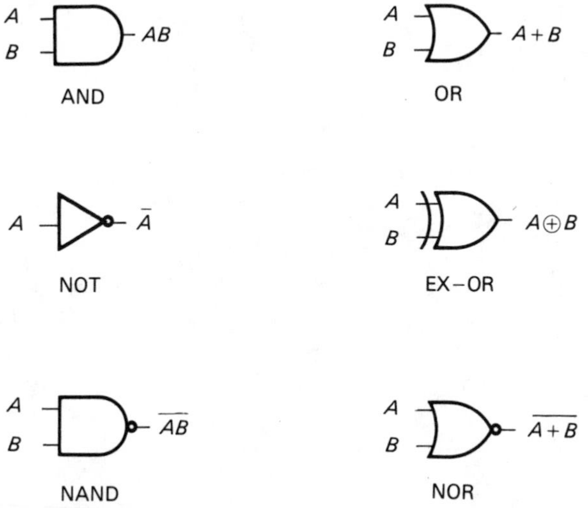

Fig. 1.1   Symbols for logic gates

The functions of the gates described so far can be summarized by means of the idealized waveform diagrams shown in Fig. 1.2. The idealized waveform assumes that the rise and fall times are negligible compared with the on and off times.

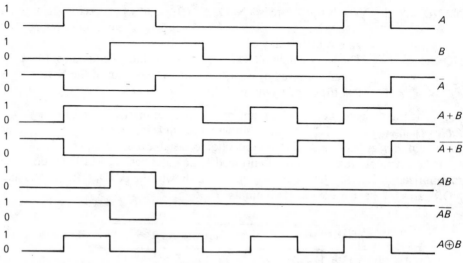

Fig. 1.2   Idealized waveform diagrams for two-input gates using positive logic

1.3   BOOLEAN ALGEBRA

From the definitions of AND, OR and NOT, the following can be stated:

$$\bar{1} \quad = 0$$
$$\bar{0} \quad = 1$$
$$0 \cdot 0 \quad = 0$$
$$0 \cdot 1 \quad = 0$$
$$1 \cdot 1 \quad = 1$$

$$0 + 0 \quad = 0$$
$$0 + 1 \quad = 1$$
$$1 + 1 \quad = 1$$

$$A \cdot A \quad = A$$
$$A \cdot 1 \quad = A$$
$$A \cdot 0 \quad = 0$$
$$A \cdot \bar{A} \quad = 0$$
$$\bar{\bar{A}} \quad = A$$

$$A + A \quad = A$$
$$A + 1 \quad = 1$$
$$A + 0 \quad = A$$
$$A + \bar{A} \quad = 1$$

   Though most of the above postulates are obvious, they can be verified by logical reasoning. Consider for example the AND laws: A can only be 0 or 1, but we must consider all possible combinations.

This verifies that $A \cdot A = A$

This verifies that $A \cdot 1 = A$

This verifies that $A \cdot 0 = 0$

Theorems
$$AB = BA$$
$$A+B = B+A$$
$$A(BC) = (AB)C$$
$$A+(B+C) = (A+B)+C$$
$$(A+B)(A+C) = A+BC$$
$$AB+AC = A(B+C)$$
$$A+AB = A$$
$$A(A+B) = A$$
$$A+\bar{A}B = A+B$$
$$A(\bar{A}+B) = AB$$

DE MORGAN'S THEOREMS:
$$\overline{A+B} = \bar{A} \cdot \bar{B}$$
$$\overline{A \cdot B} = \bar{A}+\bar{B}$$

This means that the complement of a function is obtained by complementing the variables and changing $(\cdot)$ to $(+)$ and $(+)$ to $(\cdot)$.
   The above theorems can be proved algebraically, by using the *Venn diagram* [Example (1.11)] or by using truth tables, as shown in the following examples.

Examples
Prove the following:

(1.1) $A+AB = A$
(1.2) $A+\bar{A}B = A+B$
(1.3) $A(A+B) = A$
(1.4) $A(\bar{A}+B) = AB$
(1.5) $\overline{AB+\bar{A}\bar{B}} = A\bar{B}+\bar{A}B$
(1.6) $A\bar{B}+\bar{A}B = (A+B)\overline{AB}$
(1.7) $(A+B)(A+C) = A+BC$
(1.8) $(A+B)(A+\bar{B}) = A$
(1.9) $AB+\bar{A}C = AB+\bar{A}C+BC$
(1.10) Using a truth table, verify De Morgan's theorem
(1.11) Using the Venn diagram, show that
         $A+\bar{A}B = A+B$

SOLUTIONS TO EXAMPLES:

(1.1) $\quad A + AB \quad = A(1 + B)$
$\qquad\qquad\qquad = A$

(1.2) $\quad A + B \quad\;\; = A + B(A + \bar{A})$
$\qquad\qquad\qquad = A + AB + \bar{A}B$
$\qquad\qquad\qquad = A(1 + B) + \bar{A}B$
$\qquad\qquad\qquad = A + \bar{A}B$

(1.3) $\quad A(A + B) \; = AA + AB$
$\qquad\qquad\qquad = A + AB$
$\qquad\qquad\qquad = A(1 + B)$
$\qquad\qquad\qquad = A$

(1.4) $A(A + B) \; = A\bar{A} + AB$
$\qquad\qquad\qquad = AB$

(1.5) $\overline{AB + \bar{A}\bar{B}} = (\bar{A} + \bar{B})(A + B)$
$\qquad\qquad\qquad = \bar{A}A + \bar{A}B + A\bar{B} + B\bar{B}$
$\qquad\qquad\qquad = A\bar{B} + \bar{A}B$

(1.6) $A\bar{B} + \bar{A}B \; = A\bar{B} + A\bar{A} + \bar{A}B + B\bar{B}$
$\qquad\qquad\qquad = A(\bar{B} + \bar{A}) \; + \; B(\bar{A} + \bar{B})$
$\qquad\qquad\qquad = \; (\bar{A} + \bar{B}) \, (A + B) \; = \; \overline{AB}(A + B)$

(1.7) $\quad (A + B)(A + C) = AA + AC + AB + BC$
$\qquad\qquad\qquad\qquad\; = A(1 + B + C) + BC$
$\qquad\qquad\qquad\qquad\; = A + BC$
$\quad$ or $A + BC \; = A(1 + B) + BC = \; A + AB + BC$
$\qquad\qquad\qquad = A(1 + C) + AB + BC = \; A + AC + AB + BC$
$\qquad\qquad\qquad = AA + AC + AB + BC$
$\qquad\qquad\qquad = A(A + C) + B(A + C)$
$\qquad\qquad\qquad = (A + C)(A + B)$

(1.8) $(A + B)(A + \bar{B}) \; = AA + A\bar{B} + AB + B\bar{B}$
$\qquad\qquad\qquad\qquad = A(1 + \bar{B} + B)$
$\qquad\qquad\qquad\qquad = A$

(1.9) $AB + \bar{A}C \; = AB(1 + C) + \bar{A}C(1 + B)$
$\qquad\qquad\qquad = AB + \bar{A}C + ABC + \bar{A}BC$
$\qquad\qquad\qquad = AB + \bar{A}C + BC$

(1.10) Proof by perfect induction using the following truth table:

A	B	$\bar{A}$	$\bar{B}$	$\overline{A + B}$	$\bar{A} \cdot \bar{B}$	$\overline{A \cdot B}$	$\bar{A} + \bar{B}$
0	0	1	1	1	1	1	1
0	1	1	0	0	0	1	1
1	0	0	1	0	0	1	1
1	1	0	0	0	0	0	0

Columns 6 and 5 are identical for all possible combinations of A and B; hence $\overline{A+B} = \overline{A} \cdot \overline{B}$.

Columns 8 and 7 are identical for all possible combinations of A and B; hence $\overline{A \cdot B} = \overline{A} + \overline{B}$.

(1.11) The Venn diagram is a pictorial representation of the logic variables where sets are represented by areas drawn inside a rectangle. The rectangle is the universal set representing all logic variables, see Fig. 1.3.

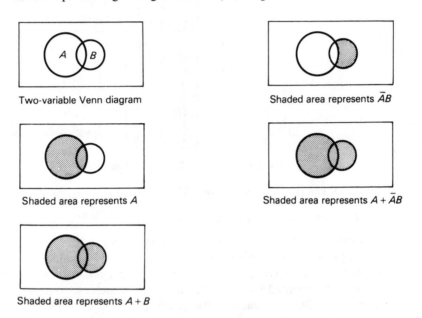

Two-variable Venn diagram   Shaded area represents $\overline{A}B$

Shaded area represents A Shaded area represents $A + \overline{A}B$

Shaded area represents $A + B$

Fig. 1.3   Venn diagram for Example (1.11)

It can be seen that $A + B$ is the same as $A + \overline{A}B$.

1.4   BOOLEAN EXPRESSIONS

In combinational logic, the output of the circuit depends only on the inputs to the circuit. Combinational logic problems are normally given in the form of logical statements or a truth table. To design and implement the problem, Boolean logical expressions (equations) are derived for the output logic function(s) in terms of the binary variables representing the inputs. The logic expressions are given either in the form of a sum of products or in the form of a product of sums. These expressions are then simplified, if possible, and implemented using suitable logic devices.

Boolean algebra can be used to simplify the logic equations or to change them from one form to another. Alternative techniques for simplification include pattern recognition techniques like the Karnaugh map and Venn diagram and tabular techniques like the Quine–McCluskey method, as will be shown later in this chapter.

Canonical Forms

Consider the truth table in Table 1.2. Since there are 3 variables we have $2^3 = 8$ possible combinations.

Decimal	Inputs			Output
	x	y	z	f
0	0	0	0	1
1	0	0	1	0
2	0	1	0	1
3	0	1	1	1
4	1	0	0	0
5	1	0	1	0
6	1	1	0	1
7	1	1	1	1

Table 1.2   Truth table for the logic function f

From the truth table we can derive the logical expression:

$$f = \bar{x}\bar{y}\bar{z} + \bar{x}y\bar{z} + \bar{x}yz + xy\bar{z} + xyz$$

This expression is called the *canonical (normal) sum of products*.

A product term which contains each of the *n*-variables as factors in either complemented or uncomplemented form is called a *minterm*. A binary variable in its true or complemented form is called a *literal*. If each minterm is represented by its decimal equivalent, the same expression can be shown as

$$f = \Sigma\,(0,2,3,6,7)$$

since $f = 1$ in rows 0,2,3,6 and 7.

A logic equation can also be expressed as a product of sums. This is done by considering the combinations for which $f = 0$. From the truth table, $f = 0$ in rows 1,4 and 5, hence

$$\bar{f} = \bar{x}\bar{y}z + x\bar{y}\bar{z} + x\bar{y}z$$

$$\bar{\bar{f}} = \overline{\bar{x}\bar{y}z + x\bar{y}\bar{z} + x\bar{y}z}$$

$$f = (x+y+\bar{z})\cdot(\bar{x}+y+z)\cdot(\bar{x}+y+\bar{z})$$

The product of sums can be expressed as

$$f = \Pi\,(1,4,5)$$

A sum which contains each of the *n* variables complemented or not is called a *maxterm*.

The expression made up of the product of all maxterms for which the function is 0 is called the *canonical product of sums*.

Examples

Using Boolean Algebra, simplify the following Boolean expressions:

(1.12) $F = \overline{\overline{(AB)}A} \cdot \overline{\overline{(AB)}B}$

(1.13) $F = \bar{A}\bar{B}\bar{D} + A\bar{B}\bar{C}\bar{D} + A\bar{B}C\bar{D}$

(1.14) $F = \overline{\overline{AB\bar{C}} + ACD} + \overline{ACD} + \overline{B\bar{C}\bar{D}} + BC\bar{D}$

SOLUTIONS TO EXAMPLES:

(1.12) $\begin{aligned} F &= \overline{\overline{(AB)}A} \cdot \overline{\overline{(AB)}B} \\ &= \overline{(AB)}A + \overline{(AB)}B \\ &= (\bar{A} + \bar{B})A + (\bar{A} + \bar{B})B \\ &= A\bar{A} + A\bar{B} + \bar{A}B + \bar{B}B \\ &= A\bar{B} + \bar{A}B \\ &= A \oplus B \end{aligned}$

The logic circuit is shown in Fig. 1.4.

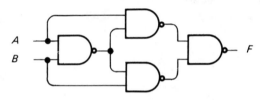

A
B

A
B
F

Fig. 1.4(b)   Logic circuit for Example (1.12) after simplification

Fig. 1.4(a)   Logic circuit for Example (1.12)

(1.13) $\begin{aligned} F &= \bar{A}\bar{B}\bar{D} + A\bar{B}\bar{C}\bar{D} + A\bar{B}C\bar{D} \\ &= \bar{A}\bar{B}\bar{D} + A\bar{B}\bar{D}(C + \bar{C}) \\ &= \bar{B}\bar{D}(\bar{A} + A) \\ &= \bar{B}\bar{D} \end{aligned}$

The logic circuit is shown in Fig. 1.5.

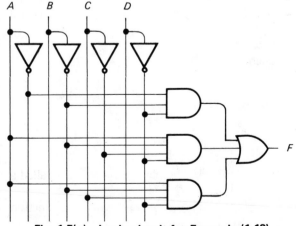

A B C D

F

Fig. 1.5(a)   Logic circuit for Example (1.13)

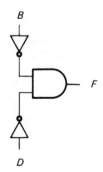

B

F

D

Fig. 1.5(b)   Logic circuit for Example (1.13) after simplification

$$
\begin{aligned}
\text{(1.14)}\ F &= \overline{\overline{\bar{A}B\bar{C}+ACD}+\overline{\overline{ACD}+\bar{B}\bar{C}\bar{D}+BC\bar{D}}} \\
&= (\bar{A}B\bar{C}+ACD)\cdot(\overline{ACD}+\bar{B}\bar{C}\bar{D}+BC\bar{D}) \\
&= (\bar{A}B\bar{C}+ACD)\cdot(\bar{A}+\bar{C}+\bar{D}+\bar{B}\bar{C}\bar{D}+BC\bar{D}) \\
&= \bar{A}\bar{A}B\bar{C}+\bar{A}ACD+\bar{A}B\bar{C}\bar{C}+AC\bar{C}D+\bar{A}B\bar{C}\bar{D}+ACD\bar{D} \\
&= \bar{A}B\bar{C}+\bar{A}B\bar{C}+\bar{A}B\bar{C}\bar{D} \\
&= \bar{A}B\bar{C}\,(1+\bar{D}) \\
&= \bar{A}B\bar{C}
\end{aligned}
$$

It is left to the reader to compare the hardware requirement for this example before and after simplification.

1.5 KARNAUGH MAPS

The Karnaugh map is the easiest method for simplifying logical expressions of up to six variables. In this method all possible combinations of the binary input variables are represented on the map. A logic function containing n variables requires 2^n squares.

1.5.1 Two-Variable Maps

Let us consider first the Karnaugh map (K-map) required for a logic function of two variables A and B. This is shown in Fig. 1.6.

	$\bar{A}$	A
$\bar{B}$	$\bar{A}\bar{B}$	$A\bar{B}$
B	$\bar{A}B$	AB

Fig. 1.6 A two-variable K-map

Example (1.15)
Draw a K-map for the logic function

$f = A\bar{B}+AB$

This is shown in Fig. 1.7.

	$\bar{A}$	A
$\bar{B}$		1
B		1

Fig. 1.7 K-map for Example (1.15)

A 1 is placed in the squares corresponding to $A\bar{B}$ and AB. Though the other two boxes are blank, they in fact contain zeros. An alternative way of drawing the map is shown in Fig. 1.8(a). This representation of the K-map is useful if the logic expression is given in a truth table form as shown in Table 1.3.

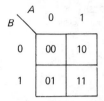

A \ B	0	1
0	00	10
1	01	11

A	B	f
0	0	0
0	1	0
1	0	1
1	1	1

Fig. 1.8(a) **Alternative representation of a two-variable K-map**

Table 1.3 **Truth table for $f = A\bar{B} + AB$**

If on the other hand the logic function is given in a decimal form as

$$f = \Sigma\ (2,3)$$

then the K-map representation in Fig. 1.8(b) is the most convenient.

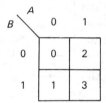

Fig. 1.8(b) **Another representation of a two-variable K-map**

Using the K-map to Simplify Logic Functions
Consider Example (1.15)

$$f = A\bar{B} + AB$$
$$= A(\bar{B} + B)$$
$$= A$$

Since the variable B changes from $\bar{B}$ to B, it is eliminated. This can be achieved on the K-map by looping the two adjacent 1s on the map (Fig. 1.9). Within the loop only variable A is not changing and we can write directly the answer as $f = A$.

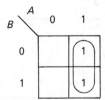

Fig. 1.9 **K-map for Example (1.15) with looped 1s**

1.5.2 Three-Variable Maps

For three variables, the K-map has $2^3 = 8$ squares; these are drawn as in Fig. 1.10(a).

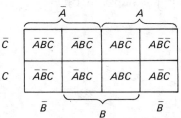

Fig. 1.10(a)   K-map for three variables

The alternative form shown in Fig. 1.10(b) will be adopted from now on, though the binary numbers and their decimal equivalents will be omitted.

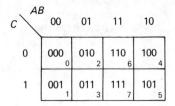

Fig. 1.10(b)   Alternative form of K-map for three variables

The main thing to remember at this stage is that only one variable can change between adjacent squares when moving in a horizontal or a vertical direction.

The terms in the logic functions are entered directly whether they are given in a truth table, logic equation, or decimal format.

Example (1.16)

Simplify the logic function $f = \overline{A}B\overline{C} + \overline{A}BC + A\overline{B}C + A\overline{B}\overline{C}$ using the K-map.

The 1s corresponding to $\overline{A}B\overline{C}$ and $\overline{A}BC$ differ by one variable and can be looped together (Fig. 1.11). Similarly the other two terms are groups together.

The solution is $f = \overline{A}B + A\overline{B}$. This means that f can be independent of the input variable C, which is redundant.

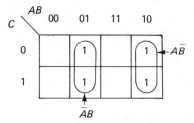

Fig. 1.11   K-map for Example (1.16)

Example (1.17)

Simplify the following logic function:

$$f = \bar{A}B\bar{C} + \bar{A}BC + AB\bar{C} + ABC$$

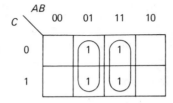

Fig. 1.12   K-map for Example (1.17)

The first two minterms can be combined as follows (Fig. 1.12):

$$\bar{A}B\bar{C} + \bar{A}BC = \bar{A}B(\bar{C} + C) = \bar{A}B \quad \text{(L.H.S. loop)}$$

The last two minterms can also be combined as follows:

$$AB\bar{C} + ABC = AB(\bar{C} + C) = AB \quad \text{(R.H.S. loop)}$$

The two loops can now be combined as follows

$$\bar{A}B + AB = B(\bar{A} + A) = B$$

It is clear then that four adjacent squares can be looped together and two variables eliminated as shown in Fig. 1.13.

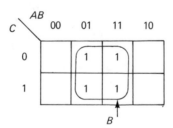

Fig. 1.13   K-map for Example (1.17) with a single loop

From the K-map, $f = B$.

Note that when moving horizontally A changes, and when moving vertically C changes; B is always 1.

Example (1.18)

Simplify the following logic function

$$f = \bar{A}\bar{B}\bar{C} + \bar{A}\bar{B}C + A\bar{B}\bar{C} + A\bar{B}C$$

The four 1s are entered on the K-map (Fig. 1.14) corresponding to the four minterms.

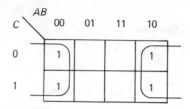

Fig. 1.14   K-map for Example (1.18)

By similar reasoning as in Example (1.17) the four minterms are looped together resulting in the following:

$$f = \bar{B}$$

1.5.3 Four-Variable Maps

A four-variable map contains $2^4 = 16$ squares as in Fig. 1.15.

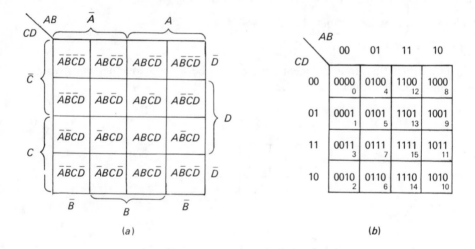

Fig. 1.15   Four-variable K-map using two different notations

Example (1.19)

Simplify the following logic function:

$$f = \bar{A}\bar{B}\bar{C}\bar{D} + \bar{A}\bar{B}\bar{C}D + \bar{A}B\bar{C}\bar{D} + \bar{A}B\bar{C}D + \bar{A}BCD + A\bar{B}C\bar{D}$$

The six minterms are entered in their corresponding squares and looped as shown in Fig. 1.16. A few points arise at this stage:

(a)   All 1s must be included in at least one loop.

(b)   A square may be used as often as necessary. In the above example the term $\bar{A}B\bar{C}D$ is used in two loops. This is possible since $\bar{A}B\bar{C}D + \bar{A}B\bar{C}D = \bar{A}B\bar{C}D$.

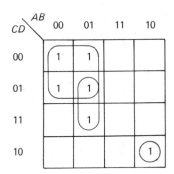

Fig. 1.16   K-map for Example (1.19)

(c)   The 1s are grouped in powers of 2 to form a horizontal row, a vertical row, larger squares or rectangles.

(d)   To minimize the number of terms, all the 1s must be looped in the minimum number of groups.

(e)   To minimize the number of variables in each term, the groups must be as large as possible.

The solution is $f = \bar{A}\bar{C} + \bar{A}BD + A\bar{B}C\bar{D}$

1.5.4 DON'T CARE and CAN'T HAPPEN

Some logic circuits do not use all of the possible input combinations; the unused input combinations are called CAN'T HAPPEN. In other circuits the result is correct whether the input is a 1 or a 0. These input combinations are called DON'T CARE. When simplifying logic expressions, the DON'T CARE and CAN'T HAPPEN terms are represented by X and may be used as 1s or 0s which ever result in a simpler answer. These and other points of interest will be illustrated by means of examples.

Example (1.20)
Simplify the following logic function:

$f = \bar{A}\bar{B}\bar{D} + A\bar{B}C\bar{D} + A\bar{B}C\bar{D}$

given that $\bar{A}BCD$ and $ABCD$ are DON'T CARE conditions.

(a)   The term $\bar{A}\bar{B}\bar{D}$ occurs in two places and contributes two 1s, namely $\bar{A}\bar{B}C\bar{D}$ and $\bar{A}\bar{B}\bar{C}\bar{D}$.

(b)   The two DON'T CARE terms marked X are ignored since they don't help in the simplification.

(c)   The four corners form one group of four.

From the K-map shown in Fig. 1.17 the solution is $f = \bar{B}\bar{D}$.

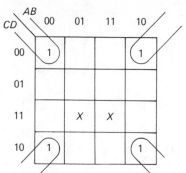

Fig. 1.17 K-map for Example (1.20)

Example (1.21)

Simplify the logic function given by

$$f = \Sigma \, (1,3,7,12,13,14,15,8,9,10)$$

with the DON'T CARE conditions

$$D = (4,5,11)$$

This numerical sum of product representation was first introduced in Section 1.4. In the K-map the boxes are also numbered in this way which makes it easier to enter the terms on the map. Since the largest digit in f is 15, a four-variable map is required. From the map (Fig. 1.18) the solution is:

$$f = A + CD + \bar{B}D$$

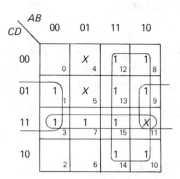

Fig. 1.18 K-map for Example 1.21

The points to note here are:

(a) The DON'T CARE at 11 is taken as a 1 to enable us to make a large loop of eight. The other two DON'T CARE conditions were taken as 0s and ignored.

(b) The two sides of the map are adjacent, i.e. (1,3) and (9,11) are adjacent. Similarly top row (0,4,12,8) and bottom row (2,6,14,10) are adjacent.

(c) A further loop of eight containing boxes 1,5,13,9,3,7,15 and 11 could be taken. This however is unnecessary since it will not include any new 1s.

(d) The answer obtained is correct, but it is not an optimum solution.

Consider the looping shown in the K-map of Fig. 1.19.

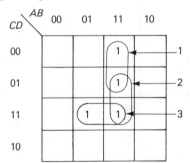

```
        AB
  CD \      00    01    11    10

  00           X    ( 1     1 )

  01     ( 1     X     1     1 )

  11     ( 1     1     1     X )

  10                 ( 1     1 )
```

Fig. 1.19   K-map for Example 1.21 with two loops

From this map

$$f = A + D$$

This brings out the fifth point:

(e)   In some cases, more than one solution is possible, and it is left to the designer's skill to choose a good solution.

Definitions

Consider the K-map of Fig. 1.20.

```
        AB
  CD \      00    01    11    10

  00                     ( 1 ) ←——— 1

  01               ( 1 ) ←——— 2

  11         ( 1    1 ) ←——— 3

  10
```

Fig. 1.20   K-map showing prime implicants

The groups shown are called *implicants*. In general any minterm or group of minterms combined as shown are called implicants.

An implicant is called a *prime implicant* if it is not a subset of another implicant. All three groups shown are prime implicants. A prime implicant is called an *essential prime implicant* if it includes a 1 that is not included in any other prime implicant. Groups 1 and 3 are essential prime implicants.

1.5.5 Five-Variable Maps

For five variables, $2^5 = 32$ squares are required. This is achieved by drawing two four-variable maps. If the variables are A, B, C, D and E then two maps are drawn as in Fig. 1.21. When groups are formed, the two maps are considered as if they were superimposed. Cells or groups from the two maps are grouped together only if they occupy the same positions on the two four-variable maps.

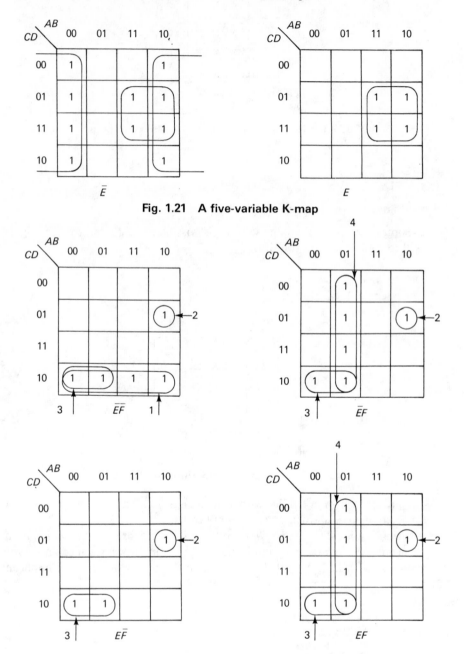

Fig. 1.21   A five-variable K-map

Fig. 1.22   A six-variable K-map

Consider the five-variable map (Fig. 1.21) for the five-variable function f:

$$f(A,B,C,D,E) = \overline{A}\overline{B}C\overline{D}\overline{E} + \overline{A}\overline{B}CD\overline{E} + \overline{A}BCD\overline{E} + \overline{A}BC\overline{D}\overline{E} + $$
$$+ AB\overline{C}\overline{D}\overline{E} + A\overline{B}C\overline{D}\overline{E} + A\overline{B}\overline{C}D + A\overline{B}CD + $$
$$+ AB\overline{C}D + ABCD$$

From the map, the large group of eight squares gives the term $\overline{B}\overline{E}$, which is in the $\overline{E}$ map only. The two four-square loops occupy the same position on the E and $\overline{E}$ maps; the E variable is therefore eliminated, giving the term AD. Thus

$$f = \overline{B}\overline{E} + AD$$

1.5.6 Six-Variable Maps

For six variables, four four-variable maps are drawn as shown in Fig. 1.22. Note that every two adjacent maps differ by one variable only. The group numbered 2 is common to all four maps and results in the term $A\overline{B}C D$, with E and F eliminated.

Group 3 is also common to all four maps and results in $\overline{A}C\overline{D}$, with E and F eliminated. Group 4 is common to the two maps on the right in which variable F does not change, thus giving the term $\overline{A}BF$. Group 1 is only in one map in which the variables E and F are in complemented form, thus giving the term $C\overline{D}\overline{E}\overline{F}$.

From the maps then we obtain:

$$f = C\overline{D}\overline{E}\overline{F} + A\overline{B}C D + \overline{A}C\overline{D} + \overline{A}BF$$

1.5.7 Looping the Zeros on the K-map

We observe that the expressions obtained from the K-map are in the sum of products form. This is usually implemented by an AND/OR circuit. It is possible to extract the expression in the product of sums form by looping the zeros instead of the ones. This results in an OR/AND circuit.

Consider the K-map of Fig. 1.23.

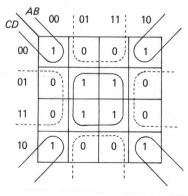

Fig. 1.23   K-map for logic function f_1

$$f_1 = BD + \bar{B}\bar{D} \text{ by looping the ones}$$
$$\bar{f_1} = B\bar{D} + \bar{B}D \text{ by looping the zeros}$$

Hence $f_1 = (\bar{B} + D)(B + \bar{D})$

Both expressions for f_1 are acceptable.

This method is particularly useful in some cases where the 1s require many loops, as in Fig. 1.24.

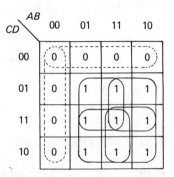

Fig. 1.24   K-map for logic function f_2

Looping the 1s, we obtain

$$f_2 = BD + AD + BC + AC$$

This can be implemented as in Fig. 1.25.

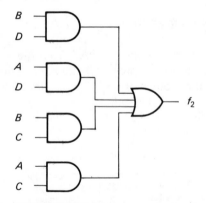

Fig. 1.25   AND/OR implementation for f_2

If the zeros are looped we get:

$$\bar{f_2} = \bar{A}\bar{B} + \bar{C}\bar{D}$$
$$f_2 = \overline{\bar{A}\bar{B} + \bar{C}\bar{D}}$$
$$= (A + B)(C + D)$$

This can be implemented as in Fig. 1.26.

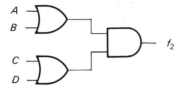

Fig. 1.26   OR/AND implementation for f_2

It can be seen that the OR/AND circuit is more economical in this case.

1.6   THE QUINE–McCLUSKEY METHOD

The K-map becomes impractical when the number of logic variables is greater than six, and the Quine – McCluskey tabular method is used. For simplicity a four-variable logic function will be used to illustrate the procedure.

> **Example (1.22)**
> Simplify the logic function f where
>
> $f = \Sigma\,(1, 2, 4, 5, 6, 10, 12, 13, 14)$

SOLUTION: The minterms are grouped according to the number of 1s as shown in Table 1.4.

		A	B	C	D
X	1	0	0	0	1
X	2	0	0	1	0
X	4	0	1	0	0
X	5	0	1	0	1
X	6	0	1	1	0
X	10	1	0	1	0
X	12	1	1	0	0
X	13	1	1	0	1
X	14	1	1	1	0

Table 1.4   Computation of prime implicants

Each term of a group is compared with each term in the group immediately below. If two terms differ in one variable only then we can use the theorem $AB + \bar{A}B = B$ and eliminate one variable. The variable eliminated is replaced by a — . The two combined terms are crossed (marked X).

Following this, the first term of the first group 0001 is compared with the first term of the second group 0101. Since they differ in B only, they can be combined resulting in $0-01$. Minterms 1 and 5 are then crossed.

This results in Table 1.5.

		A	B	C	D
	1, 5	0	—	0	1
X	2, 6	0	—	1	0
X	2, 10	—	0	1	0
X	4, 5	0	1	0	—
X	4, 6	0	1	—	0
X	4, 12	—	1	0	0
X	5, 13	—	1	0	1
X	6, 14	—	1	1	0
X	10, 14	1	—	1	0
X	12, 13	1	1	0	—
X	12, 14	1	1	—	0

Table 1.5   Computation of prime implicants

The same procedure is applied again and each term of the first group is compared with each term of the second group. The terms are combined only if they differ by one variable and the — in both coincide. This results in Table 1.6.

	A	B	C	D	
2, 6 ; 10, 14	—	—	1	0	
2, 10 ; 6, 14	—	—	1	0	Duplicate
4, 5 ; 12, 13	—	1	0	—	
4, 6 ; 12, 14	—	1	—	0	
4, 12 ; 5, 13	—	1	0	—	Duplicate
4, 12 ; 6, 14	—	1	—	0	Duplicate

Table 1.6   Computation of prime implicants

Duplicate terms are eliminated. The remaining uncrossed terms are the prime implicants of the logic function. 1, 5 in Table 1.5 is not marked X and results in $\bar{A}\bar{C}D$. Also 2, 6 ; 10, 14 in Table 1.6 is not marked X, it is not a duplicate and it results in $C\bar{D}$. Similarly 4, 5 ; 12, 13 gives $B\bar{C}$ and 4, 6 ; 12, 14 gives $B\bar{D}$.

Hence

$$f = \bar{A}\bar{C}D + C\bar{D} + B\bar{C} + B\bar{D}$$

To check, let us use the K-map since this is only a four-variable problem. From the map (Fig. 1.27), the prime implicants are exactly the same, resulting in the following:

$$f = \bar{A}\bar{C}D + C\bar{D} + B\bar{C} + B\bar{D}$$

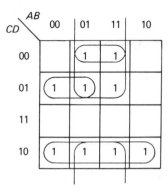

Fig. 1.27   K-map for f

Using the Quine – McCluskey method it is also possible to find the essential prime implicants. From the map it can be seen that $B\bar{D}$ is not essential. To do this using the tabular method a table is drawn as shown in Table 1.7, with the original product terms as columns and the prime implicants as rows.

	√	√	√	√	√	√	√	√	√
	1	2	4	5	6	10	12	13	14
1, 5	*			*					
2, 6 ; 10, 14		*			*	*			*
4, 5 ; 12, 13			*	*			*	*	
4, 6 ; 12, 14			*		*		*		*

Table 1.7   Finding the essential prime implicants

For each prime implicant, the original terms of the switching function contained in that prime implicant are ticked.

The first prime implicant contains 1 and 5. Therefore columns 1 and 5 are ticked (√), and an asterisk is placed under these columns in row (1, 5), and so on.

Then we select the columns with only one asterisk, namely columns 1, 2, 10 and 13. The rows corresponding to these are selected: they are (1, 5), (2, 6 ; 10, 14) and (4, 5 ; 12, 13), which are the essential prime implicants, equal to $\bar{A}\bar{C}D + C\bar{D} + B\bar{C}$.

Next we must check if any columns of the last table (terms of the original function) are not included by one of the essential prime implicants just obtained. In this case all terms are included and the minimal logic function is given by

$$f = \bar{A}\bar{C}D + C\bar{D} + B\bar{C}$$

If any terms are not included, we should have to choose other rows to cover the missing terms. Obviously we should like to cover all terms using the minimum number of rows and if possible use the rows with the fewest literals. (A *literal* is a variable in its true or complemented form as defined in Section 1.4.)

The Quine – McCluskey tabular method can be used for any number of variables, though, for large problems, it becomes time-consuming. The fact that the method is very basic and repetitive makes it suitable for computer implementation.

The recent trend, however, is to implement large logic problems using large modules like ROMs or ULMs. At this level of design, it is not always necessary to minimize logic functions. In brief, it is useful to know the method though its use is becoming less frequent, unless a computer program is available.

1.7   THE MAP-ENTERED VARIABLE TECHNIQUE

The Karnaugh map (K-map) is the most commonly used method for simplifying logic functions for up to six variables. For logic functions having more than six variables the Quine – McCluskey tabular method is more suitable. This method can be used as a hand computation technique, though it tends to become tedious for large problems. The main advantage of the tabular method is that it can be programmed for digital computers.

Using the map-entered variable technique, it is possible to use the K-map for larger number of variables if some of the variables (the seldom used variables) can be conveniently entered on the map, as shown in Fig. 1.28 for x_5 and x_6. The technique amounts to a reduction in the map size and a very useful extension to the power of the map. The map is read as follows:

(i)   Read all the 1s counting all map-entered variables as 0s.
(ii)   Take each map-entered variable separately and find an encirclement using each previously circled 1 as a DON'T CARE.

From the map in Fig. 1.28 the following expression is obtained:

$$Y = x_3 x_4 + x_1 \bar{x}_2 x_3 + x_1 x_3 x_5 + x_1 \bar{x}_2 \bar{x}_4 x_6$$

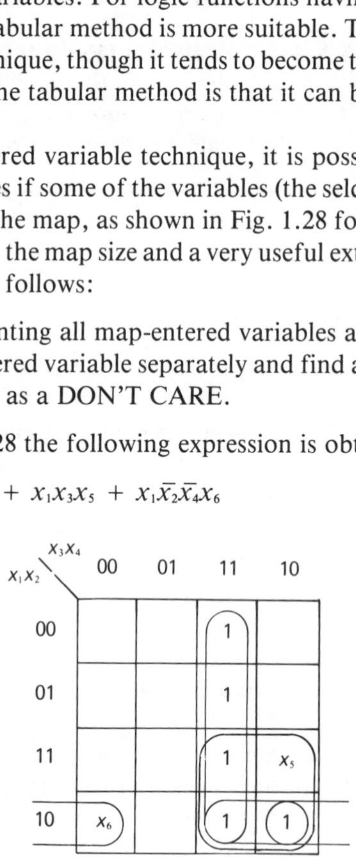

Fig. 1.28   Modified K-map for the six-variable logic function Y

This is more convenient than the usual four maps required for a six-variable function shown in Fig. 1.29.

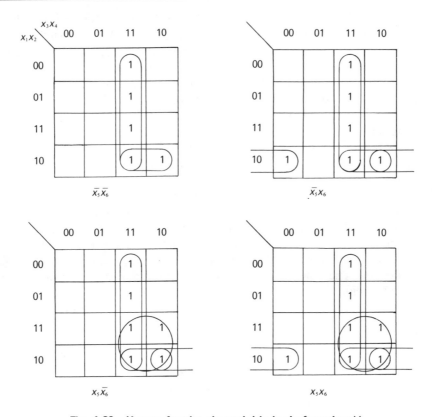

Fig. 1.29   K-map for the six-variable logic function Y

From Fig. 1.29 the same logic function can be obtained:

$$Y = x_3x_4 + x_1\bar{x}_2x_3 + x_1x_3x_5 + x_1\bar{x}_2\bar{x}_4x_6$$

1.8   LARGE AND MULTIPLE-OUTPUT FUNCTIONS

It should be obvious by now that the K-map is the most convenient method for simplifying logic functions of less than or equal to six variables. The Quine–McCluskey method is convenient for both hand computation and computer programming and can be used for small as well as large logic functions. In practice, however, it becomes cumbersome for functions with more than about twelve variables. Needless to say, researchers have investigated the minimization problem of logic functions ⟨20, 90⟩. The problems of simplifying large logic functions and multiple-output logic functions, where a number of outputs depend on the same logic variables [see Equation (3.1) and Example (3.1)], are still without a satisfactory solution.

Surely multiple outputs can be minimized separately, but this does not always lead to an optimal solution. Both the K-map and the Quine – McCluskey method can be extended to simplify a number of logic functions simultaneously by sharing common terms ⟨46⟩. This, however, soon becomes impractical if the number of functions, or if the number of variables, is large – worse still if both are large.

Some sophisticated, mostly commercial, computer packages have been developed to help in the design of certain systems, though not all these are available to the average designer. One such system, is the automatic programmable logic array synthesis (APLAS) system ⟨52, 94⟩. This system can synthesize switching functions with 72 inputs, 144 outputs and several thousand product terms and achieve up to 50 per cent reduction in chip area.

Another factor that is technology-dependent is the criterion for optimization. In other words, what is it that a designer must minimize? Is it the number of gates, the number of chips, the number of interconnections, the delay time, the number of pins, the silicon area, the cost or maybe a combination of some of these? Further, the need for optimization is dependent upon the technology employed and the volume of production. The time and effort required for optimization, with or without computer aid, is more justified if the circuit is mass-produced.

As a simple illustration, consider the following two logic functions:

$$f_1 = ABC + \bar{A}\bar{B}$$
$$f_2 = AB\bar{C} + \bar{A}\bar{B}$$

Fig. 1.30(a) shows how the common term $\bar{A}\bar{B}$ is shared by two outputs f_1 and f_2 resulting in the saving of one gate. Fig. 1.30(b) shows how another common term AB is shared in order to save gate inputs. Both are common occurrences in logic circuits.

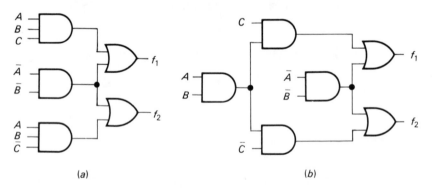

(a)   (b)

Fig. 1.30   Sharing common terms in multi-output logic

1.9   GATE CONVERSION

Though there is a large variety of gates, it is often desirable to convert logic expressions into a form suitable for whatever type of gates are available. It is desirable to use one type of gate rather than mix different gates.

Designing logic circuits using NAND gates only

Consider the following:

$$f = A \cdot B$$

This requires a two-input AND gate, but can be implemented using two NAND gates (Fig. 1.31).

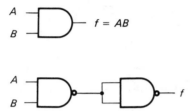

Fig. 1.31   AND/NAND conversion

Similarly an OR function can be implemented using NAND gates (Fig. 1.32).

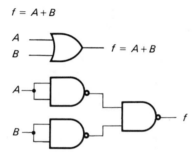

Fig. 1.32   OR/NAND conversion

It is clear that besides its normal function, a NAND gate can be used to invert and to replace the AND and OR gates.

Example (1.23)

Implement the following logic function

$$f = AB + CD$$

(i)   Using AND/OR gates (Fig. 1.33)

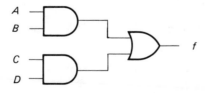

Fig. 1.33   AND/OR circuit for Example (1.23)

(ii)   Using NAND gates (Fig. 1.34)

$$f = \overline{\overline{AB} + \overline{CD}} \qquad \text{(Remember that } \overline{\overline{A}} = A\text{)}$$

$$= \overline{\overline{AB} \cdot \overline{CD}} \qquad \text{by using De Morgan's theorem}$$

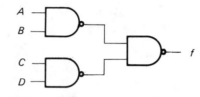

Fig. 1.34   NAND circuit for Example (1.23)

Example (1.24)
Implement the following logic function

$$f = ABC + D$$

(i)   Using AND/OR gates [Fig. 1.35(*a*)]

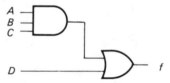

Fig. 1.35(a)   AND/OR circuit for Example (1.24)

(ii)   Using NAND gates [Fig. 1.35(*b*)]

$$f = \overline{\overline{ABC} + \overline{D}}$$

$$= \overline{\overline{ABC} \cdot \overline{D}}$$

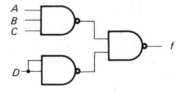

Fig. 1.35(b)   NAND circuit for Example (1.24)

Designing Logic Circuits Using NOR gates
Consider the following

$$f = A + B$$

This requires a two-input OR gate, but can be implemented using two NOR gates (Fig. 1.36).

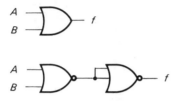

Fig. 1.36   OR/NOR conversion

This means that an OR function can be realized using NOR gates. Similarly an AND function can be realized using NOR gates. Consider $f = AB$ and the logic circuits shown in Fig. 1.37.

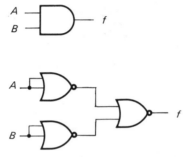

Fig. 1.37   AND/NOR conversion

Example (1.25)
Implement the following logic function

$$f = (A + B)(A + C)$$

(i)   Using OR/AND gates [Fig. 1.38(a)]

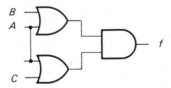

Fig. 1.38(a)   OR/AND circuit for Example (1.25)

(ii)   Using NOR gates [Fig. 1.38(b)]

$$f = \overline{\overline{(A + B)(A + C)}} \qquad \text{(Remember that } \overline{\overline{A}} = A\text{)}$$

$$= \overline{\overline{A + B} + \overline{A + C}} \qquad \text{by using De Morgan's theorem}$$

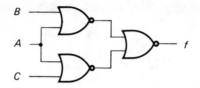

Fig. 1.38(b)   NOR circuit for Example (1.25)

Example (1.26)

Implement the EXclusive-OR function given that

$$f = A\bar{B} + \bar{A}B$$

(i)   Using a single EX-OR gate when available [Fig. 1.39(a)]

Fig. 1.39(a)   EX-OR gate

(ii)   Using AND/OR INVERTER gates [Fig. 1.39(b)]

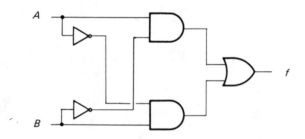

Fig. 1.39(b)   EX–OR circuit using AND/OR/NOT gates

(iii)   Using NAND gates only [Fig. 1.39(c)]

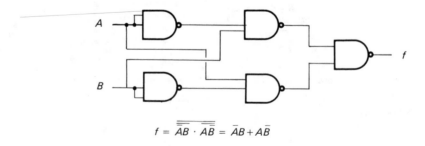

$$f = \overline{\overline{A\bar{B}} \cdot \overline{\bar{A}B}} = \bar{A}B + A\bar{B}$$

Fig. 1.39(c)   EX–OR circuit using NAND gates

(iv) Using NOR gates only [Fig. 1.39(d)]

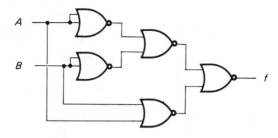

Fig. 1.39(d)   EX–OR circuit using NOR gates

$$f = \overline{\overline{\overline{A + B}} + \overline{\overline{A + B}}}$$
$$= (\overline{A} + \overline{B}) \cdot (A + B)$$
$$= \overline{A}A + \overline{A}B + A\overline{B} + B\overline{B}$$
$$= \overline{A}B + A\overline{B}$$

1.10   FACTORING AND MULTI-LEVEL CIRCUITS

Consider the following logic function

$$f = AB\overline{C}D + A\overline{B}C\overline{D} + \overline{A}BCD + \overline{A}\overline{B}C\overline{D}$$

Suppose also that only two input logic states are available; f can then be written as follows:

$$f = A\overline{C}(BD + \overline{B}\overline{D}) + \overline{A}C(BD + \overline{B}\overline{D})$$
$$= (BD + \overline{B}\overline{D})(A\overline{C} + \overline{A}C)$$
$$= (BD + \overline{B}\overline{D})(A \oplus C)$$

Assuming that the variables are available in their true and complemented forms, the function can now be implemented as in Fig. 1.40

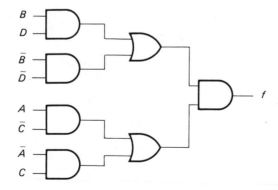

Fig. 1.40   Multi-level implementation for logic function f

Note that $A\bar{C} + \bar{A}C$ can also be generated using one EX-OR gate.

Example (1.27)

Derive the logic expression for the circuit given in Fig. 1.41 and find its two-level AND/OR equivalent assuming the variables are available in their true and complemented form.

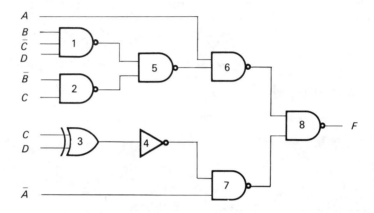

Fig. 1.41 Logic diagram for Example (1.27)

The outputs of the gates are as follows:

Gate number	Output
1	$\overline{B\bar{C}D}$
2	$\overline{\bar{B}C}$
3	$C \oplus D = C\bar{D} + \bar{C}D$
4	$\overline{C \oplus D}$
5	$\overline{\overline{B\bar{C}D} \cdot \overline{\bar{B}C}}$
6	$\overline{(\overline{B\bar{C}D} \cdot \overline{\bar{B}C}) \cdot A}$
7	$\overline{\overline{C \oplus D} \cdot \bar{A}}$
8	$F = \overline{\overline{(C \oplus D)} \cdot \bar{A} \cdot \overline{(\overline{B\bar{C}D}) \cdot (\overline{\bar{B}C})) \cdot A}}$

Using De Morgan's theorem, we have:

$$F = \overline{(C \oplus D)} \cdot \bar{A} + \overline{(\overline{B\bar{C}D} \cdot \overline{\bar{B}C})}A$$

$$= (\bar{C}\bar{D} + CD)\bar{A} + A(B\bar{C}D + \bar{B}C)$$

$$= \bar{A}\bar{C}\bar{D} + \bar{A}CD + AB\bar{C}D + A\bar{B}C$$

F can be implemented as in Fig. 1.42

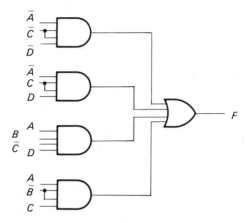

Fig. 1.42   Two-level implementation of Example (1.27)

Example (1.28)

A warning light is to glow when the mains switch is on provided that either switches *A* and *B* or *C* and *D* are turned on. Assuming that logics 1 and 0 are produced when a switch is closed and opened respectively, design a minimal logic circuit using NAND gates to produce a logic 1 signal when the light is on. Redesign the circuit using two-input NAND gates only.

SOLUTION: The logic expression for this problem can be written directly, but a K-map is drawn to emphasize that it is a five-variable problem (Fig. 1.43). The output $Z = f(A, B, C, D, S)$

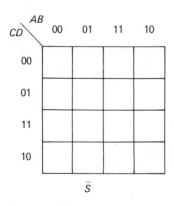

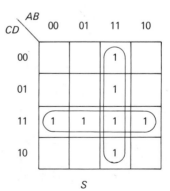

$$Z = (AB + CD) S$$
$$ = ABS + CDS$$

Fig. 1.43   K-map for Example (1.28)

If NAND gates are to be used, then De Morgan's theorem is used to convert the expression

$$Z = \overline{\overline{ABS} + \overline{CDS}}$$
$$= \overline{\overline{ABS} \cdot \overline{CDS}}$$

This is implemented as in Fig. 1.44.

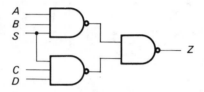

Fig. 1.44   NAND gate implementation for Example (1.28)

If only two input NAND gates are available the equation is written as follows

$$Z = \overline{(AB + CD)S}$$
$$= \overline{(\overline{AB} \cdot \overline{CD})S}$$
$$= \overline{(\overline{AB} \cdot \overline{CD}) \cdot S}$$

This can be implemented as in Fig. 1.45.

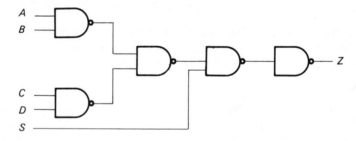

Fig. 1.45   Implementation of Example (1.28) using two-input NAND gates

Exercises

Q(1.1)   Prove the following identities:

(a) $\overline{AB(\bar{D} + \bar{C}D)} \cdot \overline{(ABCD)} \cdot \overline{A\bar{B}(CD + \bar{C}\bar{D})} = AB + A(C \oplus D)$

(b) $[(A + B)(A + \bar{B})] \ [(CD + \bar{C}\bar{D}) + (C \oplus D)] = A$

Q(1.2)   Simplify the following logical expressions and implement them using suitable logic:

(a) $f_1 = \Sigma \ (2,4,6,10,14)$

(b) $f_2 = \Sigma \ (0,2,4,6,8,10)$

(c) $f_3 = \Pi \ (2,3,6)$

Q(1.3)   Using K-maps or otherwise, simplify the following logic equation and implement it using NAND gates only:

$$f = \bar{A}\bar{B}\bar{C}D + B\bar{C}D + A\bar{B}\bar{C}D + \bar{B}CD$$

given that $ABCD$ and $AB\bar{C}\bar{D}$ are DON'T CARE states.

Q(1.4)   A chemical process is activated only if at least three out of four keys are inserted. Assuming that an inserted key produces a logic 1, design a minimal logic circuit to achieve this using AND, OR and NOT gates.

2

Logic and Memory Devices

2.1   INTRODUCTION

This chapter describes the most commonly used logic and memory devices. Logic devices are broadly divided into two categories: bipolar and MOS.

Bipolar devices include resistor transistor logic (RTL), diode transistor logic (DTL), transistor transistor logic (TTL), integrated injection logic (I^2L) and emitter coupled logic (ECL).

Metal oxide semiconductor (MOS) devices include positive-type MOS (pMOS), negative type MOS (nMOS) and complementary MOS (cMOS).

The second part of this chapter describes semiconductor memory devices. The term memory device, as used with digital computers, includes any device that can be used to store digital information or binary data, such as semiconductor memory devices (to be described here) and other devices like drums, discs, floppy discs, magnetic tape, paper tape, core memory, holographic memory, magnetic bubbles, and so on. These are normally discussed in digital computer books ⟨10, 64, 73⟩. Semiconductor memories on the other hand, though they can be used to store data just like magnetic cores or discs, can also be used to implement logic devices and it is in this context they will be described.

Semiconductor devices can be divided into two categories. The first is the read only type like read only memory (ROM), programmable ROM (PROM), erasable and programmable ROM (EPROM) and electrically alterable ROM (EAROM). The common feature of all these devices is that the information stored is non-volatile (permanent) and that after programming, one can only read from them. The second type is called random access memory (RAM) and includes static RAM and dynamic RAM devices. These are volatile and can be used for reading and writing. Programmable logic arrays (PLA), programmable array logic (PAL), gate arrays, and universal logic modules (ULM) are also described.

Definitions

NOISE IMMUNITY: this refers to the ability of a logic circuit to withstand noise superimposed on its input logic level. Referring to the graphical representation in Fig. 2.1, we can distinguish between two types of noise immunity:

Low noise immunity (LNI). This refers to the largest step function which when superimposed at the input on the maximum value of low logic level will not change the output logic level.

High noise immunity (HNI). This refers to the largest negative step function which when superimposed at the input on the minimum high logic level will not change the output logic level.

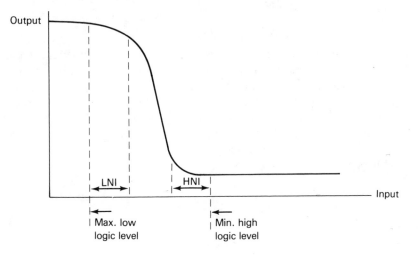

Fig. 2.1   **Representation of noise immunities**

FAN-IN: this is the number of similar logic gates that can be connected to the input without degrading the circuit.

FAN-OUT: This is the number of similar logic gates that any single gate is capable of driving.

2.2   **RESISTOR TRANSISTOR LOGIC (RTL)**

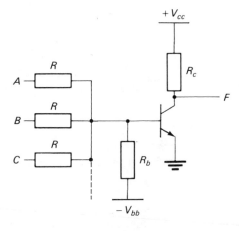

Fig. 2.2   **RTL NOR gate**

If one, or more, of the inputs *A, B, C* is high, enough base current flows to saturate the transistor and the output *F* will be low (earth). If all the inputs are low, the transistor is cut off and the output *F* is high (*Vcc*). For positive logic, the circuit shown in Fig. 2.2 behaves as a NOR gate:

$$F = \overline{A + B + C}$$

The RTL circuit is one of the earliest logic gates and was often designed using discrete components. R_b and $-V_{bb}$ are not essential to the logic requirement, but they are used for fast turn-off. Contrary to popular belief, RTL circuits are still in use in some gate array devices, though in a modified form as shown in Fig. 2.3 and Fig. 2.4. Collector diffusion isolation (CDI) has been used by Ferranti to produce RTL and current mode logic (CML) gate arrays. CDI is a manufacturing process which is known to produce faster switching speeds and packing densities that are equivalent to *n*MOS.

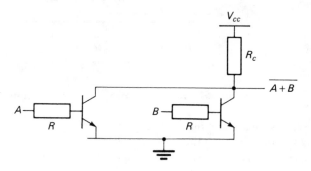

Fig. 2.3   A two-input RTL gate

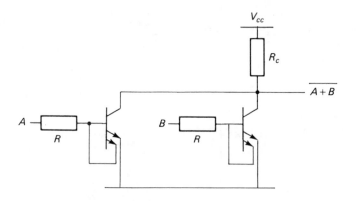

Fig. 2.4   Integrated-type RTL NOR gate

Example (2.1)

For the circuit given in Fig. 2.2, find a suitable value for *R* if the fan-out and fan-in are each four and the low d.c. noise immunity is to be better than one volt.

Assuming that noise will only occur on one input terminal at any particular time, compute the low and high noise immunities for the circuit, given that:

$$\beta = 25, \ V_{beon} = 0.7 \text{ V}, \ V_{cesat} = 0.1 \text{ V}, \ R_c = 1 \text{ k}\Omega,$$
$$R_b = 50 \text{ k}\Omega, \ V_{cc} = 10 \text{ V}, \ V_{bb} = -5 \text{ V}$$

where V_{beon} and V_{cesat} are the base and collector voltages respectively when the transistor is saturated and β is the common emitter forward current gain.

SOLUTION: To simplify the task, we shall neglect the effect of supply and resistor variations. Let the circuit have a fan-in of M and a fan-out of N. In a worst-case design, we imagine the transistor is held in the on state by a single stage that is high, all other inputs being low. Furthermore, the transistor is expected to drive all the N outputs. In the arrangement of Fig. 2.5 T_2 is held on by one high input from T_1, the other $(M-1)$ inputs being low at V_{cesat}.

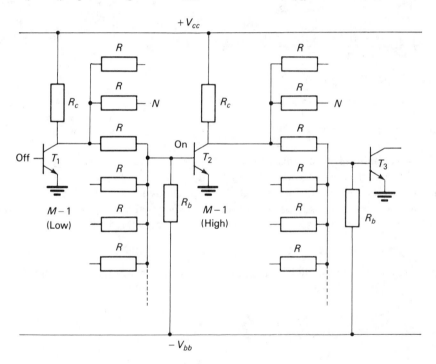

Fig. 2.5   Cascaded RTL gates

Transistor T_2 is saturated, its collector voltage is at V_{cesat} and its base is at V_{beon}. T_2 will have to sink-in current from all its N outputs. These N stages could be held on by their other inputs.

$$I_{cmax} = \frac{V_{cc} - V_{cesat}}{R_c} + N\left(\frac{V_{beon} - V_{cesat}}{R} \right) \tag{2.1}$$

If I is the current supplied by T_1 stage to its N output stages, then:

$$V_{cc} - V_{beon} = NIR_c + IR$$

$$I = \frac{V_{cc} - V_{beon}}{R + NR_c}$$

$$I_{bmin} = \frac{V_{cc} - V_{beon}}{R + NR_c} - \left(\frac{V_{beon} - V_{cesat}}{R}\right)(M-1) - \frac{V_{beon} - . V_{bb}}{R_b} \qquad (2.2)$$

This means that in worst-case situations, some of the current to T_2 is lost to the $- V_{bb}$ terminal and the $(M-1)$ low inputs.

Substituting numerical values we obtain the following:

$$I_c = \frac{10 - 0.1}{1} + \frac{4(0.7 - 0.1)}{R}$$

$$I_b = \frac{10 - 0.7}{R + 4} - \frac{3(0.7 - 0.1)}{R} - \frac{(0.7 + 5)}{50}$$

When $I_b = I_c/\beta$, we get

$$0.5 R^2 - 5.36 R + 7.6 = 0$$

This gives $8.67 \text{ k}\Omega > R > 1.64 \text{ k}\Omega$

Noise Immunity Considerations

If the transistor is turned off, all its inputs are low (V_{cesat}). A positive-going pulse could raise the input to a level that is high enough to turn the transistor on. If the low noise immunity is to be better than some specified value (LNI_s), the input has to rise from V_{cesat} to a new value V_x, say, in order to turn the device on, thus

$$V_x = LNI_s + V_{cesat}$$

In the worst case, noise might affect all inputs. Thus, from Fig. 2.6, when the transistor is just on, we have:

$$M\frac{(V_x - V_{beon})}{R} = \frac{V_{beon} - V_{bb}}{R_b} \qquad (2.3)$$

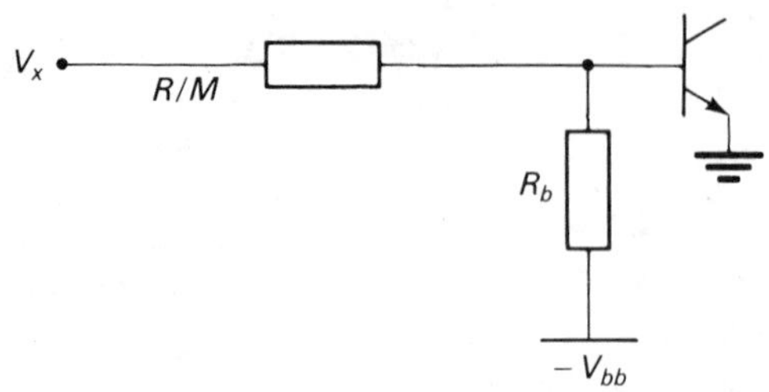

Fig. 2.6   The M inputs appear in parallel

$$R = M\frac{R_b \, (V_x - V_{beon})}{V_{beon} - V_{bb}} \tag{2.4}$$

This imposes a third limit on the value of R. The third value R_3 should be greater than R in Equation (2.4) so as to reduce the effect of noise. In this example, noise appears on one input only. The worst case occurs when the stage is driven by one input, as in Fig. 2.7.

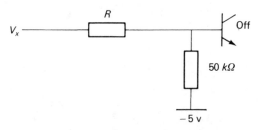

Fig. 2.7   Calculation of LNI

The input has to rise from 0.1 V to 1.1 V before it can change the state of the stage.

$$V_x = 1 + 0.1 = 1.1 \text{ V}$$

$$\frac{1.1 - 0.7}{R} = \frac{0.7 - (-5)}{50}$$

From this $R \geqslant 3.5 \text{ k}\Omega$
To satisfy all conditions

$$R = \frac{8.67 + 3.5}{2} = 6.1 \text{ k}\Omega$$

The nearest preferred value is 5.6 kΩ.

Having chosen a value for R, we now calculate the values of the LNI and HNI:

$$\frac{V_x - 0.7}{5.6} = \frac{5.7}{50}$$

$$V_x = 1.34 \text{ V}$$

Low noise immunity $= 1.34 - 0.1 = 1.24 \text{ V}$.

When the transistor is on, the voltage at the collector of the stage is V_y, as in Fig. 2.8.

$$V_y = V_{cc} - NR_cI$$

$$= 10 - 4 \times 1 \times \left(\frac{10 - 0.7}{4 \times 1 + 5.6}\right)$$

$$= 6.1 \text{ V}$$

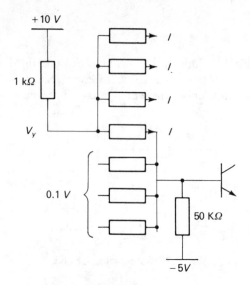

Fig. 2.8   Calculation of HNI

When the transistor is just turned off, we have:

V_y falls to V_y'

$$\frac{V_y' - V_{beon}}{R} = \frac{V_{beon} - V_{bb}}{R_b} + (M-1)\frac{(V_{beon} - V_{cesat})}{R} + I_{bmin}$$

From Equation (2.1)

$$I_c = 9.9 + \frac{4 \times 0.6}{5.6}$$

$$= 10.3 \text{ mA}$$

$$I_b = \frac{10.3}{25} = 0.41 \text{ mA}$$

$$V_y' = 5.4 \text{ V}$$

High noise immunity $= 6.1 - 5.4$

$$= 0.7 \text{ V}$$

2.3   DIODE TRANSISTOR LOGIC (DTL)

This is another early logic circuit that is less frequently used nowadays. It consists basically of a diode AND gate the output of which is used to switch a transistor between saturation and cut-off. The transistor inverts the logical function and the gate becomes a NOT AND (NAND) gate for positive logic (Fig. 2.9).

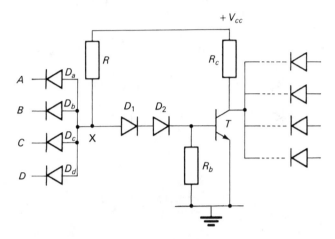

Fig. 2.9   DTL NAND gate

The inputs are connected to the outputs of similar gates. The four inputs A, B, C, and D are therefore the collectors of the input stages. In general let the fan-in be M. Similarly each output is connected to a similar stage. If the fan-out is N, then N stages are connected to the collector of the transistor T. We should remember that each of these N stages could have further $(M - 1)$ inputs.

If the collector voltage V_{cesat}, is, say, 0.1 V when the transistor is saturated. Then a logic 0 at any input A, say, will cause D_a to conduct.

If the voltage drop across a diode is 0.7 V the voltage at point X will be 0.8 V which is high enough to switch the transistor on. This explains the presence of the two diodes D_1 and D_2. The two diodes prevent the transistor from switching on with the input diodes. The transistor should be switched on only when all inputs are high. In brief, when all inputs are high (logic 1) the input diodes are reverse-biased, the voltage at X rises and the transistor conducts, making the output low (logic 0). If any input is low (logic 0), the corresponding diode conducts taking current away from the transistor. The transistor is cut off and the output goes high (logic 1). In some cases R_b is connected to a negative supply

Example (2.2)
For the DTL circuit shown in Fig. 2.10, estimate the fan-out, and noise margins, given that:

$$V_{cesat} = 0.2 \text{ V}$$
$$\beta = 30$$

The voltage drop across the diode and base emitter junction of the transistor is 0.7 V

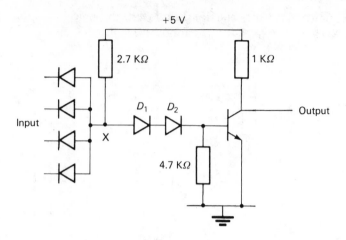

Fig. 2.10   Circuit for Example (2.2)

SOLUTION: When all inputs are high (5 V) the input diodes are reverse-biased. D_1 and D_2 are forward-biased. The transistor is conducting and the output is low at V_{cesat}.

The base voltage is 0.7 V

Voltage at X = 0.7 + 0.7 + 0.7
$$= 2.1 \text{ V}$$

Current through the 2.7 kΩ resistor is given by:

$$I_1 = \frac{5 - 2.1}{2.7} = 1.074 \text{ mA}$$

Current through the 4.7 kΩ resistor is given by:

$$I_2 = \frac{0.7 - 0}{4.7} = 0.150 \text{ mA}$$

Base current $I_b = I_1 - I_2$
$$= 0.924 \text{ mA}$$

Collector current $I_c = \beta I_b$
$$= 30 \times 0.924$$
$$= 27.72 \text{ mA}$$

Current through the 1kΩ resistor is given by: (Fig. 2.11)

$$I_3 = \frac{5 - 0.2}{1} = 4.8 \text{ mA}$$

In a worst-case situation, the saturated transistor has to sink current from all its N output stages all of which have $(M-1)$ high inputs. Each output stage supplies a current given by:

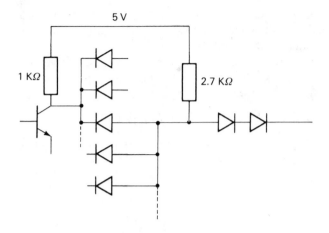

Fig. 2.11   A section of cascaded DTL circuits

$$I_4 = \frac{5 - (0.2 + 0.7)}{2.7} = 1.52 \text{ mA}$$

N output stages will sink NI_4 mA. Since maximum collector current is given by βI_b, then

$$\beta I_b = I_3 + NI_4$$

$$N = \frac{27.72 - 4.8}{1.52} = 15$$

Obviously N must be an integer.

To allow for tolerance and spread of transistor parameters, N should be taken as 10 or 12 for proper operation.

To estimate the noise margins in the high state consider the diagram in Fig. 2.12:

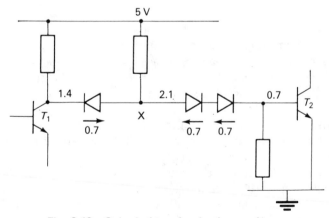

Fig. 2.12   Calculation of noise immunity

Transistor T_2 is ON as long as the voltage at X is 2.1 V or more. The input voltage at the collector of T_1 is assumed to be 5 V. If the input voltage drops and the input diode conducts, the voltage drop across it is also 0.7 V.

From the voltages given, it can be seen that as long as the voltage at the input (collector of T_1) is greater than 1.4 V, T_2 will be on. In other words, the input must drop from 5 V to 1.4 V before T_2 is brought to the edge of changing its state.

Noise immunity = 5 − 1.4 = 3.6 V (High state)

In the low state, the maximum low logic level is 0.2 V (V_{cesat}). The input has to rise from 0.2 V to 1.4 V before T_2 changes its state. Thus

Noise immunity = 1.4 − 0.2 = 1.2 V (Low state)

2.4   TRANSISTOR TRANSISTOR LOGIC (TTL)

2.4.1 Basic Transistor Transistor Logic

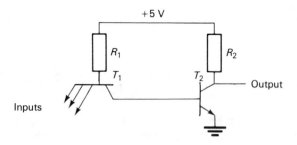

Fig. 2.13   Basic TTL logic gate

Fig. 2.13 shows the basic TTL circuit. It can be seen as a modification of the DTL circuit. Emitter base junctions replace the input diode, and the collector base junction replaces the series diode. When all inputs are high, the emitters are reverse-biased. The base voltage is high, the collector is forward-biased and current flows from the supply, through R_1 to the base of T_2. Transistor T_2 is on and saturated. The output is low at V_{cesat}.

If any inputs become low (V_{cesat}), current will be diverted through this input which is now forward-biased. Transistor T_2 will be cut off and its output will go high.

The circuit functions as a NAND gate.

Early integrated circuits were a direct translation of discrete logic, but the design techniques rapidly changed to the extent that circuit complexity is no longer a limiting factor, and the main consideration is now high performance. The standard form of integrated TTL is shown in Fig. 2.14, and discussed in Section 2.4.2.

2.4.2 Totem-Pole TTL NAND Gate

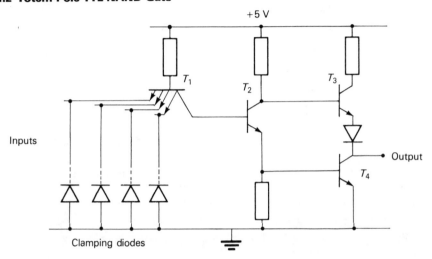

Fig. 2.14   Standard TTL logic gate

When all inputs are high ($+5$ V), the base collector junction is forward-biased. Transistors T_2 and T_4 are on. The voltage at the base of T_4 is V_{beon} (0.7 V, say). The base of T_2 is at $2V_{beon}$. The collector of T_2 is at $V_{beon} + V_{cesat}$ (0.9 V).

The voltage at the base of T_3 is not high enough to turn T_3 on. T_3 is off, and the output is low at V_{cesat}. Note that for T_3 to be on, its base voltage must be at least $V_{cesat} + V_D + V_{beon}$. V_D is the voltage drop across the diode.

The output is therefore low if all inputs are high. If we assume four inputs A, B, C, and D, the output is given by $\overline{A \cdot B \cdot C \cdot D}$.

When at least one input is low (V_{cesat}), transistor T_1 turns on. The collector of T_1 (also the base of T_2) is at 0.2 V. Transistors T_2 and T_4 turn off. The output taken from the collector of T_4 is high. The collector of T_2 is high, thus turning T_3 on (assuming an output load is connected).

Note that transistors T_3 and T_4 are connected in a *totem-pole* configuration. T_3 sits upon T_4 providing an active pull-up.

In Fig. 2.14 clamping diodes are shown at the inputs. These diodes have no effect when the inputs are high. When the inputs are low, the diodes limit any negative voltage excursions that may be present at the input.

2.4.3 Schottky TTL

The TTL logic family is the fastest of the saturated bipolar logic group, but for higher speeds the Schottky barrier diodes are employed. These allow the transistors to operate in the non-saturated mode by reducing the storage time. The Schottky barrier diode is a fast switching device and has low forward drop (about 0.25 V). The diode is used as a clamp from base to collector as shown in Fig. 2.15(a). The symbol for the resulting transistor is shown in Fig. 2.15(b).

Fig. 2.15(a)   Schottky diode used   **Fig. 2.15(b)**   The symbol for a Schottky
                          as a clamp   − clamped transistor

The Schottky-clamped transistors do not go as deeply into saturation and there-fore turn off faster than ordinary transistors.

It is possible now to select one of many TTL versions depending on applications like standard TTL, low-power TTL, low-power Schottky TTL, etc. We should remember, however, that it is usually a matter of compromise between various re-quirements. To increase the speed, smaller resistors are used at the expense of higher power dissipation. Low power TTL families employ larger resistors, which results in increased propagation delay.

2.4.4 Output Configuration

TTL families are also available in different output arrangements. The totem-pole mode is shown in Fig. 2.14. This does not allow a wired logic connection. A second arrangement is the *passive pull-up*, which employs a resistor only at the out-put and eliminates T₃ and the diode. The third arrangement is the open-collector gate, where the collector is left open as shown in Fig. 2.16.

The open-collector gate is suitable for wired-OR connection and when in use an external resistor is connected between the output and the supply. Finally we mention

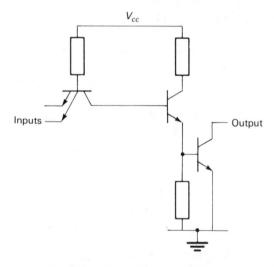

Fig. 2.16   Open-collector TTL gate

the *tri-state* arrangement, which functions as a standard TTL gate, except that a third state is added in which both output transistor T_3 and T_4 are switched off simultaneously, given a high output impedance. In this high-impedance state, the output looks like an open-circuit.

As shown in Fig. 2.17 the control level decides whether the output of a gate is normal NAND or OPEN.

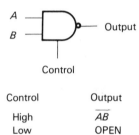

Control	Output
High	$\overline{AB}$
Low	OPEN

Fig. 2.17   Tri-state TTL gate

To illustrate how this third state may be achieved consider a typical low-power Schottky TTL gate shown in Fig. 2.18. First, the multi-emitter transistor T_1 is now replaced by Schottky diodes as in DTL circuits. This circuit is faster than the traditional multi-emitter type. Secondly, the output stage is slightly different. The transistor T_5 and two resistors replace the resistance between the base of T_4 and earth.

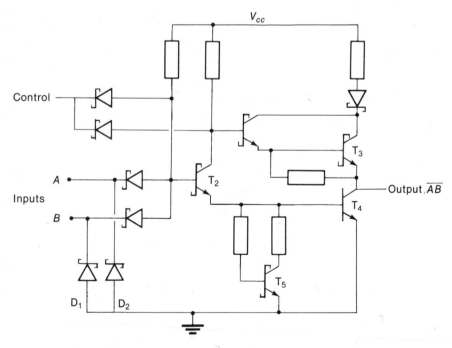

Fig. 2.18   A tri-state Schottky TTL gate

This is called a squaring circuit because it helps to square up the transfer characteristics ($V_{out} \sim V_{in}$). This is achieved by preventing T_2 from conducting until the input voltage is high enough for T_2 to supply adequate base current to T_4. The circuit also provides a low resistance to discharge the capacitance at the base of T_4 in the off conditons.

The output contains a *Darlington circuit* instead of just T_3 as in the conventional TTL. D_1 and D_2 are catching diodes to suppress negative transients. The main difference however is the control circuit added to obtain high impedance in the open state. When the level of the control line is low, transistor T_2 (hence T_4) and the Darlington pair are turned off. The output stage is non-conducting. This third state is useful when the outputs of many gates are to be connected to a common bus and the controls enable only one output at any time.

2.4.5 Other TTL Gates

The standard TTL NAND gate can easily be modified to obtain a TTL inverter. This is achieved by replacing the multi-emitter transistor T_1 with a single-emitter transistor. It takes a little bit more, however, to make a NOR gate. In Fig. 2.19, when both inputs are low, transistors T_1 and T_2 are turned on and current is taken away from transistors T_3 and T_4, which are turned off. This results in transistor T_5 being on while transistor T_6 being off. The output is high.

If both inputs are high, transistors T_1 and T_2 are off, which enables transistors T_3 and T_4 to turn on. This turns T_6 on and T_5 off, resulting in a low output. If only one input is high, the corresponding input transistor, T_1 or T_2, is off, resulting in T_3 or T_4 being on. Either arrangement will turn T_6 on and T_5 off and produce a low output. Obviously the output is low if input A or input B or both are high, which is a NOR operation.

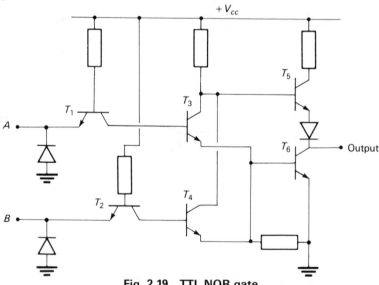

Fig. 2.19   TTL NOR gate

2.5   EMITTER COUPLED LOGIC (ECL)

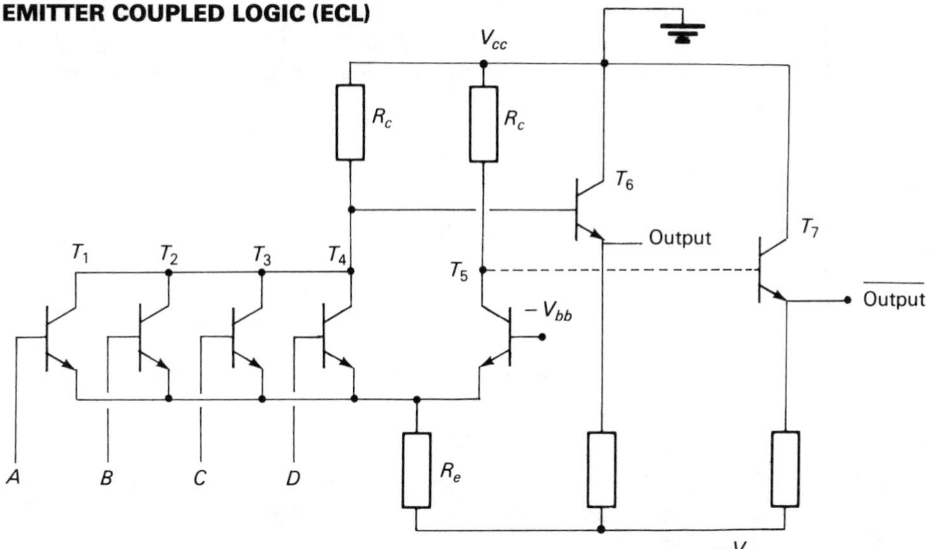

Fig. 2.20   Basic ECL logic gate

An ECL circuit (Fig. 2.20) is used when speed is a major consideration. The reason for the fast operation of these circuits is that they operate in the non-saturated mode.

The input transistors T_1 to T_4 share a common emitter resistor with the reference transistor T_5. If all inputs are low, the input transistors are cut off. Their collector terminal is high. The output taken from the emitter of T_6 follows the base potential, giving a high output (T_6 is on). If any input is more positive than the bias voltage V_{bb}, the corresponding transistor will conduct. The current previously flowing through R_e from T_5 is now supplied by the input transistor (common mode). The collector voltage will drop. The output will drop to the low logic value (T_6 is off).

The circuit functions as a NOR gate. A complementary output (OR), can be obtained from the emitter of T_7. A bias driver using a divider network is used to obtain the bias voltage V_{bb}, but this is not shown here. Vee is the most negative voltage.

	DTL	RTL	TTL	STTL	ECL	nMOS	cMOS	I²L
Positive logic function of basic gate	NAND	NOR	NAND	NAND	OR/NOR	LSI	NAND/NOR	LSI
Fan-in without expansion	10	5	8	8	5	—	8	—
Typical fan-out	8–10	4	8–10	8–10	15	—	20	—
Power dissipation per gate, mW	10	12	2–20	2–20	30	0.2	0.001	0.02
Gate delay, ns	20	15	5–30	3–10	1–2	50	20	15
Noise immunity, V	0.8	0.3	0.5–1	1	0.1	1	($V_D/3$)	0.2

Table 2.1(*a*)   Basic characteristics of logic families

The main advantage of ECL is its high speed as can be seen from Table 2.1(*a*). Its main disadvantages are the high component count and high power dissipation. Standard ECL also has a relatively small logic swing of around 800 mV and hence a low noise margin.

2.6   INTEGRATED INJECTION LOGIC (I²L)

The basic I²L inverter shown in Fig. 2.21(a) consists of a *pnp* transistor T_1 operating in the common base mode, and a multiple-collector *npn* transistor T_2. Transistor T_1 acts as a current generator and is usually drawn as such, as shown in Fig. 2.21(b).

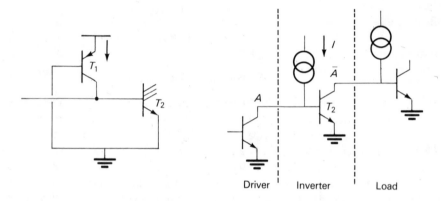

Driver   Inverter   Load

Fig. 2.21(a)   Basic I²L logic gate   **Fig. 2.21(b)   I²L gates in cascade**

When the input A is high, current I is injected through the base of transistor T_2 which is turned on. The output at the collector of transistor T_2 is low at V_{cesat}, which is about 0.1 V. If the input A is low, the driver sinks* current I from the base of transistor T_2. Transistor T_2 is turned off and its output is at V_{beon} of the load, which is about 0.7 V. This is the high level of the output. The device, therefore, acts as an inverter with output voltage swinging between 0.1 V and 0.7 V.

When this device is fabricated, the collector of the *pnp* transistor T_1 is merged with the base of the *npn* transistor T_2; also the base of transistor T_1 is merged with the emitter of transistor T_2. This can be seen from the cross-section of a typical inverter, as in Fig. 2.21(c) with three *n*-region collectors formed in the large *p*-type base region of transistor T_2. This base region also forms the *p*-type collector of transistor T_1. Similarly it can be seen that the *n*-region which forms the emitter of transistor T_2 is also used as a base for T_1, both the emitter of T_2 and the base of T_1 being earthed.

We note here that merged regions like the base of transistor T_2 and the collector of transistor T_1 require no metal interconnection. The merged structure means it requires a very small silicon area. Individual gates need not be isolated since the base of

* See Section 2.8.1

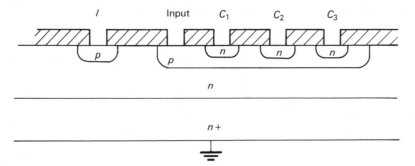

Fig 2.21(c)   Cross-section of an I^2L inverter with three collectors C_1, C_2, C_3

the *pnp* transistor and the emitter of the *npn* transistors are both earthed. Also it is the only bipolar device which require no resistors.

These devices enjoy very low power dissipation compared with other bipolar devices and can have a density which is at least as good as that of *n*MOS. Their properties make them very suitable for VLSI.

Transistor T_2 normally has multiple collectors for fan-out. Typically four collectors at present.

2.6.1   NAND/NOR Gates

Though I^2L is used mainly for LSI devices, logic gates like NAND and NOR can be made, as shown in Figs 2.21(d) and 2.21(e) respectively.

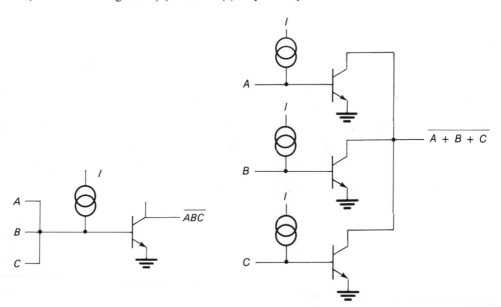

Fig. 2.21(d)   A three-input I^2L NAND gate   **Fig. 2.21(e) A three-input I^2L NOR gate**

2.7   METAL OXIDE SEMICONDUCTOR DEVICES (MOS)

2.7.1 MOS-FET Switch

Metal oxide semiconductor (MOS) devices are based on MOS field effect transistors (MOSFET). The FET is a unipolar device having one type of carrier. The carriers can be either holes for the *p*-type FET, or electrons for the *n*-type FET, but not both as in bipolar transistors. The MOSFET consists of a semiconductor substrate and a metal gate separated by an oxide insulator. In some manufacturing processes, polysilicon gates are used instead of metal gates but the overall behaviour is the same.

The FET under consideration is also known as an *insulated-gate* FET since the gate is separated from the semiconductor substrate by a thin layer of insulator. MOSFET is extensively used in digital circuit, since it requires a smaller silicon chip area than do bipolar transistors. This is a very attractive feature in MSI and LSI devices. Fig. 2.22 shows the basic structure of *n*-channel MOSFET. The substrate for the *n*MOS device may be at ground or negative (typically -4 V) potential. The substrate connection is not always shown.

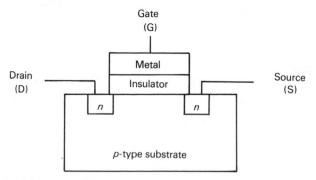

Fig. 2.22   Construction of enhancement-mode nMOSFET

The line between drain substrate and source is shown broken in Fig. 2.23(*a*) to emphasize that this path has no conduction. The gate is also shown separated, to indicate a very high resistance between gate and channel. Fig. 2.22 shows drain, source and gate physically separated. If the gate is made positive, the field from the gate draws electrons into the region between drain and source creating an *n*-channel for the electron carriers to flow. The gate voltage thus enhances the availability of electron carriers. This explains the name *enhancement-mode* MOSFET.

Fig. 2.23   Symbols for the FET In Fig. 2.22

It is possible to construct a depletion-mode MOSFET by diffusing n-type impurities between source and drain, i.e. creating a conducting channel, as in Fig. 2.24(a). Depletion-mode transistors turn on at zero gate voltage and require negative gate voltage to switch off.

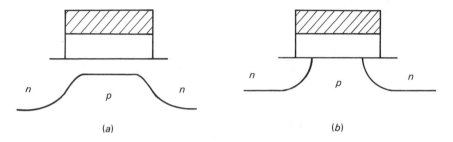

(a)   (b)

Fig. 2.24   Enhancement and depletion modes
(a) Depletion transistor
(b) Enhancement transistor

We note finally that, since electrons have greater mobility than holes, nMOS devices employing electrons as carriers are faster than p-type devices employing holes as carriers. Otherwise the p-channel and n-channel MOSFET operate in the same manner if the polarity of the supplies and the arrow on the symbol are reversed. In the n-channel shown the drain is made positive with respect to the source. To turn the transistor on, the gate is made positive. The enhancement mode nMOS is the type most commonly used in integrated circuits (ICs).

The MOSFET can be thought off as being a low resistance, 1 kΩ say, situated between drain and source when the gate voltage is high enough with respect to the source to switch the transistor on. If the transistor is off, the resistance is very high, 10^{10} Ω say, which is effectively an open-circuit. Hence the transistor can be used as a switch which is either on or off.

2.7.2   *n*MOS Logic Circuits

Logic circuits can be constructed using pMOSFETs, nMOSFETs or a combination of both. The last is known as complementary MOS (cMOS). The pMOS technology was the first to be developed. At present however nMOS technology is more widely used due to its superior qualities. As was already mentioned, nMOS devices employ electrons as carriers. Electrons have greater mobility which means that nMOS devices are faster than pMOS devices. Further, the p-channel devices, due to the low mobility have more area for the same resistance. In other words nMOS ICs can have higher packing density, which makes them more economical and attractive for LSI.

In the following, nMOS devices will be described, but it should be obvious that all these devices can be pMOS devices if the polarities of the supplies are reversed.

*n*MOS NOT Gate

The basic NOT (inverter) gate is made of two transistors, as shown in Fig. 2.25.

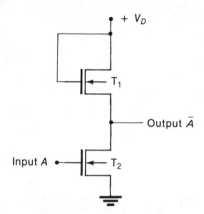

Fig. 2.25 *n*MOS NOT gate

T_1 acts as a the load for T_2 (pull-up transistor). T_1 is on since its gate is positive. If the channel width of T_1 is made narrower than that of T_2, the resistance of T_1 in the on condition becomes higher than that of T_2. By careful adjustment, it is possible to make the on resistance of T_1 about 10–100 times the on resistance of T_2.

If the input is low (logic 0), T_2 is turned off (open-circuit). The output is the same as that of V_D (usually 5 V), and the output is high (logic 1). If the input is high (logic 1), T_2 is turned on. The two transistors act as a potential divider. If the ratio of the two resistors is 100:1, then the output is 1/101 of V_D, i.e. the output is low (logic 0). The circuit is therefore a NOT gate.

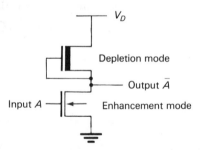

Fig. 2.26 **nMOS inverter employing depletion and enhancement techniques**

We note also that in some manufacturing processes the load transistor T_1 is a depletion type, as shown in Fig. 2.26. Depletion-mode transistors turn on at zero gate voltage (threshold voltage), and turn off at about -4 V. T_2 is still of the enhancement type and requires at least 1 V to turn on for a V_D of 5 V. This technique is widely used in industry. The depletion mode pull-up inverter (Fig. 2.26) is known to have better rising transients than the enhancement mode pull-up inverter (Fig. 2.25) ⟨78⟩.

nMOS NOR Gate

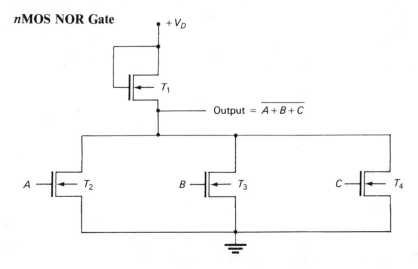

Fig. 2.27   A three-input NOR gate

Fig. 2.27 shows a three-input NOR gate using nMOSFETs. The three parallel transistors act as switches, with T_1 as the load resistor. If A,B or C is high (also if two or all the three) the corresponding transistor is on and the output is shorted to earth, i.e. the output is low. The output is therefore $= \overline{A+B+C}$

When all inputs are low (zero volts), the input transistors are off. Transistor T_1 is always on since its gate is positive, making the output high.

nMOS NAND Gate

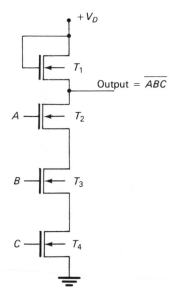

Fig. 2.28   A three-input NAND gate

The circuit shown in Fig. 2.28 is a three-input NAND gate. If A and B and C are all, high, the corresponding transistors T_2, T_3 and T_4 are on, the output is low. The output is high only if at least one input is low. If A is low, say, then T_2 is off, which causes the output to rise to V_D since T_1 is always on.

*n*MOS AND/OR Gates

AND gates can be obtained by inverting the output of NAND gates. Similarly OR gates can be obtained by inverting the output of NOR gates.

2.7.3 Complementary MOS Devices

Another type of MOS logic device is the complementary MOS (*c*MOS), which employs both *p*MOS and *n*MOS. Like other logic devices, the *c*MOS has its own advantages and disadvantages and these will be considered later. First, however, we shall look at the basic logic devices, the NOT, NAND and NOR gates.

*c*MOS NOT Gate

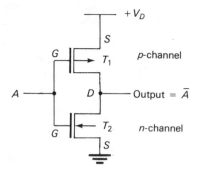

Fig. 2.29   A cMOS NOT gate

In the NOT gate of Fig. 2.29 we should remember that a positive (logic 1) input at A will turn T_2 on, while a low voltage (logic 0) will turn T_1 on. T_1 and T_2 cannot both be on simultaneously.

We need to remember also that the on resistance, of the transistor, is low (1 kΩ say) while the off resistance is virtually open-circuit (about 10^{10} Ω).

When A is high (logic 1) and equals V_D (typically 5 V) T_1 is off and T_2 on. The output is at earth potential (logic 0). If the input A is low (logic 0), T_1 is on and T_2 is off. The output is very nearly equal to V_D (logic 1).

The circuit inverts the input, which is a NOT function.

*c*MOS NOR Gate

The NOR gate in Fig. 2.30 is a simple extension of the inverter shown in Fig. 2.29.

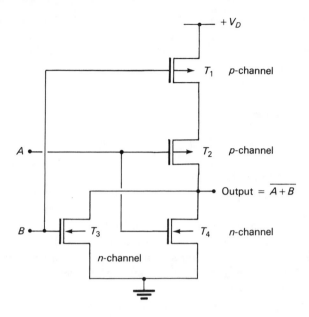

Fig. 2.30   A two-input NOR gate

If input A or input B (or both) is high (logic 1) the corresponding n-channel transistor is on and the p-channel transistor is off. The on transistor presents a low resistance between output and earth. The output voltage is low. The output is therefore $= \overline{A+B}$, which is a NOR operation. The output becomes high if the two inputs are low, since this causes the two p-channel transistors to turn on and the two n-channel transistors to turn off.

*c*MOS NAND Gate

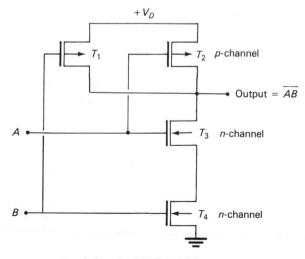

Fig. 2.31   A cMOS NAND gate

To see the function of the circuit in Fig. 2.31, consider the case when the two inputs A and B are both high (logic 1). The two n-channel transistors T_3 and T_4 are on while the two p-channel transistors T_1 and T_2 are off, causing the output to be low (logic 0). This is a NAND function.

2.7.4 Other Technologies

Besides the common devices already described, others are available to satisfy various requirements. Amongst these is the high performance MOS (hMOS), which is a scaled version of the nMOS. The improved performance is obtained by scaling down the dimensions, and this results in improved speed and power consumption and reduced parasitic capacitances. The smaller channel length reduces carrier transit time, provided that the channel is not too small for the gate to lose control. Lines of 1.2 and 1 micron are now — in 1984 — considered practical. Among the new products that make use of this new technology is the complementary high-performance MOS (chMOS). This is a new device which combines the low power consumption of cMOS with the high density and high speed of hMOS.

Many large semiconductor companies are investing in these fast devices, also known as very high speed integrated circuits (VHSIC), with military and space applications in mind. Recent technological developments and improved fabrication processes are helping this trend in scaling reach its physical limitations. Amongst these we mention the following:

(i)   The use of electron beams, ion beam and X-ray lithography which helps the production of smaller dimensions compared to the usual photolithography. Lithography decides the definition of circuit details on masks and wafers.

(ii)   The use of ion implant doping helps to produce small geometry devices by using low temperature processing and giving good junction definition compared to diffusion.

(iii)   The use of dry etching rather than wet etching also helps produce fine geometries.

In the field of material technology, gallium arsenide (GaAs) is making remarkable progress. GaAs is not expected to replace silicon, but it seems set to get a good share of the IC market by the end of this decade especially for high frequency (gigahertz) applications. The fastest gate array reported to date is an LSI GaAs 3.75×3.75 mm chip with 1050 gates (equivalent to 5000 transistors) with propagation time in use of 210 ps. The search will continue for the material and/or technology that is capable of producing the sub-micron, ultra fast, super VLSI chip of the 1990s.

Another type is the vMOS (V-groove MOS), where the transistor is formed on the side of a V-shaped groove. These devices have small channel length, high density and high input impedance. They are known to be capable of handling heavy current with high switching speeds, which makes them suitable for applications like switched-mode power supplies.

Finally we mention the silicon on sapphire (SOS) technology, in which cMOS devices are built on a sapphire (aluminium oxide) substrate instead of silicon substrate. Sapphire is a good insulator and cMOS/SOS devices have reduced parasitic capacitance. The pMOS and nMOS gates can be closer, resulting in higher density and better speed than with ordinary cMOS. SOS substrates have been the most successful of the silicon on insulator (SOI) family. They have not been applied to VLSI due to technological difficulties which result in low yield and high cost. Other SOI techniques employ buried dielectrics such as SiO_2 and Si_3N_4 by implanting oxygen or nitrogen ions. Researchers in the field of SOI claim that it has the following advantages over bulk single-crystal silicon:

1   increased circuit speed;
2   lower power dissipation;
3   reduced parasitic capacitance;
4   higher packing density for cMOS;
5   elimination of latch-up in cMOS.

2.7.5 Characteristics of nMOS Devices

Table 2.1 shows that the nMOS is slow compared with any bipolar device. The reason for this is that the output resistance in the high state is high and the input capacitance of the next stage (assuming nMOS as well) is also high. The combination of high output resistance and high load capacitance results in a slow switching time. This is really the main disadvantage compared with bipolar logic devices. The speed disadvantage is exaggerated for pMOS, which is even slower because of the low mobility of the carriers. Other properties of nMOS devices are their good noise immunity, which is in the region of 1 V, and their high fan-out capability.

Another important property, which is common to all MOS devices, is their very low power dissipation. This, together with the fact that they use no resistance, makes them very suitable for large-scale integration. In MOS technology a resistor can take five to ten times the area of a transistor in integrated circuits. Thus nMOS devices are very popular for LSI but their low speed makes it difficult for them to compete with fast bipolar devices like STTL or ECL in SSI. This explains why in Table 2.1(a) LSI was inserted to describe their function. At present, nMOS technology represents nearly 50 percent of the world consumption. Table 2.1(b) compares the properties of the three main technologies currently employed for the fabrication of LSI devices.

Properties	Technology		
	nMOS	cMOS	Bipolar
Speed	+	+	+ +
Power dissipation	+	+ +	−
Noise margin	+	+ +	−
Drive capability	−	+	+ +
Logic swing	−	+ +	−
Density	+ +	+	−
Suitability for scaling down	+	+ +	−

Table 2.1(b)   Relative merits of LSI technologies

2.7.6 Characteristics of cMOS Devices

All cMOS employs both p- and n-channel transistors, which makes it more complex and have a lower packing density than nMOS. The devices do however have good characteristics which makes them suitable for SSI and comparable with small-scale bipolar logic devices. In fact they are manufactured to be pin-for-pin TTL compatible. Nowadays cMOS and TTL are the most widely used SSI devices.

Amongst the main advantages of cMOS over nMOS devices in the SSI scale is their speed. This is achieved due to the low output resistance in the high state. In cMOS the output resistance in the high state is that of the pMOS transistor, which is about 1 kΩ. With nMOS the resistance is made about 10 to 100 times higher, as mentioned earlier. A low output resistance means faster switching, assuming the input capacitance of the driven stage is the same. At the time of writing (1984), the speed of cMOS devices is expected to fall below 2 ns soon.

A typical value of fan-out is given as 20 in Table 2.1(a); this could go up to 50 if speed is not important. Propagation delay is about 20 ns for say two loads, but this increases by about 5 ns for each 5 pF load. The power dissipation of cMOS devices is very low, being even lower than equivalent nMOS devices. The reason for the very low power dissipation can be understood by looking at any of the cMOS circuits already considered. We see that in any logical condition there is at least one transistor which is off, and hence no large current can flow from the supply to earth. Another advantage of cMOS devices is that they accept a wide variation in power supply, a useful feature if low-cost power supplies are used. The cMOS devices also enjoy good noise immunity, which makes them suitable for noisy industrial environment. They are, however, suceptible to static charge build-up and must be handled with care. New devices are protected against static charge by diodes on each input.

Large investments by major companies and improvements in manufacturing techniques suggest that cMOS will be the technology for both small- and large-scale integration for the rest of the decade.

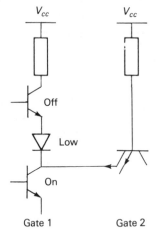

Fig. 2.32(a)   Current sinking

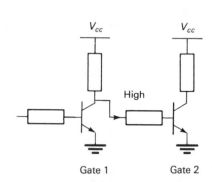

Fig. 2.32(b)   Current sourcing

2.8   INTERFACING OF LOGIC MODULES

2.8.1 Current Sinking and Sourcing

With two gates connected as shown in Fig. 2.32(*a*), gate 1 acts as a driver. When the driver provides a low input to the driven gate (gate 2; i.e. *the output of the driver is low*) gate 1 is said to be *sinking* current from gate 2. Gate 1 should be capable of sinking this current without being damaged or changing the logic level.

In Fig. 2.32(*b*) the first gate is supplying current to drive gate 2. The output of gate 1 is high and current flows from gate 1 to the input of gate 2. This is known as current sourcing. Care must be taken to ensure that a gate used as a driver can provide enough current to operate the driven gate.

2.8.2 cMOS Driving TTL

TTL logic is accepted as a standard, and is still the most commonly used. But *c*MOS, as was pointed out, is a new technology that has many advantages such as low power dissipation and good noise immunity. It is also suitable for small- and medium-scale integration and many types of gate are produced using *c*MOS. In some applications, it is desirable to combine the properties of both types of logic, using TTL, where speed is important, and *c*MOS, where power dissipation must be low. Manufacturers are aware of this situation and many *c*MOS devices are compatible with TTL. Still it is useful to look at the problems one might face when interfacing TTL and *c*MOS. A *c*MOS device accepts power supplies from 3 to 15 V, and TTL requires 5 V. Therefore a 5 V power supply can be used for both.

If the output of the *c*MOS device is high, its value is about $+5$ V. This is acceptable as a high input for the TTL stage. The TTL current is then a maximum of

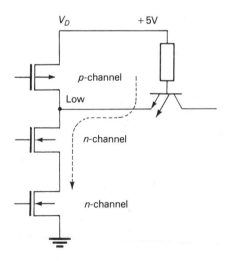

Fig. 2.33   cMOS driving TTL

40 μA, which can be supplied by the cMOS gate through the on resistance of the p-channel transistor. When the output of the cMOS device is low as shown in Fig. 2.33 the cMOS device has to sink a current of up to 1.6 mA from the TTL gate. This means that the cMOS device must sink 1.6 mA through the on resistance of the n-channel FET. (The on resistance varies from 100 ohms to 5 kilohms.) The voltage drop could make the output voltage of the driver too high to satisfy the low input voltage of the TTL gate, which is about 0.8 V.

If a 0.4 V noise margin is allowed, the output of cMOS must be below 0.4 V when 1.6 mA flows in. Special devices called *buffer* cMOS can be used between cMOS and TTL to overcome this problem.

2.8.3 TTL Driving cMOS

First consider the case when the output of the TTL driver is low. A low output is about 0.2 V (0.4 V maximum); cMOS gates accept up to 1.5 V as a low input. Hence there is no problem.

If, on the other hand, the TTL gate has a high output, the high voltage is not 5 V as one might expect, but about 3.6 V or maybe as low as 2.4 V. This is due to the voltage drop across the transistor T_3 and the diode, as explained earlier (Fig. 2.34). The cMOS gates require about 3.5 V as a high input. Therefore even though 3.6 V is high enough, the noise margin is reduced to 0.1 V, which is not very satisfactory. Normally, an external pull-up resistor is used to raise the output of the TTL to about 5 V. This explains why open-collector TTL gates are better than totem-pole outputs for driving cMOS gates.

ECL gates are not compatible with other types of gates, but nMOS and pMOS are compatible and can be interfaced with cMOS. As a rule, one should check if interfacing circuits are required before connecting gates of various types.

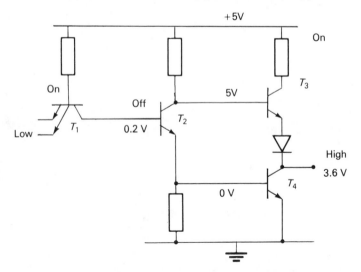

Fig. 2.34   Showing the high output of a TTL driver

2.8.4 Loading Requirements

The manufacturers' data books give adequate information and they should be consulted. For logic gates the following definitions are given in the Texas Instruments TTL data book (reproduced by courtesy of Texas Instruments).

High-level Input Current I_{IH}
The current into an input when a high-level voltage is applied to that input.

High-level Output Current I_{OH}
The current into an output with input conditions applied that according to the product specification will establish a high level at the output.

Low-level Input Current I_{IL}
The current into an input when a low-level voltage is applied to that input.

Low-level Output Current I_{OL}
The current into an output with input conditions applied that according to the product specification will establish a low level at the output.
Note The current out of a terminal is given as a negative value.

High-level Input Voltage V_{IH}
An input voltage within the more positive (less negative) of the two ranges of values used to represent the binary variables.

High-level Output Voltage V_{OH}
The voltage at a output terminal with input conditions applied that according to the product specification will establish a high level at the output.

Low-level Input Voltage V_{IL}
An input voltage level within the less positive (more negative) of the two ranges of values used to represent the binary variables.

Low-level Output Voltage V_{OL}
The voltage at an output terminal with input conditions applied that according to the product specification will establish a low level at the output.
For the TTL 7400, 7404, 7410, 7420, and 7430 the following data are given

$$
\begin{aligned}
I_{IH} &= 40 \ \mu A & I_{OH} &= -400 \ \mu A \\
I_{IL} &= -1.6 \ mA & I_{OL} &= 16 \ mA \\
V_{IH} &= 2 \ V(min.) & V_{OH} &= 2.4 \ V(min.) \\
V_{IL} &= 0.8 \ V(max.) & V_{OL} &= 0.4 \ V(max.)
\end{aligned}
$$

Typical values for the output voltages (supply voltage is 5 V), are given as V_{OH} = 3.4 V and V_{OL} = 0.2 V. These are better than the worst-case conditions given above but they show that the output in the high state is not the ideal 5 V or even 3.6 V as calculated in Fig. 2.34. For a reliable design the designer must take the worst possible values that are guaranteed as acceptable high or low logic levels.

If two gates are connected the output of the driver is connected to the input of the driven gate. For reliable operation then V_{OL} of the driver must be less than V_{IL} of the driven gate. If they are equal the circuit might function correctly, but the low d.c. noise immunity is reduced to zero.

In the high state V_{OH} of the driver must be greater than the V_{IH} of the driven gate. Again, if they are just equal, the high d.c. noise immunity is reduced to zero.

In addition to the above-mentioned voltage requirements, attention must be paid to the flow of currents. In brief, the driver should be capable of sinking (sourcing) the appropriate amount of current from (to) the driven gate (or gates) without changing the corresponding voltage levels.

If these requirements cannot be satisfied the designer must consult the interface circuits data book for a suitable buffer.

2.8.5 Unused Inputs

In Chapter 1 unused inputs were connected to a used input. This is a common practice and can be used for all gates, though the fan-out capability of the driving gate must be kept in mind. Another possibility is to connect the unused inputs of AND and NAND gates to a high logic level. For TTL gates, for example, the unused inputs are connected to the 5 V supply through a 1 kΩ resistor. Similarly the unused inputs of OR and NOR gates can be connected to the low logic level (normally ground). It is not desirable to leave unused inputs of gates floating, for stray signals may be picked up which may cause improper operation.

2.9   WIRED-OR CONNECTION

In two-level gate circuits, the output gate can, in some cases, be removed and the outputs of the first level connected directly. This result in what is commonly called *wired-OR connection*.

Consider the logic diagram in Fig. 2.35(*a*).

With the output gate removed, the logic diagram is shown in Fig. 2.35(*b*).

It can be seen that if the output of one of the two NAND gates becomes low, it will pull the other output low. The output is high only if the outputs of both gates are high, which is equivalent to the AND function. (In a DTL NAND circuit the output is high when the output transistor is off. See Section 2.3.)

Similarly if two NOR gates are connected as shown, the output is determined by the AND function of the two NOR gates, which results in a simple input expansion. (See RTL NOR circuits, where the output is high when the transistor is off.)

We note however, that when circuits are connected in this way their collector resistors are effectively in parallel; therefore, when a large number of gates are con-

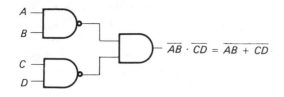

$$\overline{AB} \cdot \overline{CD} = \overline{AB + CD}$$

Fig. 2.35(a)   Two levels of logic gates

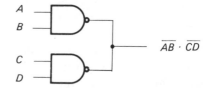

$$\overline{AB} \cdot \overline{CD}$$

Fig. 2.35(b)   Wired–OR connection of the NAND gates in Fig. 2.35(a)

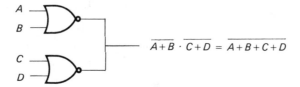

$$\overline{A+B} \cdot \overline{C+D} = \overline{A+B+C+D}$$

Fig. 2.35(c)   Wired–OR connection of NOR gates

nected in parallel, their effective collector resistance may become two low. The answer to this problem is that open collector gates are used in these situations with an external resistor.

Now let us consider connecting the outputs of ECL gates, remembering that an emitter follower stage is used. The output is high when the transistor is on (in Fig. 2.20 T_6 is on and its emitter is high). The output is low when the transistor is off. Therefore if the outputs are tied together, it will go high if any output is high. Thus when ECLs are connected in a wired-OR configuration, the result is an OR connection. Hence if two 2-input OR gates are connected the output is $(A + B) + (C + D)$. If two 2-input NOR gates are connected, the output is $\overline{(A + B)} + \overline{(C + D)}$.

It should be emphasized, however, that not all gates can be connected this way. Consider the totem-pole* outputs shown in Fig. 2.36, where transistor T_4 can sink

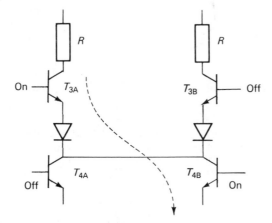

Fig. 2.36   The outputs of two totem-pole TTL gates tied together

* See Section 2.4.2

about 16 mA. If the output of gate A is high while that of gate B is low, a current as high as 50 mA can flow as shown from the supply through T_{3A} and T_{4B}, and T_{4B} could be damaged. This can be worse if more gates are connected.

However open collector gates allow the connection. In this case T_3 and its collector resistor R are eliminated (the output is open), and an external resistor is connected as shown in Fig. 2.37. But despite what has been said above, the wired-OR connection is not considered good practice and should be avoided where possible.

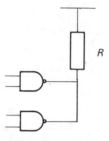

Fig. 2.37   Two open-collector gates wired to give an AND function

2.10   TRI-STATE GATES

A tri-state TTL gate was described earlier. These devices are available as bipolar and MOS and are widely used in microprocessors and computers. Fig. 2.38 shows a tri-state buffer gate, whose output can be 0, 1, or open-circuit depending on the control line, as shown in Table 2.2. Note that the control line is negated compared with Fig. 2.17.

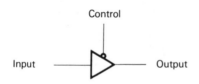

Fig. 2.38   A tri-state buffer

Input	Control	Output
0	1	Open
1	1	Open
0	0	0
1	0	1

Table 2.2   Truth table for the tri-state buffer

One common application for these devices is for the transfer of data between various parts of a computer over a common bus line, or even from one computer to another. Suppose that in Fig. 2.39 the control lines are:

$$a = 1 \qquad a' = 1$$
$$b = 0 \qquad b' = 1$$
$$c = 1 \qquad c' = 0$$

Then only registers B and C′ are connected to the bus system and data can pass from register B to C′. Other registers are effectively disconnected (their output is neither high nor low but open-circuit).

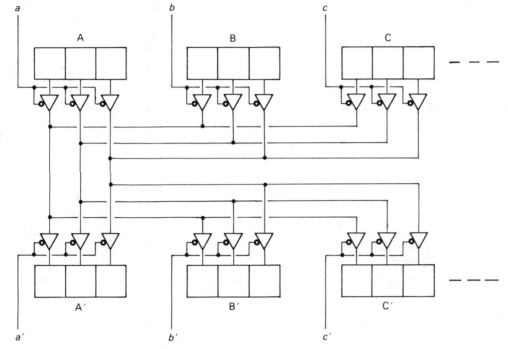

Fig. 2.39 Tri-state buffer gates used in a data transfer arrangement across a commom bus

2.11   READ ONLY MEMORY (ROM)

A semiconductor read only memory (ROM) is an LSI device consisting of two main parts, the address decoder with n inputs to decode the inputs into 2^n address lines, and a memory containing 2^n words of m bits each for an m output ROM. The total memory is $m \cdot 2^n$ bits. Beside being able to store information as a memory, a ROM is also used as a code convertor, character generator, look-up table and for the implementation of logic functions.

The ROM shown in Fig. 2.40 is an array of diodes (or transistors). The information contained in the ROM is specified by the interconnection pattern. The data are programmed by including or omitting a small conducting jumper between the diode and the bit line. The first word in this case contains the pattern 0110, which can be read by applying 000 at the address lines A_1, A_2 and A_3 (make line 0 high).

ROMs may be fabricated by either bipolar or MOS technology. In MOS ROMs the basic storage element is an MOS transistor. The physical presence of a transistor in a given location indicates a logical 1 and its absence indicates a logical 0. This is achieved by changing the photomask in the manufacturing process. In the bipolar ROM, each bit is represented by a bipolar transistor and connecting or disconnecting the emitter lead indicates logic 1 or 0 respectively.

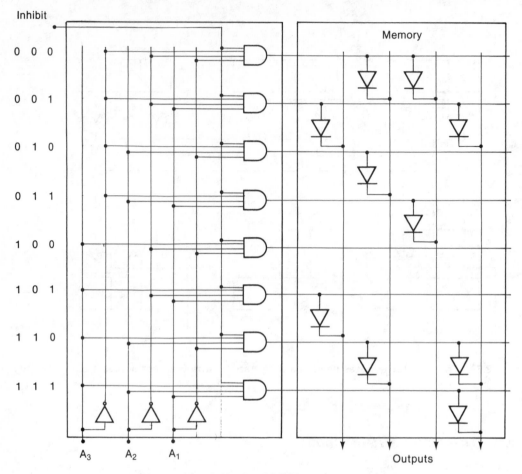

Fig. 2.40   8 x 4 ROM store

The user is required to fill in a truth table as shown in Table 2.3. The manufacturer prepares a mask pattern to program the ROM. They are available in standard sizes like 256×4 (1K) bits, 2048×8 (16K) bits, 256K bits, up to 1M bits.

ROMs can be connected in order to increase the number of words, the word length or both, as illustrated in Fig. 2.41. Memory expansion also applies to other devices like RAMs. Speed (access time) is in the region of 50–90 ns. Power dissipation is in the region of 200–500 mW/chip.

The fact that ROMs are programmed by the manufacturer, who creates a special mask for each program, means that ROMs can be expensive unless they are ordered in large quantities. Further, the user must make sure that his program is correct since further changes can be very costly. One way of achieving this is by trying the program on an EPROM (to be described in Section 2.13), before designing the final mask.

Apart from the fact that the program is permanent and cannot be modified, the ROMs are non-volatile devices, which simply means that the program, or data, stored is not lost when power is switched off. ROMs also have non-destructive read-

Word	Inputs							Outputs			
	A_5	A_4	A_3	A_2	A_1	E		O_4	O_3	O_2	O_1
0	0	0	0	0	0	0					
1	0	0	0	0	1	0					
2	0	0	0	1	0	0					
3	0	0	0	1	1	0					
4	0	0	1	0	0	0					
5	0	0	1	0	1	0					
6	0	0	1	1	0	0					
7	0	0	1	1	1	0					
8	0	1	0	0	0	0					
9	0	1	0	0	1	0					
10	0	1	0	1	0	0					
11	0	1	0	1	1	0					
12	0	1	1	0	0	0					
13	0	1	1	0	1	0					
14	0	1	1	1	0	0					
15	0	1	1	1	1	0					
16	1	0	0	0	0	0					
17	1	0	0	0	1	0					
18	1	0	0	1	0	0					
19	1	0	0	1	1	0					
20	1	0	1	0	0	0					
21	1	0	1	0	1	0					
22	1	0	1	1	0	0					
23	1	0	1	1	1	0					
24	1	1	0	0	0	0					
25	1	1	0	0	1	0					
26	1	1	0	1	0	0					
27	1	1	0	1	1	0					
28	1	1	1	0	0	0					
29	1	1	1	0	1	0					
30	1	1	1	1	0	0					
31	1	1	1	1	1	0					
All	x	x	x	x	x	1		1	1	1	1

(x represents DON'T CARE)

Table 2.3   Truth-table/order-bland for a 32 × 4 ROM

out and random access; as the name implies, once they are programmed one can only read from them.

Timing Waveforms

When using memory devices with other electronic devices one must take into account timing problems, and allow for signal delays even though they are measured in nano seconds.

Fig. 2.41 shows that different memory modules may be selected and de-selected at different times to enable and disable the data output lines. These introduce chip select time delay t_s and chip deselect time delay t_d. Another useful parameter is the

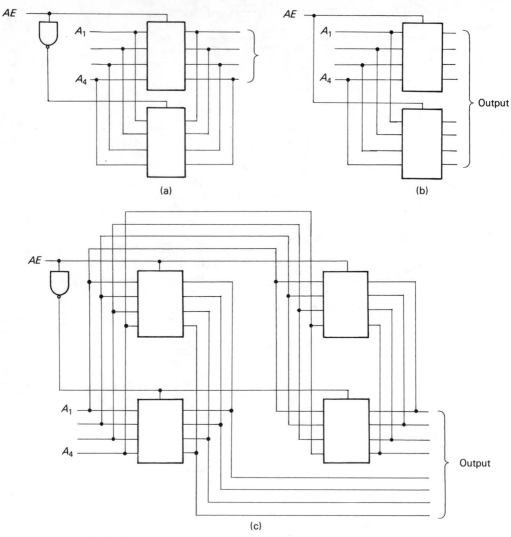

Fig. 2.41   Cascading of ROM modules

(a) Two **16** × **4** devices connected to form a **32** × **4** memory
(b) Two **16** × **4** devices connected to form a **16** × **8** memory
(c) Four **16** × **4** devices connected to form a **32** × **8** memory

access time t_a, which is the time between the application of the address and the presence of the valid data at the output. Fig. 2.42(a) shows that the output data will be available at the output of the memory device at a time t_s after the chip is selected and a time t_a after the address is applied. if the chip is de-selected, it takes a time t_d before the date disappears.

The access time is usually larger than the chip select time, and Fig. 2.42(b) shows what happens if the chip is selected first. The previously addressed date will simply appear at the output, and at a time t_a later the new data will appear.

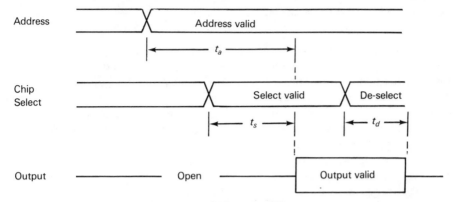

Fig. 2.42(a)   ROM timing diagram

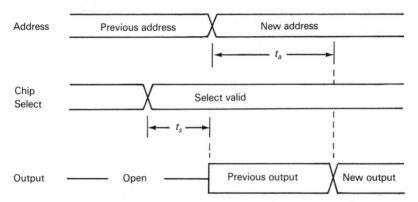

Fig. 2.42(b)   Timing diagram showing early chip selection

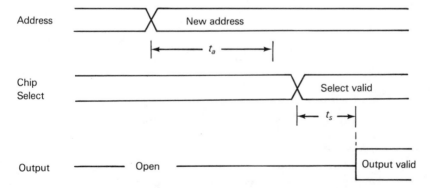

Fig. 2.42(c)   Timing diagram showing late chip selection

Fig. 2.42(c) on the other hand shows that the address may be valid, but no data can be seen until a time t_s after the chip is selected.

2.12   PROGRAMMABLE READ ONLY MEMORY (PROM)

Fig. 2.43 shows a simplified block diagram of a programmable read only memory (PROM). These are similar to ROMs and the memory may be a diode or a transistor array, but the memory includes a fusible link of nichrome or polysilicon and programming is achieved by blowing the links by passing high current pulses (typically 20–30 mA). A blown link represents a logic 0. This is not always so, however, since PROMs are available with platinum silicide fusible links in which the intact fuse represents logic 0, and 1s are programmed by applying appropriate voltages. This facility allows programming to be done by the user instead of the manufacturer. As in the ROM, the program cannot be changed. PROMs are usually bipolar and like ROMs, they are available in standard sizes like 1 K × 4, 1 K × 8, 4 K × 8, and are usually exchangeable with ROMs (same pin-out and packaging configuration).

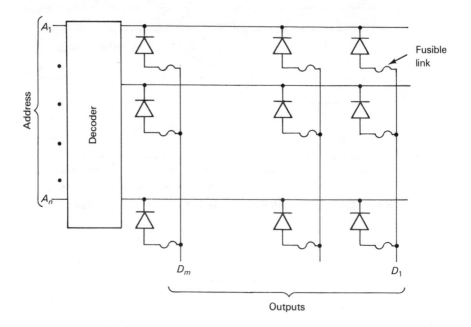

Fig. 2.43   A simplified block diagram of a PROM

Access to the memory is achieved via the address decoder as in the ROM. Power dissipation is about 600 mW, and access time varies from 35 ns for bipolar PROMs up to 450 ns for cMOS PROMs.

MOS PROMs are susceptible to static discharge and special care must be taken when handling these devices. This includes grounding all equipment, touching ground before handling the devices, and using specially designed carriers.

2.13   ERASABLE AND PROGRAMMABLE ROM (EPROM AND EEPROM)

Like a ROM and PROM, the device consists of a memory and an address decoder. EPROM offers greater flexibility in that the program can be erased and the device reprogrammed. The storage cell is a FET with a floating gate. The programming pulse has the effect of injecting electrons in the gate, thus turning the transistor on (a drain source channel is established). The program is non-volatile, but can be erased by exposing the floating gate to ultraviolet radiation which clears the charge and hence the written data. This procedure erases all the program. The device can be reprogrammed.

EPROMs are available in standard sizes and, like other memory devices, are getting faster, larger and cheaper. Sizes vary from $\frac{1}{4}$ K $\times$ 8 up to 512 K with access time of the order 150–450 ns, though one supplier reported a 1 K $\times$ 8 device with 40 ns access time using silicon on sapphire cMOS. A new addition to this family is the electrically erasable PROM (EEPROM), some of which can be powered by a single 5 V supply. Its main attraction is that the user can erase and program selected locations instead of erasing all the memory. They are available as 82 $\times$ 1, 32 $\times$ 16, 64 $\times$ 8, 1 K $\times$ 4, 2 K $\times$ 8, up to 8 K $\times$ 8. Erase plus write time is about 10 ms, access time is 250 ns (typically). The device is also known as Electrically Alterable ROM (EAROM).

2.14   RANDOM ACCESS MEMORY (RAM)

Read-write random access memories are usually divided into two types, according to the storage mechanism used in the memory cells. Dynamic cells store information in the form of a charge on capacitors. Because of charge leakage, dynamic cells are periodically refreshed in order to maintain the stored data. The other type of RAM is static, in which the storage device is a flip-flop, and hence needs no refreshing. RAMs may replace ROMs in most applications. They have the advantage that information can be read from them and written into them, as compared to reading only in ROMs. Their main disadvantage is that they lose their stored data when power is switched off. Back-up batteries can be used when volatility of memory data is a problem.

RAMs can be used for temporary storage of programs and data in computers. Fig. 2.44 shows a block diagram of a memory device with three address lines (8 words) and four outputs (4 bits per word). Each word can be addressed separately. Data can be written into or read from the addressed location by applying the appropriate signal at the Read-Write line (R/W). If the address lines A_1 A_2 A_3 are 000, word '0' will be selected. A '0' on the R/W line enables the input buffer and disables the output buffer (tri-state devices). The data at the input will be written into the device. If a 1 is now applied to the R/W line the output buffer is enabled, while the input buffer is disabled, and the content of word '0' appears at the output. Fig. 2.45 is the logic diagram of a commercial 32 $\times$ 2 RAM. Words are selected through the five-input decoder when the read–write enable input CE is at logic 1. $\overline{WS_0}$ and

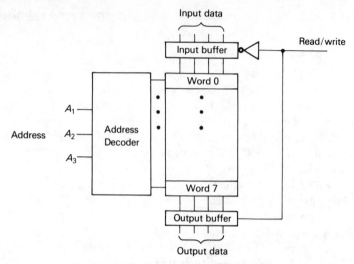

Fig. 2.44   A block diagram of a RAM

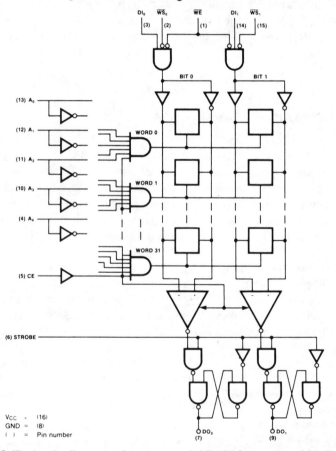

Fig. 2.45   Logic diagram of a commercial RAM (courtesy of Mullard)

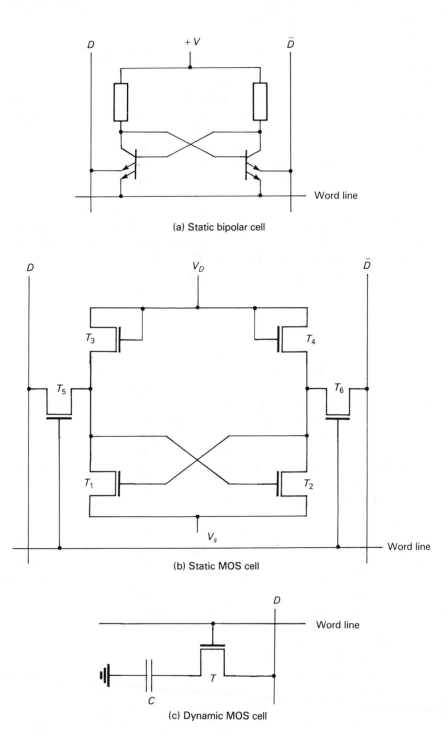

(a) Static bipolar cell

(b) Static MOS cell

(c) Dynamic MOS cell

Fig. 2.46   Three different types of RAM cell

$\overline{WS_1}$ are the write inputs for bit 0 and bit 1 of the word selected. $\overline{WE}$ is the write control input.

With $\overline{WS_N}$ and $\overline{WE}$ at logic 0, data on the DI_0 and DI_1 data lines are written into the addressed word. The read function is enabled when either $\overline{WS_N}$ or $\overline{WE}$ is at logic 1.

In the RAM shown, it is possible to write while reading. An internal latch is on the chip to provide the write-while-read facility. When the latch control (strobe) is logic 1 and the data are being read from the device, the latch is effectively bypassed. The data at the output will be those of the addressed word. When the strobe goes to logic 0 the outputs are latched and will remain latched regardless of the state of any other address or control line. When the strobe goes to logic 1 the outputs unlatch and the outputs will be that of the present address word.

Dynamic RAMs employ MOS technology, while static RAMs are made using either MOS or bipolar techniques, as shown in Fig. 2.46. Access time is of the order of 30−300 ns for static RAMs and 100−300 ns for dynamic RAMs. Typical sizes for dynamic RAMs are 16K × 1, 32K × 1 and 64K × 1; larger devices like 8K × 8 and 16K × 4 are becoming available. Static devices are available as 1K × 1, up to 16K × 1; 1K × 4, 1K × 8, 4K × 4, 2K × 8, etc.

2.14.1 RAM Cells

(a) Static Bipolar Memory cell

Fig. 2.46(a) shows a simple cell for a static RAM employing bipolar transistors. All cells in a column share the same digit line and all cells in a row share a common word line. Digit lines are normally held at a higher potential than word lines. Reading is achieved by raising the potential of the word line causing one of the digit line emitters to conduct current to the digit line. The information stored in the cell can then be detected by the sense amplifier (not shown). When writing, the word line potential is raised and the potential, of one digit line is lowered causing the flip-flop to assume the appropriate state.

(b) Static MOS Memory cell

Fig. 2.46(b) shows a static MOS cell using six FETs. T_1 and T_2 form the basic storage cell. T_3 and T_4 replace the load resistors used in bipolar devices. T_5 and T_6 are used to connect the outputs of T_1 and T_2 to the digit lines. To read the cell, the word line is used to turn on T_5 and T_6. The stored data are transfered to the digit line. To write into devices, T_5 and T_6 are turned on by the word line, enabling T_1 and T_2 to assume the logic states on the digit lines.

(c) Dynamic MOS Memory Cell

Fig. 2.46(c) shows a simple cell using one transistor. Some dynamic RAMs use more complicated cells but the idea is the same. The simplicity of the cell and the fact that fewer FETs are used compared with static devices means that higher densities are possible. Dynamic cells must be refreshed periodically (every few milliseconds) to replace charge leakage.

Writing is achieved by applying an appropriate signal on the word line to activate the transistor. The data on the bit line are now transfered to the capacitor by means of the gate transistor. Reading is achieved by activating the transistor using the word line. This connects the capacitor to the bit line. The charge on the line is then sensed.

The name 'random access' is rather misleading since ROMs and PROMs are also random access devices. It is probably better to call them read/write devices to distinguish them from the read only memories. RAMs are volatile devices, and the program can be erased by switching the power off. They can, therefore, be used as scratch pads and for temporary storage.

2.14.2 New RAMs and the non-volatile RAM

The largest of the RAM family, at the time of writing (1984), are the 64 K static RAM and the 256 K dynamic RAM. The 256 K dynamic RAMs usually operate from a five-volt supply, have a typical access time of 80 ns and 4 ms refresh cycle. The full speed power consumption varies from 170 mW to 350 mW. In terms of speed, the fastest reported 1 K memory is a 1 K × 1 static RAM made using gallium arsenide (GaAs) technology. This has a six-transistor cell with access time of 2 ns for 500 mW power consumption. The fastest silicon device is the 4 K ECL RAM with access time of 3.2 ns for 3000 mW power consumption. Another fast device is the 64-bit ECL RAM with 3 ns access time and 700 mW power consumption. A 4 K × 1 *n*MOS RAM has 5 ns access time for 350 mW power consumption. The fastest reported 64 K × 1 static I²L RAM has 25 ns access time for 270 mW power consumption. The fastest reported 64 K × 1 static *n*MOS RAM has 40 ns access time for 400 mW power consumption. The fastest reported 64 K × 1 static *c*MOS RAM has 50 *n*s access time for 100 mW power consumption.

A new useful device available now is the non-volatile RAM (NV-RAM). In the NV-RAM, the cell has a shadow non-volatile storage transistor. The data are written in the cell as usual but can be transferred to the non-volatile storage transistors when an enable signal is received. The main disadvantage of the NV-RAMs is that they require about five times the chip area of the normal RAM. Currently they are available as $\frac{1}{4}$ K × 8 and $\frac{1}{2}$ K × 8.

Another device which combines the high performance of *h*MOS and low power consumption of *c*MOS is the recently announced *ch*MOS dynamic RAM. Intel claim that their 256 K *ch*MOS dynamic RAMs (*ch*MOS DRAMs) consumes 90 percent less power than previous 64 K *n*MOS dynamic RAMs.

2.14.3 Timing Diagrams

The idea of the timing diagram for RAMs is similar to that of a ROM except that the RAM has both a read and a write cycle while the ROM has only a read cycle.

In Fig. 2.47 it is assumed that the read/write (R/W) input is at logic 1 for the read state and logic 0 for the write state. No chip select line is shown in Fig. 2.47; it is

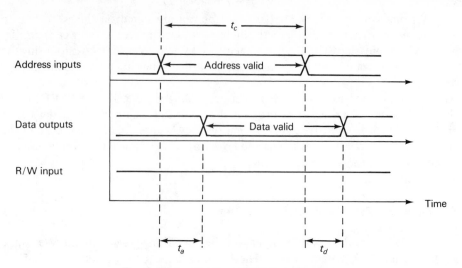

Fig. 2.47(a)   Read-cycle timing diagram

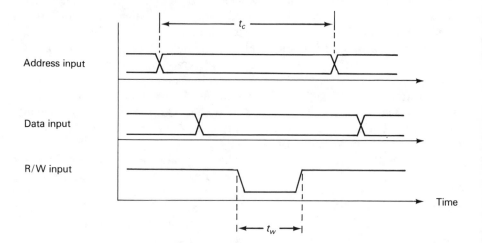

Fig. 2.47(b)   Write-cycle timing diagram

assumed that the chip is selected if multi-chips are used or that the R/W line is acting as a select line for either writing or reading.

In the read cycle of Fig. 2.47(*a*):

t_c is the minimum cycle time.

t_a is the access time.

t_d is the time old data remain valid after changing the address.

In Fig. 2.47(*b*) the R/W line is lowered to zero for writing a period t_w, which is the write-pulse width. The access time was defined earlier as the time between the application of the address and the presence of valid data at the output. The read cycle is the time between successive read cycles. The cycle time is defined as the minimum duration between successive read or write operations. For writing, both address and data must be stable before applying the write pulse.

2.15   PROGRAMMABLE LOGIC ARRAY (PLA)

The PLA consists of three main sections, as shown in Fig. 2.48. The decoding array is simply a matrix to form the AND products. The memory array forms the OR of these product terms. The connections at the cross-points of the matrix are usually diodes or transistors. The PLA can be considered as a ROM with programmable address. The difference is that in a ROM all combinations of inputs produce an output, whereas in a PLA some combinations may have no effect, and some groups of combinations may be indistinguishable. In other words, it is not necessary to program all the possible states.

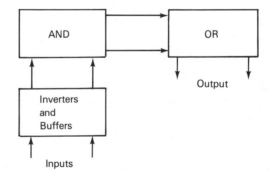

Fig. 2.48   A block diagram of a PLA

PLAs can be used to implement logic functions, code conversion and storing of microprogram instructions. The programming is done by the manufacturer according to a truth table supplied by the designer. Appendix 3.2 shows a typical mask-programmable logic array.

Table 2.4 shows a truth table for a module of 14 inputs, 8 outputs and 48 unique product terms. A ROM with 14 inputs will have 2^{14} product terms. Typical access time is 50 ns, and typical power dissipation is about 500 mW.

Field programmable logic arrays (FPLA) are also available, and can be programmed by the user. As an example, suppose a particular product term P_i contains inputs I_j, then the $\bar{I}_j$ are fused by passing a certain current. A FPLA with 16 inputs, 8 outputs and 48 product terms is shown in Fig. 2.49. This module employs fusible links as the programming elements, FPLAs are also available with 12 inputs, 6

Product Terms	Inputs															Output							
	I_{14}	I_{13}	I_{12}	I_{11}	I_{10}	I_9	I_8	I_7	I_6	I_5	I_4	I_3	I_2	I_1	O_8	O_7	O_6	O_5	O_3	O_3	O_2	O_1
1																						
2																						
3																						
4																						
5																						
.																						
47																						
48																						

Table 2.4   A truth table for a PLA

outputs and 50 product terms; 14 inputs, 8 outputs and 48 product terms; and 16 inputs, 8 outputs and 48 product terms. The output can be programmed to be active low or active high. This is useful if the complement of the output contains fewer product terms. The fuse at the input to the EX-OR in Fig. 2.49 is blown for the complemented output.

Besides their use for random logic, PLAs are also used as macro-cells for VLSI applications. The main advantage of PLAs is their regular structure, which results in fast design turnaround and low cost. Their main disadvantage, however, is their poor silicon utilization due to the large percentage of unused devices when the circuit is personalized for a particular application. This can be partially remedied by using Boolean algebra and other classical minimization techniques. A more powerful approach is the segmentation and sharing of the product term lines by several product terms and/or folding and sharing the input output lines. That is outside the scope of this book, but the interested reader can find further details in the appropriate references ⟨52,94⟩.

2.16   PROGRAMMABLE ARRAY LOGIC (PAL)

This is the latest of the programmable logic family and aims to replace conventional TTL logic. The basic PAL implements the logic function as a sum of products by using a programmable AND array which feeds a fixed OR array.

The accepted notation is to use an 'X' to represent an intact fuse. Using this notation, the AND of two inputs is shown in Fig. 2.50(c). Though one common line is used, the input terms with Xs are not physically connected. In other words, Figs 2.50(a), 2.50(b) and 2.50(c) are equivalent.

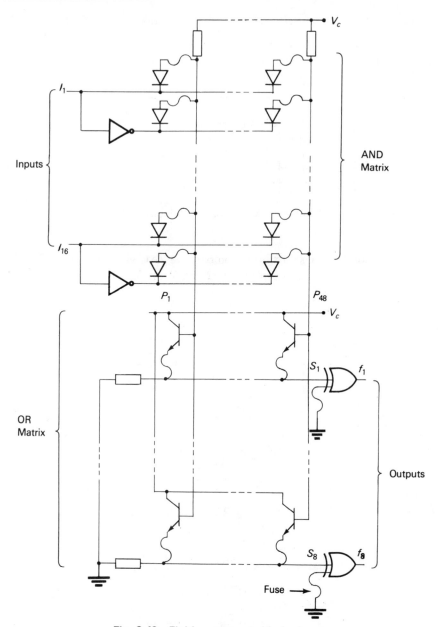

Fig. 2.49   Field programmable logic array

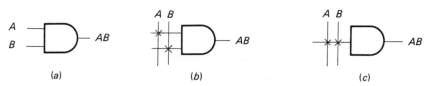

Fig. 2.50   Symbols for a two-input AND

For a logic function $f = AB + \bar{A}\bar{B}$, the PAL Logic diagram is shown in Fig. 2.51.

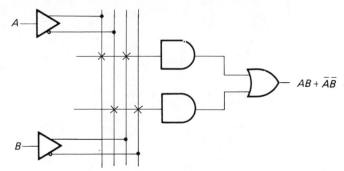

$$AB + \bar{A}\bar{B}$$

Fig. 2.51   PAL implementation of a two-input logic function

Using this notation, it is possible to compare the PAL with other programmable logic devices. This can be seen from Fig. 2.52 and Table 2.5.

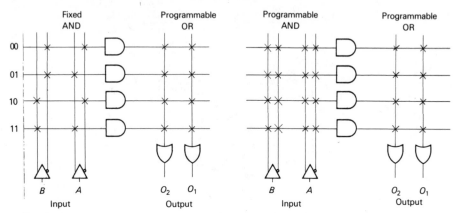

Fig. 2.52(a)   4-word, 2-bit PROM   Fig. 2.52(b)   2-input, 2-output, 4-product FPLA

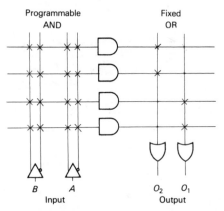

Fig. 2.52(c)   2-input, 2-output, 4-product PAL

Device	AND	OR	Output
PAL	Programmable	Fixed	TS,R,FB, I/O
FPLA	Programmable	Programmable	TS,OC,FP
PROM	Fixed	Programmable	TS,OC

TS	= Tri-state	I/O	= Input/Output
R	= Registered	OC	= Open collector
FB	= Feedback	FP	= Fusible polarity

Table 2.5   Comparison of programmable logic characteristics

PAL devices are available in different configurations to suit different types of applications. These are:

2.16.1 AND-OR/NOR Cell

This is for the implementation of the sum of products: the basic cell is shown in Fig. 2.53

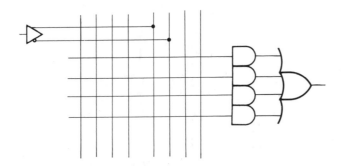

Fig. 2.53   Basic combinational PAL cell

2.16.2 Programmable Input/Output

This has the added capability of a programmable enable line. It includes three-state drivers as shown in Fig. 2.54, and this allows a pin to be used as an input, an output or dynamic input/output. The input/output pin is an input when the three-state gate is disabled. The tri-state buffer is controlled by one of the product terms.

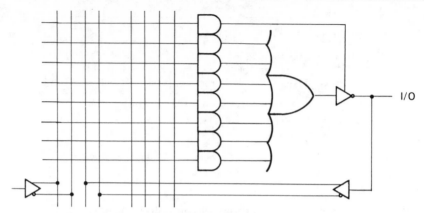

Fig. 2.54   Basic programmable input/output PAL cell

2.16.3 Registered Output with Feedback

The product term is stored in a flip-flop. The Q output of the flip-flop can be gated to the output pin by enabling the active low three-state buffer. The feedback facilitates the use of PAL as a state sequencer (Fig. 2.55). Another version includes an EX-OR at the D input for the HOLD operation in counters.

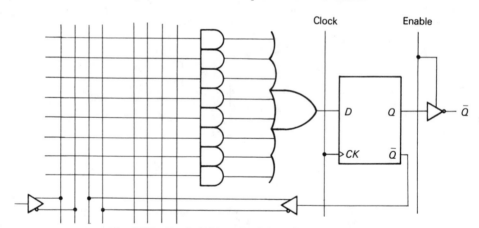

Fig. 2.55   Basic PAL cell with registered output

2.16.4 Arithmetic PAL Cell

This incorporates an EX-OR and gated feedback and is used for basic arithmetic operations such as add and subtract. The EX-OR allows carries to be 'EX-ORed' with current sums generated by the array. The output of the flip-flop is 'ored' with an input I to form $\bar{I}+\bar{Q}$, $\bar{I}+Q$, $I+\bar{Q}$ and $I+Q$, which are fed into the PAL matrix as shown in Fig. 2.56. Fig. 2.57 shows how a PAL is programmed to produce 16 logical functions for an arithmetic logic unit (ALU).

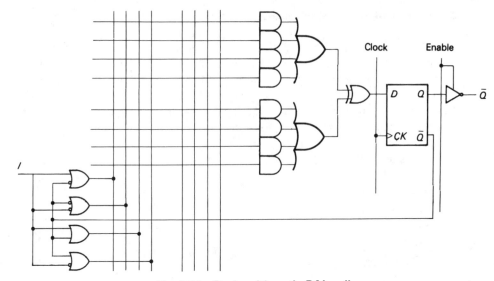

Fig. 2.56   Basic arithmetic PAL cell

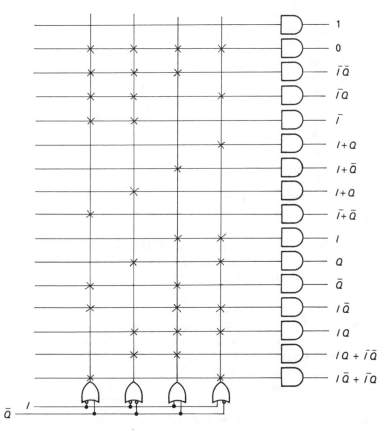

Fig. 2.57   PAL generation of ALU functions

2.16.5 Uses of PALs and Associated Devices

After selecting the most suitable device for the type of logic under consideration, which may be, say, combinational or sequential the designer then selects the most suitable size. Like other logic devices, PALs are available with different combinations of inputs, outputs and product terms. It is not unusual, however, to find that the chip is only partially used. The use of a PAL for random logic will be illustrated in Chapter 3. The number code for PAL devices can give a good idea about the chip size, and suitability for a particular application.

Consider a PAL with the number code PAL10H8CJ:

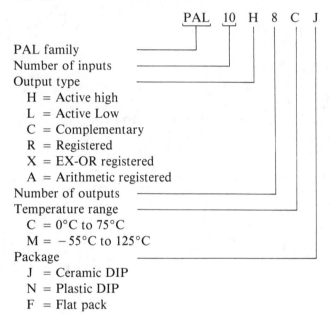

$$\text{PAL} \quad 10 \quad \text{H} \quad 8 \quad \text{C} \quad \text{J}$$

PAL family
Number of inputs
Output type
   H = Active high
   L = Active Low
   C = Complementary
   R = Registered
   X = EX-OR registered
   A = Arithmetic registered
Number of outputs
Temperature range
   C = 0°C to 75°C
   M = −55°C to 125°C
Package
   J = Ceramic DIP
   N = Plastic DIP
   F = Flat pack

The convention used in this section is that used by Monolithic Memories who invented the PAL. (See MMI PAL Handbook).

At present PAL ICs are available with 20, 24 and 28 pins. They can be programmed using PROM programmers with the addition of a personality card. The propagation delay for the combinational cells and the set-up time for the registered cell are, at present, about 15 ns. The use of PALS is discussed in Section 3.7.

Hard Array Logic (HAL)

The similarity between a HAL and a PAL is equivalent to the similarity between a ROM and a PROM. The HAL is a mask-programmed version of a PAL and is made to the customer's requirements in the last stage of fabrication. For large volumes, the designer might use a PAL as a fusible prototype and mass-produce the circuit by making a metal mask to 'customize' the HAL.

HAL and PAL Medium Scale Integration

HAL Medium Scale Integration (HMSI) devices are made using HAL technology and they perform commonly used functions such as counting and multiplexing. PAL Medium Scale Integration (PMSI) devices are derived from PAL for trial, and a HAL mask is generated if the demand warrants high volume production. Present HMSI devices include octal counters, octal shift registers, multifunction registers, 10-bit up/down counters, dual 8:1 multiplexers, quad 4:1 multiplexers and 16:1 multiplexers.

PAL versus Other Programmable Devices

A single PAL can replace from five to ten SSI/MSI chips with the obvious advantages of reduced board size and labour, and increased reliability. Further, PAL devices have a security fuse which can be blown to prevent unauthorized copying.

PALs are comparable with FPLAs except that in FPLAs both the AND and OR arrays are programmed. Both PALs and FPLAs have a limited number of product terms. A PROM, on the other hand, has all possible product terms for a given number of inputs and is more suitable for applications like binary multiplications than random logic, where only a limited number of product terms may be required.

PALs can also be compared with gate arrays (see Section 2.17). Gate arrays, however, have higher gate density and a longer turn-around time, which is measured in months as compared with hours for PALs.

2.17   UNCOMMITTED LOGIC ARRAY (ULA)

Designers who want a complete system on a specially tailored chip can still obtain one using full custom design. This however could be costly and could take one or even two years. The ULA, which is also known as a gate array, offers a solution called *semi-custom design*.

The ULA is an LSI chip made of an array of uncommitted active and passive components that are fully processed except for the final layer of aluminium interconnection. The interconnection pattern is generated from the problem specifications either by the customer, using ULA–CAD facilities, or by the manufacturer. The logic is effectively programmed in the device in a manner not unlike programming data in a ROM. Because the ULA is mass-produced and only one interconnection mask is required to convert the ULA chip into the required customer circuit, the development cost can be low and the production time short compared with full custom design.

ULA automation software is available for layout; design rule checking; simulation; testing; and verification. The ULA devices are available in many technologies like ECL, cMOS and bipolar. They are currently available in system complexities varying from a few hundred gates up to 10 000 gates. Clock rates from 6 to 60 MHz and gate delays from 2.5 to 25 ns.

A typical example of these devices is the UK 5000 cMOS array. As the name suggests, it has the equivalent of 5000 gates and has been developed in the UK by a consortium of seven partners. Ferranti, on the other hand, produces a wide range of bipolar ULAs for systems of 100 to 10 000 gates. The basic Ferranti cell may be RTL, current mode logic (CML) or buffered CML.

A number of variations are possible in the overall organization of gate arrays as well as in the structure of the cells. Generally speaking, the cells are arranged as a matrix with special peripheral cells around the edge of the chip. The channels between the cells are used to lay the metal tracks for the interconnections. The logic cells contain components like transistors and resistors that can be connected together to form the required functional elements. The designer has the task of mapping the logic diagram into an interconnection pattern to commit the device to the required logic.

In single-metal-layer devices, the pattern is used to create a single mask which determines the connections of the components within the cells as well as the connections between the cells. In modern gate arrays two layers of metal are employed. The latter devices normally require one mask for each layer and a third to produce the

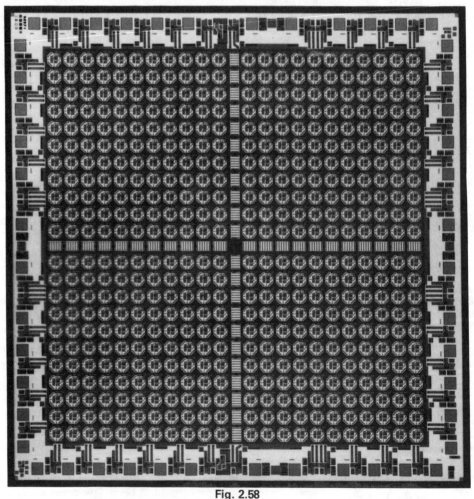

Fig. 2.58
(Courtesy of Ferranti Electronics Ltd.)

contact holes through the insulating layer that separates the two levels of metal. The use of two layers of metal obviously requires more complicated processing, but results in higher on-chip packing density and considerably simplifies routing.

Fig. 2.58 shows a Ferranti ULA chip. The cell in Fig. 2.59 is taken from a similar 440 bipolar ULA. Because this is a single-metal-layer device three diffused crossunders are provided in each cell to simplify the layout of the interconnecting tracks. The layout can be performed manually. CAD tools however are available to simplify this task.

To illustrate the use of these devices, Fig. 2.60 shows the connections required

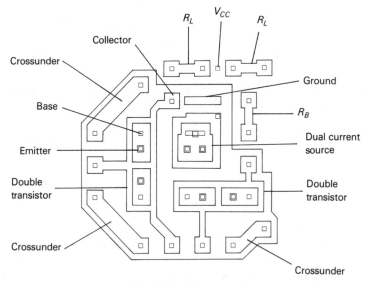

Fig. 2.59 Matrix cell
(Courtesy of Ferranti Electronics Ltd.)

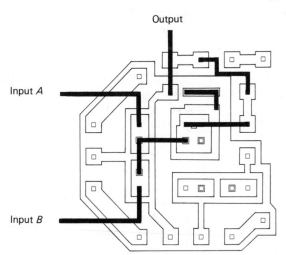

Fig. 2.60 Layout for a two-input NOR gate

to create a two-input NOR gate. The resulting circuit diagram shown in Fig. 2.61 is a current-mode logic (CML) NOR gate. Transistor T_3 prevents the input transistors from going into saturation by limiting the current to ground. Note that enough components are left on the cell to create a second two-input NOR gate. Alternatively a four-input NOR gate can be created. Other circuits are also possible.

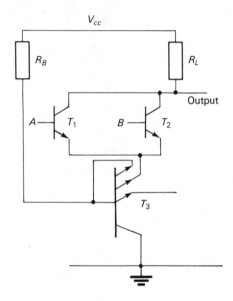

Fig. 2.61   The two-input CML NOR gate of Fig. 2.60

As another example, consider the nMOS ED500 gate array developed at Edinburgh University. This device has 552 logic cells and 40 peripheral cells. The logic cell in Fig. 2.62 has four transistors. T_1 is a depletion mode transistor while T_2, T_3 and T_4 are enhancement mode transistors. Transistors T_1 and T_2 are prewired as an inverter (see Fig. 2.26). A number of cross-unders are also available to facilitate inter-connections. These are buried within the structure to enable tracks to cross over them.

Fig. 2.63 shows a number of adjacent cells. It can be seen that the cells repeat horizontally along the rows but are alternately mirror imaged in the vertical direction to enable adjacent rows to share the supply lines. Connections between cells in adjacent rows are performed using inter-cell cross-unders which pass under the supply lines.

Transistors T_2, T_3 and T_4 in the logic cell can be connected in different ways to create the required logic. Fig. 2.64 shows a three-input NOR gate with the three enhancement mode transistors connected in parallel and transistor T_1 acting as a common load. The resulting circuit is shown in Fig. 2.65. Two-input NOR or NAND gates are also possible. If only T_3 and T_4 are used together with T_1, transistor T_2 is disabled by connecting its gate to ground. The gate of the three enhancement transistors are marked G in Fig. 2.62. The gate of the depletion transistor is prewired to keep the transistor on. Connection tracks run along the grid lines shown dotted.

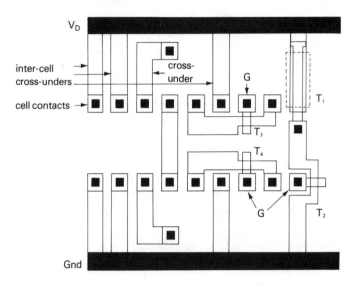

Fig. 2.62   Logic cell for the ED500

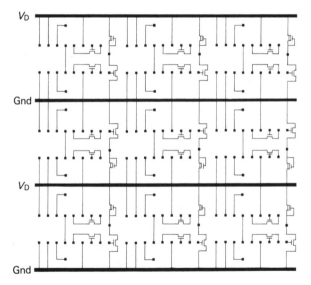

Fig. 2.63   A 3 × 3 array of cells
(Courtesy of University of Edinburgh)

It remains to say that the placement and layout of the interconnecting tracks within each cell and between cells can become a problem for large designs. It is also likely that some cells may be sacrificed to ensure a successful routing. Good designs, however, tend to have some degree of regularity and some patterns may occur many times within a design. These can be used to advantage, especially when CAD tools are used, by defining macros or layout subroutines for these repetitive patterns. The

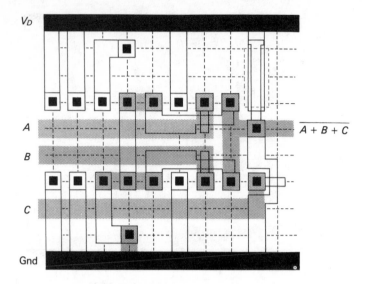

Fig. 2.64   Layout of a 3-input NOR gate

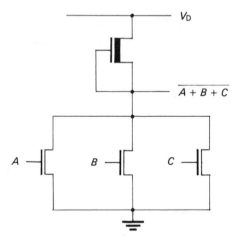

Fig. 2.65   The 3-input NOR gate of Fig. 2.64

macros, which can be simple gates, flipflops, adders etc., once defined can be called and placed in the required positions. Gate arrays or uncommitted logic arrays, as they are sometimes called, are becoming popular and their share of the IC market is expected to grow as more CAD software becomes available at reasonable cost.

2.18   UNIVERSAL LOGIC MODULE (ULM)

Another useful MSI device is the digital multiplexer, which is also called a selector because it acts like a selector switch and selects one of the inputs to appear at the output by means of the control inputs. The device is also called a universal logic module (ULM) because an n-variable ULM can be used to implement any logic function of n or fewer variables. If the number of variables of the logic function is greater than that of the module, then two or more modules can be connected in a multi-level array.

Fig. 2.66 shows a three-variable ULM and the way the inputs are addressed by the control inputs. Two of the variables are used as control inputs (address selector). The third variable is connected to the input. The input can also be the complement of the variable, logic 0 or logic 1, as will be shown in Chapter 3. If used as a selector switch this device is called '4 lines to 1', since the two control lines can connect any of the four input lines to the output. (Two control inputs can have four possible combinations, namely 00, 01, 10, and 11.)

ULMs are available as 2-, 3-, 4- or 5-variable modules and packaged as single, quad two-input, dual four-input, etc. Typical power dissipation is from 20 to 200 mW and typical propagation delay is from 5 to 50 ns.

Another member of this family is the demultiplexer, which performs the opposite function, namely that of connecting a single input to any one of many outputs by means of the control inputs.

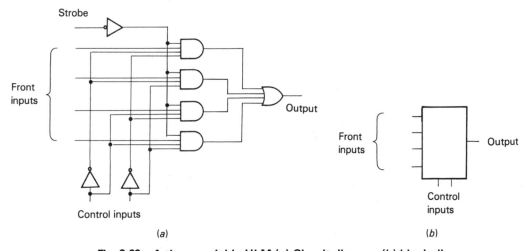

(a)   (b)

Fig. 2.66   A three-variable ULM (a) Circuit diagram (b) block diagram

A common application for digital multiplexers, which is not discussed in this book, is in time division multiplexing where bits (characters) from many signal sources are interleaved and transmitted over a common line. Digital demultiplexers are used at the receiving end to direct the bits (characters) to their respective destinations, as shown in Fig. 2.67. For such applications, the device is also known as a selector switch.

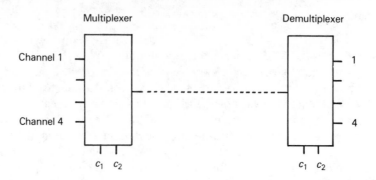

Fig. 2.67   Time division multiplexing of four signals

In the rest of this book, multiplexers will be used as logic building blocks and will normally be called universal logic modules (ULMs). The reader should be aware that analogue multiplexer chips are also available, but these are outside the scope of this book.

3

Combinational Logic at Different
Levels of Integration

3.1   INTRODUCTION

Logic circuits can still be implemented with discrete components like transistors, diodes
and resistors. The usual practice, however, is to use standard off-the-shelf SSI/MSI
digital integrated circuits (ICs). ICs are available as gates, adders, coders, shift
registers, flip-flops, etc., and can be connected to realize combinational and sequential
logic. Other useful devices available to the logic designer are the programmable ICs.
These are normally of LSI complexity and capable of performing a variety of tasks.
Some, like PROMs, FPLAs and PALs, can be programmed by the user (normally by
blowing selected fuses). Others, like ROMs and PLAs, are programmed by the
manufacturer and are sometimes called *mask programmable devices*.

A recent development in logic design is the semi-custom approach. The term
covers uncommitted logic arrays and standard cells, although some designers include
programmable logic in this category. The uncommitted logic arrays, also known as
gate arrays, are made of a large number of cells fully designed except for the last level of
metallization. The metallization determines the connection of the components within
the cell (gate connections) as well as the connections between the cells which commit the
device to a particular design. Though CAD tools are available successful routing
without sacrificing component density can still be a problem for large gate arrays. For
standard cells the designer uses custom-designed cells stored in a computer library. The
cells are called and placed in rows on the chip with spaces between the rows for interconnections. Commonly CAD tools are employed for automatic placement and
routing. This approach is more economical in terms of silicon than gate arrays since
there are no unused gates. The prototype, however, is more expensive to develop and
requires a full set of masks.

In some situations a specially tailored IC, using full custom design, may be
required. This is perfectly feasible and very economical in terms of silicon, but may be
expensive unless a large volume is required. Typical designs take one to two years to
complete with the full custom-design approach, though this will improve as more efficient CAD tools become available. The various design approaches are summarized in
Fig. 3.1.

It should be pointed out at this stage that most logic problems can be realized
using SSI/MSI digital devices. It is common practice, however, to mix these with
LSI/VLSI devices to implement complex problems. That may be done for convenience, speed, economy of space, security, cost, or to satisfy other requirements. When

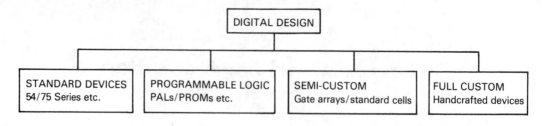

Fig. 3.1   Categories of digital logic design

there is such a mix, interfacing may become difficult especially if the devices are made using different technologies, as was mentioned in Chapter 2. The reader should also be aware that interfacing with the outside analogue world is another problem that requires analog/digital and digital/analog convertors.

In this chapter a representative sample of available devices will be used to illustrate their use for logic design. This includes standard SSI/MSI digital devices, and programmable devices. Standard cells and full custom design relate to the design and fabrication of ICs, which are outside the scope of this book. Gate arrays are described in Chapter 2.

3.2   DISCRETE COMPONENT IMPLEMENTATION

To see how the availability and variety of modern logic devices have influenced the design strategy, the following three logic functions f_1, f_2 and f_3 are considered. Each function depends on four input variables x_1, x_2, x_3 and x_4, and are given in full canonical form in Equation (3.1).

$$
\begin{aligned}
f_1 &= \bar{x}_1\bar{x}_2 x_3\bar{x}_4 + \bar{x}_1 x_2\bar{x}_3\bar{x}_4 + x_1\bar{x}_2 x_3\bar{x}_4 + \\
&\quad x_1\bar{x}_2 x_3 x_4 + x_1 x_2\bar{x}_3\bar{x}_4 + x_1 x_2\bar{x}_3 x_4 \\
f_2 &= \bar{x}_1 x_2\bar{x}_3\bar{x}_4 + \bar{x}_1 x_2\bar{x}_3 x_4 + x_1\bar{x}_2 x_3\bar{x}_4 + \\
&\quad x_1\bar{x}_2 x_3 x_4 + x_1 x_2\bar{x}_3 x_4 \\
f_3 &= \bar{x}_1\bar{x}_2\bar{x}_3 x_4 + \bar{x}_1\bar{x}_2 x_3\bar{x}_4 + \bar{x}_1\bar{x}_2 x_3 x_4 + \\
&\quad x_1\bar{x}_2 x_3\bar{x}_4 + x_1\bar{x}_2 x_3 x_4 + x_1 x_2\bar{x}_3\bar{x}_4
\end{aligned}
\qquad (3.1)
$$

If we minimize the three expressions individually by using Karnaugh maps, as shown in Fig. 3.2, we obtain the expressions of Equation (3.2).

$$
\begin{aligned}
f_1 &= x_2\bar{x}_3\bar{x}_4 + x_1 x_2\bar{x}_3 + \bar{x}_2 x_3\bar{x}_4 + x_1\bar{x}_2 x_3 \\
f_2 &= \bar{x}_1 x_2\bar{x}_3 + x_2\bar{x}_3 x_4 + x_1\bar{x}_2 x_3 \\
f_3 &= x_1 x_2\bar{x}_3\bar{x}_4 + \bar{x}_1\bar{x}_2 x_4 + \bar{x}_2 x_3
\end{aligned}
\qquad (3.2)
$$

If, on the other hand, a modified version of the Quine–McCluskey method for multiple-output functions is used, the expressions in Equation (3.3) are obtained.

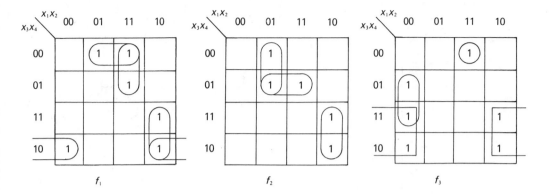

Fig. 3.2   Karnaugh maps for f_1, f_2, f_3

$$\left.\begin{array}{l} f_1 = x_1x_2\bar{x}_3 + \bar{x}_2x_3\bar{x}_4 + x_1\bar{x}_2x_3 + \bar{x}_1x_2\bar{x}_3\bar{x}_4 \\ f_2 = x_1\bar{x}_2x_3 + x_2\bar{x}_3x_4 + \bar{x}_1x_2\bar{x}_3\bar{x}_4 \\ f_3 = \bar{x}_1\bar{x}_2x_4 + \bar{x}_2x_3\bar{x}_4 + x_1\bar{x}_2x_3 + x_1x_2\bar{x}_3\bar{x}_4 \end{array}\right\} \tag{3.3}$$

The number D of diodes required to implement any group of E logic functions $f_1, f_2, \ldots, f_E$ having $\Omega_1, \Omega_2, \ldots, \Omega_\mu$ product terms with no product term repeated, is given by Equation (3.4).

$$D = \sum_{j=1}^{\mu} K_j + \sum_{i=1}^{E} K_i \tag{3.4}$$

K_i is the number of product terms in the expression f_i.
$K_i = 0$ if there is only one product term in f_i.
K_j is the number of literals in a product term Ω_j.
$K_j = 0$ if product term Ω_j contains only one literal.

Using Equation (3.4), the three functions in their original form in Equation (3.1) require a total number D_1 of diodes given by

$D_1 = 36 + 17 = 53$

The minimized expressions in Equation (3.2) require a total D_2 given by

$D_2 = 27 + 10 = 37$

From Equation (3.3), the same three functions, when minimized using a method more suitable for multiple-output functions, require a total D_3 given by

$D_3 = 23 + 11 = 34$

This confirms that the procedure is very suitable for discrete components. The final implementation is shown in Fig. 3.3.

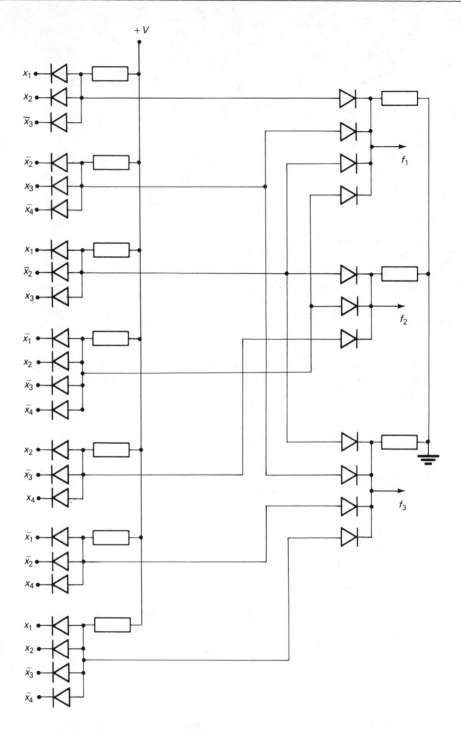

Fig. 3.3   Diode implementation of Equation (3.3)

3.3   GATE LEVEL IMPLEMENTATION

Though Equation (3.2) looks simpler than Equation (3.1) and contains fewer product terms, both in fact require nine AND gates plus three OR gates. Equation (3.3) requires seven AND gates plus three OR gates. Further the OR gates in Equation (3.1) are of five or six inputs each while those in Equation (3.3) are of three or four inputs each. Some of the AND gates in Equation (3.3) are also smaller. The expressions in Equation (3.3) will be implemented.

    In Chapter 1, the gate level design was discussed and it was shown that AND/OR gates can be replaced by NAND or NOR gates.

    The logic expression in Equation (3.3) show that the maximum number of inputs required is four. If four-input gates are available, AND/OR implementation may be adopted. If only dual four-input NAND gates, say, are available, (e.g. the 7420 IC) then De Morgan's theorem can be used to rearrange the expressions as follows:

$$f_1 = \overline{\overline{x_1 x_2 \bar{x}_3} \cdot \overline{\bar{x}_2 x_3 \bar{x}_4} \cdot \overline{x_1 \bar{x}_2 x_3} \cdot \overline{\bar{x}_1 x_2 \bar{x}_3 \bar{x}_4}}$$

$$f_2 = \overline{\overline{x_1 \bar{x}_2 x_3} \cdot \overline{\bar{x}_2 x_3 x_4} \cdot \overline{\bar{x}_1 x_2 \bar{x}_3 \bar{x}_4}}$$

$$f_3 = \overline{\overline{x_1 \bar{x}_2 x_4} \cdot \overline{\bar{x}_2 x_3 \bar{x}_4} \cdot \overline{x_1 \bar{x}_2 x_3} \cdot \overline{x_1 x_2 \bar{x}_3 x_4}}$$

These are implemented as in Fig. 3.4(a) and Fig. 3.4(b).

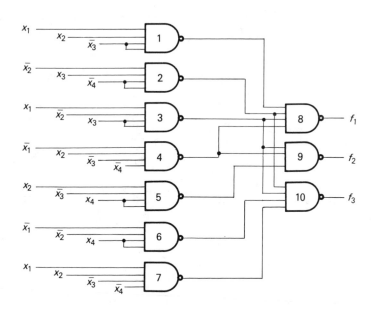

Fig. 3.4(a)   Logic diagram for Equation (3.3)

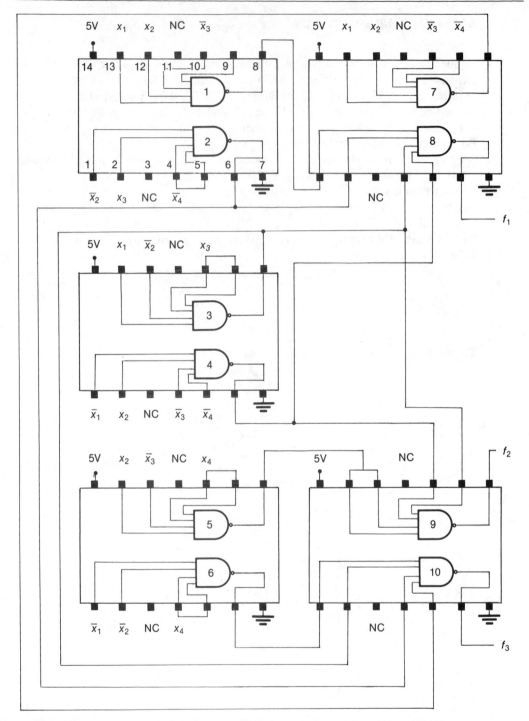

Fig. 3.4(b) IC connection for equation (3.3) using five dual four-input NAND gates
NC = No connection

3.4   UNIVERSAL LOGIC MODULES (MULTIPLEXER) IMPLEMENTATION

What is a good technique for discrete components is not necessarily good or useful when large- or medium-scale integrated devices are used. The first such device to be considered is the universal logic module (ULM).

A universal logic module of a given number of variables is a module capable of implementing any logic function of up to that number of variables. If the number of variables of the logic function is greater than that of the module, the modules are connected in an array of two or more levels.

Consider a general n-variable logic function $f(x_1, x_2, \ldots, x_n)$.

If this function is to be realized using three-variable modules, and if $n \geqslant 3$, then by Shannon's expansion theorem the function is expanded about the variables, say x_1 and x_2, used at the control inputs which gives:

$$f(x_1, x_2, \ldots, x_n) = \bar{x}_1\bar{x}_2\, f(0, 0, x_3, \ldots, x_n) + \\ \bar{x}_1 x_2\, f(0, 1, x_3, \ldots, x_n) + \\ x_1\bar{x}_2\, f(1, 0, x_3, \ldots, x_n) + \\ x_1 x_2\, f(1, 1, x_3, \ldots, x_n) \tag{3.5}$$

If the expansion is continued, the residues $f(0, 0, x_3, \ldots, x_n), \ldots, f(1, 1, x_3, \ldots, x_n)$ are expanded with respect to the variables used at the control inputs of the second level, third level, and so on, until the residue terms are functions only of one variable, say x_n. This term can take the values $\bar{x}_n, x_n, 0$, or 1 and is connected to the front input of the last level.

Consider the function

$$f = \bar{x}_1\bar{x}_2\bar{x}_3 x_4 + \bar{x}_1\bar{x}_2 x_3\bar{x}_4 + \bar{x}_1 x_2 x_3 x_4 + \\ \bar{x}_1 x_2 x_3\bar{x}_4 + x_1 x_2\bar{x}_3 x_4 + x_1 x_2 x_3\bar{x}_4 \tag{3.5(a)}$$

(i)   If a four-variable ULM is available, the logic function can be implemented directly as shown in Fig. 3.5(a); x_4 is arbitrarily chosen as the front input.

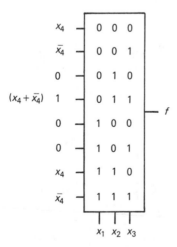

Fig. 3.5(a)   Implementation of Expression [3.5(a)] using a single four-variable ULM

(ii)   If two- and three-variable ULMs are available, the expression can be expanded with respect to the variables chosen for the first and second levels. The residue variable is used as the front input. Once again the choice is arbitrary.

$$f = \bar{x}_1 \left[\bar{x}_2\bar{x}_3(x_4) + \bar{x}_2x_3(\bar{x}_4) + x_2\bar{x}_3(0) + x_2x_3(1) \right] +$$

$$x_1 \left[\bar{x}_2\bar{x}_3(0) + \bar{x}_2x_3(0) + x_2\bar{x}_3(x_4) + x_2x_3(\bar{x}_4) \right]$$

This can be implemented as in Fig. 3.5(*b*).

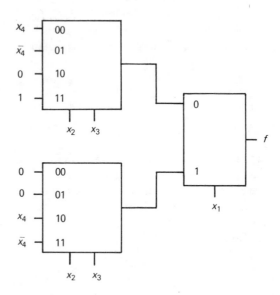

Fig. 3.5(b)   ULM implementation of Expression [3.5(a)] using two- and three-variable modules

(iii)   It is also possible to use all the variables as control inputs and only connect logic (0) and logic (1) to the inputs. This expression is expanded with respect to the variables used at the two levels as shown.

$$f = \bar{x}_1\bar{x}_2 \left[\bar{x}_3\bar{x}_4(0) + \bar{x}_3x_4(1) + x_3\bar{x}_4(1) + x_3x_4(0) \right] +$$

$$\bar{x}_1x_2 \left[\bar{x}_3\bar{x}_4(0) + \bar{x}_3x_4(0) + x_1\bar{x}_4(1) + x_3x_4(1) \right] +$$

$$x_1\bar{x}_2 \left[\bar{x}_1\bar{x}_4(0) + \bar{x}_3x_4(0) + x_3\bar{x}_4(0) + x_3x_4(0) \right] +$$

$$x_1x_2 \left[\bar{x}_3\bar{x}_4(0) + \bar{x}_3x_4(1) + x_3\bar{x}_4(1) + x_3x_4(0) \right]$$

It can be seen that there is a direct correspondence between the positions of the variables and logic constants in the expansion and their position in the circuit shown in Fig. 3.5(*c*).

With experience, it should be possible to draw the circuit directly from the logic expression or the truth table without the need for the above expansion.

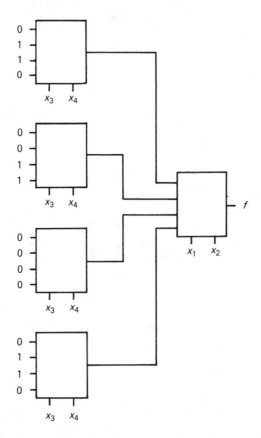

Fig. 3.5(c)   Alternative ULM implementation of Expression [3.5(a)]

Now consider the three functions in Equation (3.1). Since all functions depend on four variables, readily available four-variable ULMs will be used. It is clear that there is no need for the lengthy Quine–McCluskey procedure, or even a K-map, since we would still have to expand the functions to their full canonical form.

Starting from Equation (3.1), we choose to use x_1x_2 and x_3 as the control variables. The functions are first expanded as shown below in the table, and are implemented as shown in Fig. 3.6.

$f_1 = \bar{x}_1\bar{x}_2\bar{x}_3(0)$	$f_2 = \bar{x}_1\bar{x}_2\bar{x}_3(0)$	$f_3 = \bar{x}_1\bar{x}_2\bar{x}_3(x_4)$
$+ \; \bar{x}_1\bar{x}_2x_3(\bar{x}_4)$	$+ \; \bar{x}_1\bar{x}_2x_3(0)$	$+ \; \bar{x}_1\bar{x}_2x_3(1)$
$+ \; \bar{x}_1x_2\bar{x}_3(\bar{x}_4)$	$+ \; \bar{x}_1x_2\bar{x}_3(1)$	$+ \; \bar{x}_1x_2\bar{x}_3(0)$
$+ \; \bar{x}_1x_2x_3(0)$	$+ \; \bar{x}_1x_2x_3(0)$	$+ \; \bar{x}_1x_2x_3(0)$
$+ \; x_1\bar{x}_2\bar{x}_3(0)$	$+ \; x_1\bar{x}_2\bar{x}_3(0)$	$+ \; x_1\bar{x}_2\bar{x}_3(0)$
$+ \; x_1\bar{x}_2x_3(1)$	$+ \; x_1\bar{x}_2x_3(1)$	$+ \; x_1\bar{x}_2x_3(1)$
$+ \; x_1x_2\bar{x}_3(1)$	$+ \; x_1x_2\bar{x}_3(x_4)$	$+ \; x_1x_2\bar{x}_3(\bar{x}_4)$
$+ \; x_1x_2x_3(0)$	$+ \; x_1x_2x_3(0)$	$+ \; x_1x_2x_3(0)$

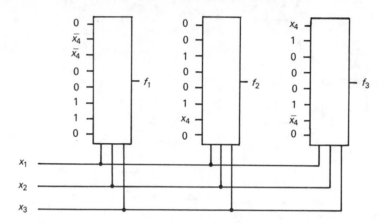

Fig. 3.6 ULM implementation of Equation (3.1)

If a smaller module was chosen, one would have to use more than one level as in the last example. Then the major problems facing the designer would be:

(*a*) How to reduce the number of levels and/or modules.
(*b*) How to decide on what variables to allocate for the control of every level.

The choice of control variables also affects the total number of modules. These problems will be discussed in Chapter 7, but we note now that both questions find no answer in either the K-map or the Quine–McCluskey method. For the time being we assume that no simplification is possible, and try in the following to predict the maximum number of modules that might be required to implement any *n*-variable logic function.

The maximum number of modules Mo, required to implement a given function, and the distribution of modules in various levels are of the form shown in Table 3.1, where:

J = number of variables of the module.
$c = J - 1$ is the number of control variables of the module.
$I = 2^c$ is the number of front inputs to a module.
v = total number of the variables of the function.

The number of levels l required to implement a function of v variables is given by Equation (3.6).*

$$l = \left| \frac{v-1}{c} \right| \tag{3.6}$$

Where $|g|$ is the smallest integer equal to or greater than g.

$*l = \left| \dfrac{v}{c} \right|$ if only constants are used at the front input as in Fig. 3.5(*c*)

No. of variables v	1st level	2nd level	3rd level	4th level	5th level	Max. No. of modules Mo
J	1					I^0
$J + (J - 1)$	1	I				$I^0 + I^1$
$J + 2(J - 1)$	1	I	I^2			$I^0 + I^1 + I^2$
$J + 3(J - 1)$	1	I	I^2	I^3		$I^0 + I^1 + I^2 + I^3$
$J + 4(J - 1)$	1	I	I^2	I^3	I^4	$I^0 + I^1 + I^2 + I^3 + I^4$

Table 3.1   The number of levels and the maximum number of modules required to
implement a given function of v variables

From Table 3.1 it can be seen that the maximum number Mo of modules required
to implement a given function is of the form of a geometric progression, which may
be represented by Equation (3.7).

$$Mo = \sum_{i=0}^{i=l-1} (I^i) \tag{3.7}$$

This summation may be rewritten in the following form:

$$Mo = \frac{1 - I^l}{1 - I} = \frac{I^l - 1}{I - 1} \tag{3.8}$$

There are cases however, where the number of variables is smaller than the
maximum that can be accommodated, i.e.

$$v < J + (l-1)(J-1)$$

In such a case, it is advisable to use the redundancy at the first level, since one unused
control variable in any level reduces the number of modules in succeeding levels by a
factor of two. Equation (3.9) gives the number of modules if the control inputs in the
first level are only partially used, or if a smaller module is used in the first level.

$$Mo = 1 + 2^{c1} \left(\frac{1 - I^{l-1}}{1 - I} \right) \tag{3.9}$$

where c_w is the number of control variables used in level w.

Similarly Equation (3.10) gives the number of modules if different numbers of
control variables are used in different levels, or possibly different modules are used in
different levels.

$$Mo = 1 + 2^{c1} + 2^{c1 + c2} + \ldots + 2^{c1 + c2 + \ldots + c_{l-1}} \tag{3.10}$$

Finally, Equation (3.11) is derived to calculate the number of modules if
different numbers of control variables are used for different modules within one
level. This, however, is a situation rarely encountered in practice.

$$Mo = 1 + 2^{c1} + \dots + 2^{c1 + \dots + c_{b-1}} +$$

$$\left[\frac{(a_1)2^{c1 + \dots + c_{b-1} + r1} + \dots + (a_g)2^{c1 + \dots + c_{b-1} + rg}}{a_1 + \dots + a_g} \right] +$$

$$\dots + \left[\frac{(a_1)2^{c1 + \dots + c_{b-1} + r1 + \dots + c_{l-1}} + \dots + (a_g)2^{c1 + \dots + c_{b-1} + rg + \dots + c_{l-1}}}{a_1 + \dots + a_g} \right]$$

$$(3.11)$$

Level b has $a_1, \dots, a_g$ units, each having $r_1, \dots, r_g$ control inputs respectively. From these results, the following may be generalized:

(i)   To implement a function with a large number of variables using small modules, many levels are required. Using large modules does help in reducing the number of levels and also the effect of propagation delay. Otherwise, some technique must be found to reduce the number of levels and/or modules to a minimum, which will generally be less than the maximum predicted by Equations (3.7–3.11).

(ii)   Advantage should be taken of redundant control inputs at an early stage. This leads to a larger saving in modules. This fact will be made use of in future circuits.

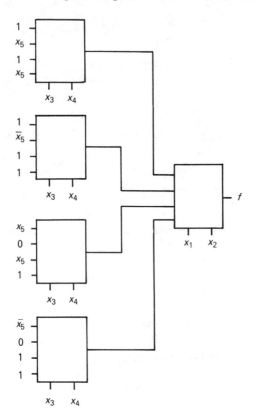

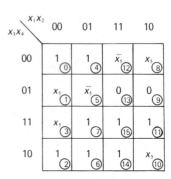

Fig. 3.7(a)   ULM implementation of Equation (3.12)

Fig. 3.7(b)   Map-entered variable for Equation (3.12)

(iii)   Methods like the K-map or the Quine–McCluskey method are not useful in the way they were before.

(iv)   The map-entered variable technique mentioned in Chapter 1 could be useful, not for minimization but for displaying the logic function in a form suitable for direct implementation.

Consider the following logic function:

$$f = \Sigma \,(0, 1, 3, 4, 5, 7, 8, 9, 10, 12, 13, 14, 15, 17, 21, 22, 23, 24, 28, 29, 30, 31)$$

This is a function of five variables and can be expanded as shown and implemented using two levels of three-variable ULMs as in Fig. 3.7(a).

$$
\begin{aligned}
f = \bar{x}_1\bar{x}_2\,[\bar{x}_3\bar{x}_4(1) + \bar{x}_3x_4(x_5) + x_3\bar{x}_4(1) + x_3x_4(x_5)] \;+ \\
\bar{x}_1x_2\,[\bar{x}_3\bar{x}_4(1) + \bar{x}_3x_4(\bar{x}_5) + x_3\bar{x}_4(1) + x_3x_4(1)] \;+ \\
x_1\bar{x}_2\,[\bar{x}_3\bar{x}_4(x_5) + \bar{x}_3x_4(0) + x_3\bar{x}_4(x_5) + x_3x_4(1)] \;+ \\
x_1x_2\,[\bar{x}_3\bar{x}_4(\bar{x}_5) + \bar{x}_3x_4(0) + x_3\bar{x}_4(1) + x_3x_4(1)]
\end{aligned}
\qquad (3.12)
$$

The map-entered variable method can be used effectively by using the input variable as the map-entered variable. If this is done, the inputs to the second level of modules can then be read directly from the map [Fig. 3.7(b)].

The sixteen positions ⓪ to ⑮ correspond to the 16 inputs (one column per module). A more economical implementation (in terms of modules), will be discussed in Chapter 7. [Example (7.5)]

3.5   SEMICONDUCTOR MEMORIES IMPLEMENTATION

A read only memory (ROM) as described in Chapter 2 is a device consisting of two main parts; one is a decoder to decode the inputs I into 2^I address lines that are connected to the 2^I words in the other part, which is the memory matrix.

If the memory matrix has m' outputs corresponding to m' bits of storage locations for each word then the total memory is given by

$$M = m' \cdot 2^I \text{ bits}$$

To illustrate how a ROM (or similar semiconductor memory device) is used to implement logic functions, consider the following simple Boolean equations:

$$f_1 = A\bar{B} + AB$$

$$f_2 = \bar{A}\bar{B} + AB$$

The inputs A and B are decoded into four address lines and can be used to select any of the four words in the memory. Two simplified representations of the memory section are shown in Fig. 3.8. In some cases the diodes (or transistors) defining the bit pattern are shown as in Fig. 3.9. Note that even if f_1 is simplified to $f_1 = A$, it will still occupy two-bit positions on the memory section.

Let us return to the three logic expressions given by Equation (3.1); since these are functions of four input variables, they would need a module having at least four

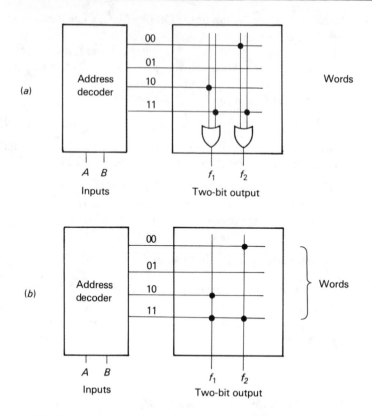

Fig. 3.8   ROM implementation of two logic functions

inputs and three outputs if we are to implement the three functions using one memory module.

Assume we have at our disposal a ROM module with four inputs and four outputs; such a module will have 4.2^4 bits.

Here again, there is no need for the traditional minimization techniques since the actual storage required to implement the three functions is 3.2^4 bits whichever set of equations (3.1), (3.2) or (3.3) is used. The starting point is therefore the truth table given in Table 3.2. The final implementation is shown in Fig. 3.9.

To implement the logic functions given in Equation (3.1), reproduced in Table 3.2, one module is required. Larger logic functions may require more than one module. This may be done by connecting ROMs to extend the number of words, word length (bits), or both, as shown in Chapter 2.

In general, if v is the number of input variables of the logic function, and I is the number of inputs of the module, the number of modules required when connected for word extension is given by:

$$Mo' = 2^{(v-I)}$$

where $I \leqslant v$

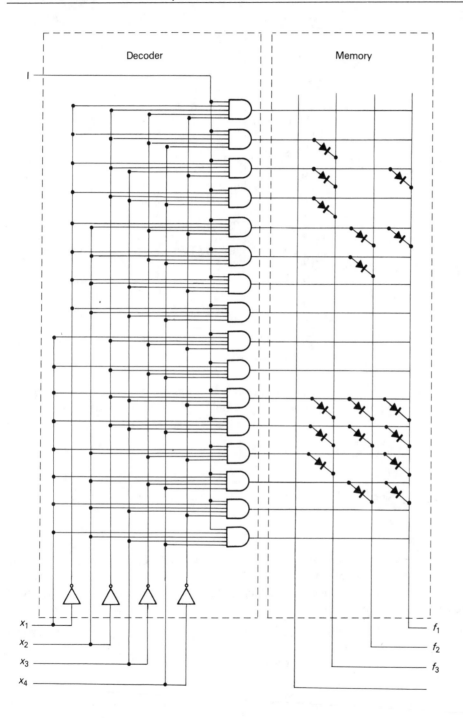

Fig. 3.9   ROM implementation of the three functions in Equation (3.1)
The memory matrix is represented by a diode matrix

Words	Inputs Binary select I_4 I_3 I_2 I_1				Enable	Outputs			
	x_1	x_2	x_3	x_4	E	f_4	f_3	f_2	f_1
0	0	0	0	0	0	0	0	0	0
1	0	0	0	1	0	0	1	0	0
2	0	0	1	0	0	0	1	0	1
3	0	0	1	1	0	0	1	0	0
4	0	1	0	0	0	0	0	1	1
5	0	1	0	1	0	0	0	1	0
6	0	1	1	0	0	0	0	0	0
7	0	1	1	1	0	0	0	0	0
8	1	0	0	0	0	0	0	0	0
9	1	0	0	1	0	0	0	0	0
10	1	0	1	0	0	0	1	1	1
11	1	0	1	1	0	0	1	1	1
12	1	1	0	0	0	0	1	0	1
13	1	1	0	1	0	0	0	1	1
14	1	1	1	0	0	0	0	0	0
15	1	1	1	1	0	0	0	0	0
All	X	X	X	X	1	1	1	1	1

Table 3.2   Read only memory truth table

(X represents DON'T CARE)

If on the other hand $I \geqslant v$, but the number of outputs is greater than that of the modules, the word length can be extended and the number of modules is then given by:

$$Mo'' = \left\lceil \frac{\text{No. of outputs required}}{m'} \right\rceil$$

If both words and word lengths are to be extended, the number of modules is given by:

$$Mo = Mo' \cdot Mo''$$

So far, one module is sufficient for the example used, requiring no minimization. For larger logic functions, the designer will find himself facing some problems that have no answer in the K-map, or the Quine–McCluskey method. We may therefore generalize the following two points:

(i)   The memory increases rapidly when the number of inputs increases; in fact the memory doubles for every additional input.

(ii) If small modules are used, many modules have to be cascaded as already described.

The criterion for minimization will therefore centre on two main points:

(i)   The minimization of storage locations, thus enabling the designer to make a better use of the storage available and in some cases make it possible to use smaller modules, or a fewer number of modules and still to fulfill the design requirements.

(ii)   The minimization directly of the number of modules.

Suitable techniques will be discussed in Chapter 7.

3.6   PROGRAMMABLE LOGIC ARRAY (PLA) IMPLEMENTATION

A PLA, already described in Chapter 2, is a programmable logic device similar to a ROM. It consists of two main parts; the first receives the input variables and their complements, and produces the desired product terms. These are then summed as required in the second part to produce the final output in the form of a sum of product terms.

The PLA can take more inputs than a ROM; this is because all combinations of inputs cause an output in a ROM, whereas in a PLA only selected combinations cause an output. As such, a PLA can be more efficient than a ROM in many situations, since, having its own programmed addresser, it is attractive for many applications where the number of inputs is large, but only a limited number of words are actually needed to accommodate the product terms of the logic function. The PLA in Appendix (3.2) is a 16-input 8-output module with 48 product terms. If this was a ROM it would require 2^{16} words, and if the number of unique product terms is, say 48 or less, then $2^{16} - 48$ words would be unused. The phrase 'product terms' is used here to emphasize that the terms may contain DON'T CARE states as opposed to minterms where all the variables are present in their true or complemented form.

Looking back at the example used in Section 3.2, and assuming that the number of inputs and outputs do not exceed that of the module at hand, we need worry only about the number of product terms. If the number of unique product terms (no product term repeated) in the three functions exceeds that of the module, then the application of K-maps or the Quine–McCluskey method could be helpful in reducing the number of product terms. We should emphasize that this step is not necessary if the number of unique product terms is less than or equal to the number which can be accommodated by the module.

In this case the original functions given in Equation (3.1) contain nine product terms. To illustrate the design procedure, we assume a module with four inputs and four outputs, but only eight product terms.

The use of the K-map results in a simplified expression as given in Equation (3.2), but the number of unique product terms is still nine. Using the Quine–McCluskey method resulted in only seven different product terms, as shown in Equation (3.3). The three functions can now be implemented with just one module, starting from Table 3.3 and ending with programming the device, as shown in Fig. 3.10.

It is interesting to note that, unlike the case with discrete components, we need only to reduce the product terms to the point where their number, with no repetition, is less than or equal to that of the module. In our example above, we used the Quine–McCluskey method. This method is very lengthy, especially for large functions and, in the example one could do without it by simply combining two minterms,

Product terms	Inputs $I_4\ I_3\ I_2\ I_1$ $x_1\ x_2\ x_3\ x_4$	Outputs $f_4\ f_3\ f_2\ f_1$
1	1 1 0 X	0 0 1
2	X 0 1 0	1 0 1
3	1 0 1 X	1 1 1
4	0 1 0 0	0 1 1
5	X 1 0 1	0 1 0
6	0 0 X 1	1 0 0
7	1 1 0 0	1 0 0

Table 3.3   Truth table for the PLA for the example of Equation (3.3) (X represents DON'T CARE)

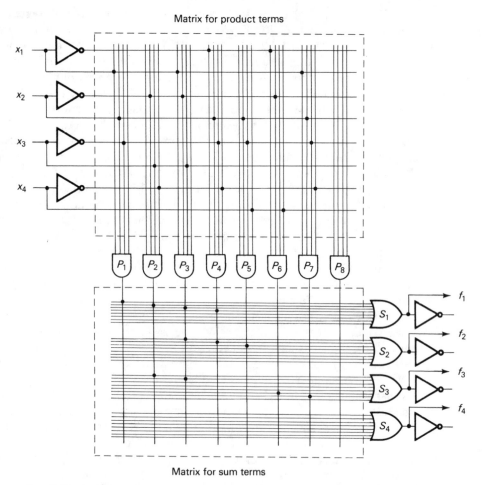

Fig. 3.10   **Implementation of the three functions in Equation (3.3) using one PLA module**

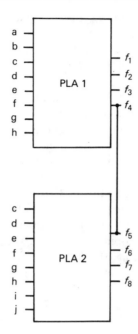

Fig. 3.11   Two small PLA modules implementing large logic functions

say the first and third in f_3, that is:

$$\bar{x}_1\bar{x}_2\bar{x}_3x_4 + \bar{x}_1\bar{x}_2x_3x_4 = \bar{x}_1\bar{x}_2x_4$$

This simple step reduces the number of unique product terms from nine to eight, which is good enough for our module. If no simplification is found useful, then more than one module might be used in cascade, as was the case with ROMs.

The designer's intuition could still play an important part in making a good use of special cases and available modules. In Fig. 3.11, two eight-input four-output PLAs are used to realise eight expressions, each a function of up to ten variables. It is a special case because no product term (P.T.) contains all variables. The P.T.s are therefore separated into two groups, one generated by PLA1, the other by PLA2.

In this case f_1, f_2 and f_3 contain P.T.s having $a, b, \ldots, h$ as input variables, while f_6, f_7 and f_8 contain P.T.s having $c, d, \ldots, j$ as input variables. Only f_4 and f_5 contain P.T.s from both modules, and are connected in parallel (virtual OR). This special but rather common case could also be made use of in other memory devices like ROMs and RAMs.

Whether a FPLA or a PLA is used depends largely on the type of application. With a protype design, for example, a FPLA, which is programmed by the user, is very convenient and economical. PLAs, which require mask design, are used for large volumes. Finally we note that the PLA shown in Appendix 3.2 shows a tri-state output controlled by $\overline{CE}$. Though this is not shown in Fig. 2.49, FPLAs, like PLAs are available with tri-state or open-collector outputs.

3.7   PROGRAMMABLE ARRAY LOGIC (PAL) IMPLEMENTATION

PALs are described in Chapter 2. To illustrate the difference between PAL and other devices, the example of Section 3.2, is implemented using a commercially available device (PAL 14H4). The device has 14 inputs and 4 outputs and is obviously not the best way to implement this example, which only has four inputs, but it serves as an illustration. The other thing to note is that the traditional minimization techniques are still relevant and useful if they help to reduce the number and/or size of the product terms. This might enable designers to make the best use of available components.

From Equation (3.3), the inputs x_1, x_2, x_3, and x_4 are assigned to pins 1, 2, 3 and 4. The outputs f_1, f_2 and f_3 are taken from pins 17, 16 and 15 respectively. The others have no connections (NC), as shown in Fig. 3.12. This single IC replaces the five chips in Fig. 3.4(b) with plenty to spare. The benefits become more obvious when the number of inputs is large but the number of product terms is limited.

Programmable logic development systems (PLDS) are available to simplify the design, programming, and testing of PAL and similar programmable logic devices. The PLDS is basically a programmer, like an EPROM programmer, plus suitable adapters. One such a system employs a PAL assembler and simulator (PALASM) design adapter and a programming testing (P/T) adapter. The data are entered to the PALASM via a terminal in the form of Boolean equations. The PALASM design adapter translates the equations into a fuse pattern which is stored in a RAM. The P/T adapter is then used to program the device by blowing the appropriate fuses according to the fuse pattern.

Programming a PAL device like the one given in Fig. 3.12 takes about 10–15 minutes. Using the above mentioned PLDS, the designer must specify the family pin code, which is given in the manual, and enter the input data. The input data for this example consist of the device number, the variables assigned to the device pins and the source equations, given in the following format.

```
0 0 0 1  PAL14H4
0 0 0 2
0 0 0 3
0 0 0 4
0 0 0 5  X1 X2 X3 X4 NC NC NC NC NC GND NC NC NC NC F3 F2 F1 NC NC VCC
0 0 0 6  F1 = X1*X2*/X3 + /X2*X3*/X4 + X1*/X2*X3 + /X1*X2*/X3*/X4
0 0 0 7  F2 = X1*/X2*X3 + X2*/X3*X4 + /X1*X2*/X3*/X4
0 0 0 8  F3 = /X1*/X2*X4 + /X2*X3*/X4 + X1*/X2*X3 + X1*X2*/X3*/X4
```

The PALASM generates the fuse pattern. This may be displayed on the terminal screen. For this example the fuse pattern is displayed in the following form:

```
            0  0                           1  0                         2  0
0 0 0 0   X  -  X  -  -  X  -  -  -  -  -   -  -  -  -  -  -  -  -  -  -   -  -  -  -  -  -  -  -  -  -
0 0 2 8   -  X  -  -  X  -  -  -  -  X       -  -  -  -  -  -  -  -  -  -   -  -  -  -  -  -  -  -  -  -
0 0 5 6   -  X  X  -  X  -  -  -  -  -       -  -  -  -  -  -  -  -  -  -   -  -  -  -  -  -  -  -  -  -
0 0 8 4   X  -  -  X  -  X  -  -  -  X       -  -  -  -  -  -  -  -  -  -   -  -  -  -  -  -  -  -  -  -
0 1 1 2   -  X  X  -  X  -  -  -  -  -       -  -  -  -  -  -  -  -  -  -   -  -  -  -  -  -  -  -  -  -
0 1 4 0   X  -  -  -  -  X  -  -  X  -       -  -  -  -  -  -  -  -  -  -   -  -  -  -  -  -  -  -  -  -
0 1 6 8   X  -  -  X  -  X  -  -  -  X       -  -  -  -  -  -  -  -  -  -   -  -  -  -  -  -  -  -  -  -
0 1 9 6   X  X  X  X  X  X  X  X  X  X  X    X  X  X  X  X  X  X  X  X  X   X  X  X  X  X  X  X  X  X  X
0 2 2 4   -  X  -  X  -  -  -  -  X  -       -  -  -  -  -  -  -  -  -  -   -  -  -  -  -  -  -  -  -  -
0 2 5 2   -  X  -  -  X  -  -  -  -  .X      -  -  -  -  -  -  -  -  -  -   -  -  -  -  -  -  -  -  -  -
0 2 8 0   -  X  X  -  X  -  -  -  -  -       -  -  -  -  -  -  -  -  -  -   -  -  -  -  -  -  -  -  -  -
0 3 0 8   X  -  X  -  -  X  -  -  -  X       -  -  -  -  -  -  -  -  -  -   -  -  -  -  -  -  -  -  -  -
0 3 3 6   X  X  X  X  X  X  X  X  X  X  X    X  X  X  X  X  X  X  X  X  X   X  X  X  X  X  X  X  X  X  X
0 3 6 4   X  X  X  X  X  X  X  X  X  X  X    X  X  X  X  X  X  X  X  X  X   X  X  X  X  X  X  X  X  X  X
0 3 9 2   X  X  X  X  X  X  X  X  X  X  X    X  X  X  X  X  X  X  X  X  X   X  X  X  X  X  X  X  X  X  X
0 4 2 0   X  X  X  X  X  X  X  X  X  X  X    X  X  X  X  X  X  X  X  X  X   X  X  X  X  X  X  X  X  X  X
```

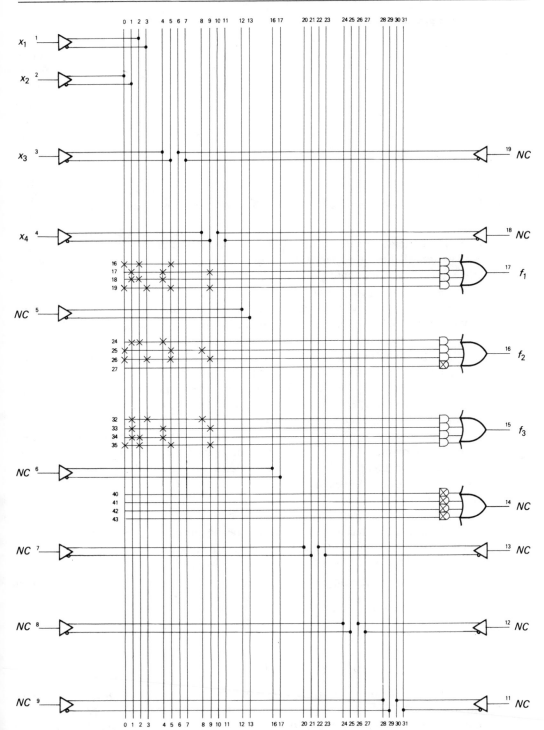

Fig. 3.12   PAL implementation of Equation (3.3) using PAL 14H4.
NC = No connection

3.8   STATIC HAZARDS

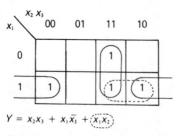

$$Y = x_2 x_3 + x_1 \bar{x_3} + (x_1 x_2)$$

Fig. 3.13(a)   K-map for logic function Y

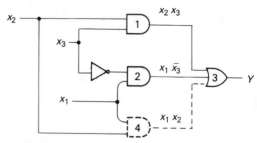

Fig. 13.13(b)   Circuit diagram for logic function Y

Consider the logic function Y and its associated circuit diagram shown in Fig. 3.13. First we ignore the dotted part of the diagrams and the logic expression. When all inputs are 1 the output of gate 1 is 1, the output of gate 2 is 0. The overall output of gate 3 is 1, which is the steady-state output.

When x_3 goes to 0, the input to gate 1 will immediately sense the change, but because of the delay in the inverter, gate 2 does not sense the change till some time later. As a consequence when the output of gate 1 changes to 0, the output of gate 2 is still 0, giving a 0 output at gate 3; when the change in x_3 propagates to gate 2 its output will change to 1, thereby changing the output of gate 3 to its steady-state value of 1.

This shows that an output of 0 is indicated at the output terminal Y momentarily until it reaches its proper steady-state value of 1. This is called a *static hazard*, which is simply a false output caused by the gate configuration. Hazards of this kind can be prevented logically by adding a third encirclement to the K-map as shown dotted. That is done by including all 1 to 1 transitions which are a unit distance apart in at least one encirclement. This results in the addition of gate 4 which holds the output during the transition. By extending this procedure, hazards can be eliminated from any two-level gate network.

When a PLA module is used, it has been stated that the K-map, or other methods, might be used to reduce the number of product terms. If this is done, a situation similar to that encountered in Fig. 3.13(b) could arise. Consider implementing the same logic function using a PLA as shown in Fig. 3.14(a).
Ignore p_3 for the time being. We have:

$$x_1 = x_2 = x_3 = 1$$
$$p_1 = 1$$
$$p_2 = 0$$
$$Y = 1$$

If x_3 changes to 0, p_1 changes to 0.

Now p_2 remains 0 for a short time until the change in x_3 propagates through the inverter, thus giving a temporary output of 0. After a short time, the steady-state output of 1 will be established. If p_3 is included, the output will be maintained at 1 as in the gate circuit.

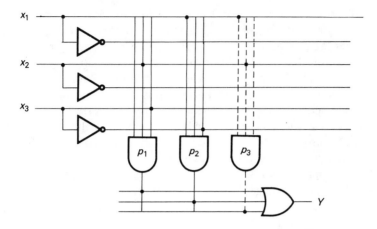

Fig. 3.14(a)   Possibility of hazard in a PLA device

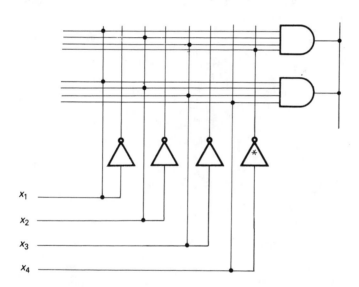

Fig. 3.14(b)   Possibility of hazard in a ROM device

Having said that, we must remember that we are dealing with a single-chip module, and the delay in the inverter is only a very small part of the total access time.

In the case of a ULM or a ROM module, all the basic switching terms are used; even then hazards could still arise due to the variations in the path lengths of the various signals.

Consider the addressing of the last two words in Fig. 3.9 reproduced in Fig. 3.14(b).

Suppose f_1 is programmed as 1 at words 14 and 15; therefore when x_1, x_2, x_3 and x_4 are 1111 or 1110, the output is 1. Assume we address word 15 first, i.e. all inputs are 1; then we change x_4 to 0 so as to address word 14. The change in x_4 will be detected by position 15 before position 14 because of the gate marked *. For a short time δt, no word is addressed and no output wil be shown. Once again we must realize that we are dealing with a single chip and that this delay is very small compared with the propagation delay specified by the manufacturer. This effect is not very serious in combinational logic where there is no feedback to aggravate the problem, but in some situations the hazard must be accounted for. This hazard is overcome by the designer by making sure that the ROM does not have to drive fast switching logic and carefully reading the appropriate data sheets.

The effect of this hazard on asynchronous systems will be considered in Chapter 5.

Example (3.1)
Design the necessary logic to drive a seven-segment display.

SOLUTION USING DISCRETE LOGIC GATES: The display consists of seven segments as shown in Fig. 3.15.

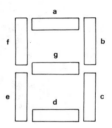

Fig. 3.15   Seven-segment display

To display the digits 0 to 9, Table 3.4 is constructed.

Digit (decimal)	Input A B C D	Output a b c d e f g	Display
0	0 0 0 0	1 1 1 1 1 1 0	0
1	0 0 0 1	0 1 1 0 0 0 0	1
2	0 0 1 0	1 1 0 1 1 0 1	2
3	0 0 1 1	1 1 1 1 0 0 1	3
4	0 1 0 0	0 1 1 0 0 1 1	4
5	0 1 0 1	1 0 1 1 0 1 1	5
6	0 1 1 0	0 0 1 1 1 1 1	6
7	0 1 1 1	1 1 1 0 0 0 0	7
8	1 0 0 0	1 1 1 1 1 1 1	8
9	1 0 0 1	1 1 1 0 0 1 1	9

Table 3.4   Table for the seven-segment display

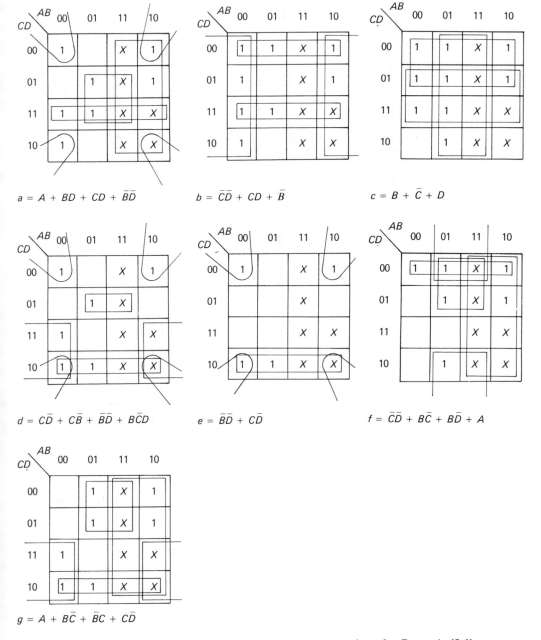

$a = A + BD + CD + \bar{B}\bar{D}$

$b = \bar{C}\bar{D} + CD + \bar{B}$

$c = B + \bar{C} + D$

$d = C\bar{D} + C\bar{B} + \bar{B}\bar{D} + \bar{B}CD$

$e = \bar{B}\bar{D} + C\bar{D}$

$f = \bar{C}\bar{D} + B\bar{C} + B\bar{D} + A$

$g = A + B\bar{C} + \bar{B}C + C\bar{D}$

Fig. 3.16 K-maps for simplifying the seven-segment expressions for Example (3.1)

The expressions for $a, b, \ldots, g$ are entered on the K-maps as shown in Fig. 3.16. The expressions can be implemented using suitable logic like AND, OR and NOT gates and used to drive the segments. This is shown in Fig. 3.17.

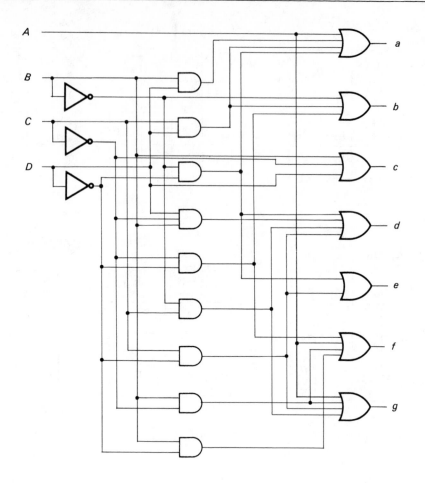

Fig. 3.17 Logic diagram for the problem of Example (3.1)
Note. Seven-segment display driver chips are available as shown in Appendix 3.1

ROM IMPLEMENTATION FOR EXAMPLE (3.1): To implement this problem using a ROM, or similar memory device ten words only are required each of 7 bits.

Assume that a four-input, eight-output device is available, though part of a larger device can be used.

The inputs A_0, ..., A_3 give a total of 16 address locations (words). The outputs are D_0, ..., D_7, only seven of which will be used.

Six words (words 10, ..., 15) can be left empty; in this case, however, some alphabets will be stored and can be displayed on the seven-segment display, just to illustrate that alphabets can be stored as well as decimal numbers. For other alphabets and symbols more complex displays containing more segments are required.

The truth table for the ROM then becomes as shown in Table 3.5. The seven outputs of the ROM are then used to drive the seven-segment display.

Address A_3 A_2 A_1 A_0	Output D_7 D_6 D_5 D_4 D_3 D_2 D_1 D_0	Display
0 0 0 0	X 1 1 1 1 1 1 0	0
0 0 0 1	X 0 1 1 0 0 0 0	1
0 0 1 0	X 1 1 0 1 1 0 1	2
0 0 1 1	X 1 1 1 1 0 0 1	3
0 1 0 0	X 0 1 1 0 0 1 1	4
0 1 0 1	X 1 0 1 1 0 1 1	5
0 1 1 0	X 0 0 1 1 1 1 1	6
0 1 1 1	X 1 1 1 0 0 0 0	7
1 0 0 0	X 1 1 1 1 1 1 1	8
1 0 0 1	X 1 1 1 0 0 1 1	9
1 0 1 0	X 1 1 1 0 1 1 1	A
1 0 1 1	X 1 0 0 1 1 1 0	C
1 1 0 0	X 1 0 0 1 1 1 1	E
1 1 0 1	X 1 0 0 0 1 1 1	F
1 1 1 0	X 0 1 1 0 1 1 1	H
1 1 1 1	X 0 0 0 1 1 1 0	L

Table 3.5   ROM table for Example 3.1 (X represents DON'T CARE)

Example (3.2)

Use a memory device to simulate the function of three-input AND, OR, NAND and NOR gates.

Input			Output			
A	B	C	AND f_1	OR f_2	NAND f_3	NOR f_4
0	0	0	0	0	1	1
0	0	1	0	1	1	0
0	1	0	0	1	1	0
0	1	1	0	1	1	0
1	0	0	0	1	1	0
1	0	1	0	1	1	0
1	1	0	0	1	1	0
1	1	1	1	1	0	0

Table 3.6   The truth table for Example (3.2)

SOLUTION: A memory with three address inputs is required. The memory device may be ROM, RAM, etc.

The input variables are used to select one of the eight words stored. Each word has four bits representing the four logic functions f_1 for AND, f_2 for OR, f_3 for NAND and f_4 for NOR. The function selector will select the required function and display it at the output terminal as shown in Fig. 3.18.

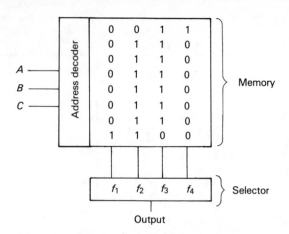

Fig. 3.18   Implementation of Example (3.2)

APPENDIX 3.1   A SAMPLE OF AVAILABLE ICs

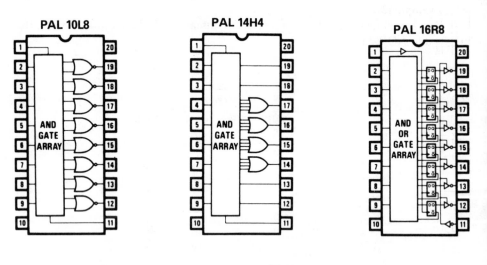

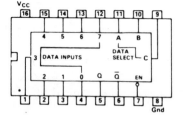

151 1 of 8 Data Selector/Multiplexer

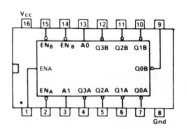

155 Dual 1 of 4
Decoder/Demultiplexer

02 Quadruple 2-input NOR gate

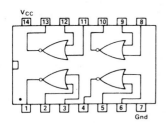

03 Quadruple 2-input NAND gate — open collector inputs

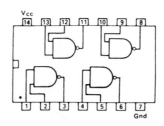

04 Hex inverter

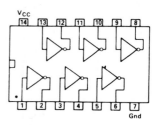

05 Hex inverter-open collector outputs

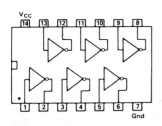

09 Quad 2-input AND gate-open collector outputs

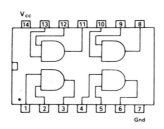

21 Dual 4-input AND gate

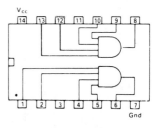

22 Dual 4-input NAND gate — open collector outputs

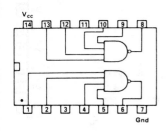

30 8-input NAND gate

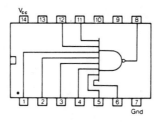

32 Quadruple 2-input OR gate

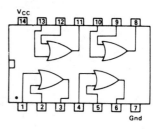

47 BCD-to-7 segment decoder/driver — open collector outputs

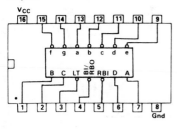

51 Dual 2-wide 2-input/3-input AND-OR-INVERT gate

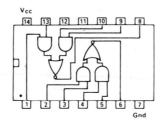

86 Quadruple 2-input exclusive OR gate

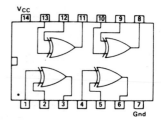

Bipolar FPLA (16X48X8)
82S100

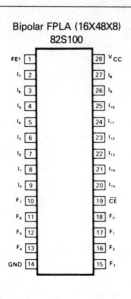

FE₁	1	28	V_CC
I₇	2	27	I₈
I₆	3	26	I₉
I₅	4	25	I₁₀
I₄	5	24	I₁₁
I₃	6	23	I₁₂
I₂	7	22	I₁₃
I₁	8	21	I₁₄
I₀	9	20	I₁₅
F₇	10	19	CE
F₆	11	18	F₀
F₅	12	17	F₁
F₄	13	16	F₂
GND	14	15	F₃

4096-Bit Dynamic RAM (4096X1)
2680

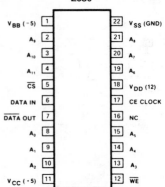

V_BB (-5)	1	22	V_SS (GND)
A₉	2	21	A₈
A₁₀	3	20	A₇
A₁₁	4	19	A₆
CS	5	18	V_DD (12)
DATA IN	6	17	CE CLOCK
DATA OUT	7	16	NC
A₀	8	15	A₅
A₁	9	14	A₄
A₂	10	13	A₃
V_CC (+5)	11	12	WE

1024-Bit Static MOS RAM (1024X1)
2115

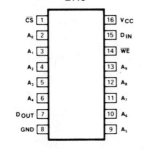

CS	1	16	V_CC
A₀	2	15	D_IN
A₁	3	14	WE
A₂	4	13	A₉
A₃	5	12	A₈
A₄	6	11	A₇
D_OUT	7	10	A₆
GND	8	9	A₅

4096-Bit Bipolar ROM (1024X4)
8228

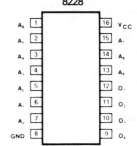

A₆	1	16	V_CC
A₅	2	15	A₇
A₄	3	14	A₈
A₁	4	13	A₉
A_C	5	12	O₁
A₃	6	11	O₂
A₂	7	10	O₃
GND	8	9	O₄

2048-Bit Electrically Programmable
MOS ROM (256X8)
1702A

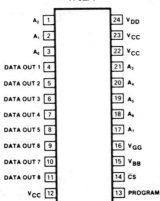

A₂	1	24	V_DD
A₁	2	23	V_CC
A₀	3	22	V_CC
DATA OUT 1	4	21	A₃
DATA OUT 2	5	20	A₄
DATA OUT 3	6	19	A₅
DATA OUT 4	7	18	A₆
DATA OUT 5	8	17	A₇
DATA OUT 6	9	16	V_GG
DATA OUT 7	10	15	V_BB
DATA OUT 8	11	14	CS
V_CC	12	13	PROGRAM

2048-Bit Bipolar PROM (256X8)
82S114

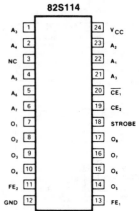

A₃	1	24	V_CC
A₄	2	23	A₂
NC	3	22	A₁
A₅	4	21	A₃
A₆	5	20	CE₁
A₇	6	19	CE₂
O₁	7	18	STROBE
O₂	8	17	O₈
O₃	9	16	O₇
O₄	10	15	O₆
FE₂	11	14	O₅
GND	12	13	FE₁

APPENDIX 3.2   BIPOLAR MASK PROGRAMMABLE LOGIC ARRAY

DESCRIPTION

The 82S200 (tri-state outputs) and the 82S201 (open collector outputs) are Bipolar Programmable Logic Arrays, containing 48 product terms (AND terms), and 8 sum terms (OR tems). Each OR term controls an output function which can be programmed either true active-high (F_p), or true active-low ($F_p{}^*$). The true state of each output function is activated by any logical combination of 16-input variables, or their complements, up to 48 terms. Both devices are mask programmable by supplying to Signetics Program Table data in one of the formats specified in this data sheet.

The 82S200 and 82S201 are fully TTL compatible, and include chip enable control for expansion of input variables, and output inhibit. They feature either open collector or tri-state outputs for ease of expansion of product terms and application in bus-organized systems.

Both devices are available in commercial and military temperature ranges. For the commercial temperature range (0°C to +75°C) specify N82S200/201, I or N, and for the military temperature range (–55°C to +125°C) specify S82S200/201, I.

PLA EQUIVALENT LOGIC PATH

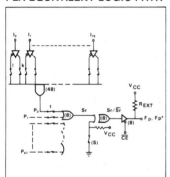

LOGIC FUNCTION

Typical Product Term:
$P_0 = I_0 \cdot I_1 \cdot \overline{I_2} \cdot I_5 \cdot \overline{I_{13}}$

Typical Output Functions:
$F_0 = (\overline{CE}) + (P_0 + P_1 + P_2)$ @ S = Closed
$F_0^* = (\overline{CE}) + (\overline{P_0} \cdot \overline{P_1} \cdot \overline{P_2})$ @ S = Open

NOTE

For each of the 8 outputs, either the function Fp (active-high) or F$_p^*$ (active low) is available, but not both. The required function polarity is programmed via link (S).

APPLICATIONS

- **CRT display systems**
- **Random logic**
- **Code conversion**
- **Peripheral controllers**
- **Function generators**
- **Look-up and decision tables**
- **Microprogramming**
- **Address mapping**
- **Character generators**
- **Sequential controllers**
- **Data security encoders**
- **Fault detectors**
- **Frequency synthesizers**

TRUTH TABLE

MODE	Pn	$\overline{CE}$	Sr $\overset{?}{=}$ f(Pn)	Fp	F$_p^*$
Disabled (82S201)	X	1	X	1	1
Disabled (82S200)	X	1	X	Hi-Z	Hi-Z
Read	1	0	Yes	1	0
	0	0		0	1
	X	0	No	0	1

PIN CONFIGURATION

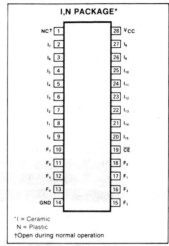

THERMAL RATINGS

TEMPERATURE	MILITARY	COMMERCIAL
Maximum junction	175°C	150°C
Maximum ambient	125°C	75°C
Allowable thermal rise ambient to junction	50°C	75°C

LOGIC DIAGRAM

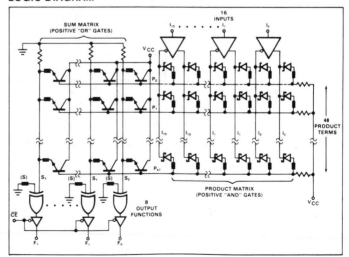

Courtesy of Mullard Limited

4

Synchronous Sequential Circuits

4.1   INTRODUCTION

In the last few chapters, we have considered combinational logic circuits, where the outputs are functions only of the present inputs. Now we consider sequential circuits, in which the outputs at any given time are functions of the external inputs as well as some stored information determined by previous inputs.

The block diagram of Fig. 4.1 shows that a sequential system may be described as a combinational circuit with a memory section to remember past inputs and feedback. The variables that represent past inputs or the state of the circuit before the present input is applied, are called the *state variables* or *secondary variables*. The clock is used in synchronous circuits only.

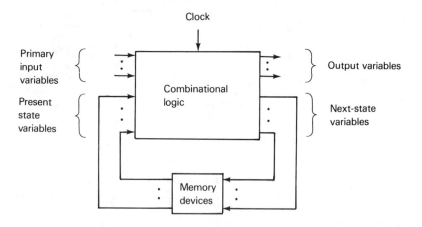

Fig. 4.1   A block diagram of a sequential system

4.2   STORAGE DEVICES

The devices used to remember the state of the circuit are called bistables or flip-flops. These are capable of storing one bit of information which may be logic 1 or logic 0.

Different types of flip-flop are available nowadays. They may be designed using discrete components or standard gates or they may be obtained on chips. A few widely used chips are noted in Appendix 4.1 but it will be useful to look now at the construction and behaviour of the most common types.

4.2.1 Set-Reset Flip-Flop (SR/FF)

This device has two inputs, the set S and the reset R, as shown in Fig. 4.2. When a logic 1 is applied to S, the flip-flop is set to 1 ($Q = 1$). When a logic 1 is applied to R, the flip-flop is reset to 0 ($Q = 0$).

Fig. 4.2   A block diagram of an SR/FF

If Q and Q_+ represent the present and next states of the output, then the truth table can be drawn as in Table 4.1(a). Remember that the complement of the output, $\bar{Q}$, is also available.

| Inputs | | Outputs | |
S	R	Q	Q_+
0	0	0	0
0	1	0	0
1	0	0	1
1	1	0	X
0	0	1	1
0	1	1	0
1	0	1	1
1	1	1	X

(*X* represents DON'T CARE/CAN'T HAPPEN)

Table 4.1(a)   Truth table for SR/FF

| Outputs | | Inputs | |
Q	Q_+	S	R
0	0	0	X
0	1	1	0
1	0	0	1
1	1	X	0

(*X* represents DON'T CARE/CAN'T HAPEN)

Table 4.1(b)   Excitation table for SR/FF

The truth table shows that if the output is low (logic 0) it will stay low unless set by a 1 on the set input. If the output is high (logic 1) it will stay high unless reset by a 1 on the reset input.

The x shows that the device cannot be set and reset simultaneously. In other words $S = R = 1$ is not allowed.

Another useful table is the excitation table [Table 4.1(b)] which indicates the inputs required to achieve a particular transition. This is obtained from Table 4.1(a).

Table 4.2 shows a condensed version of the truth table which is easy to remember.

It is also useful to develop an expression for the next state Q_+ in terms of the present state Q and the inputs S and R. This is shown in Fig. 4.3.

| Inputs | | Output |
S	R	Q
0	0	No change
1	0	Set ($Q=1$)
0	1	Reset ($Q=0$)
1	1	Not allowed

Table 4.2 Condensed truth table for *SR*/FF

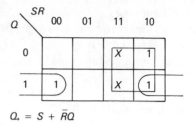

$Q_+ = S + \bar{R}Q$

Fig. 4.3 Derivation of the next state equation

The *SR*/FF can be implemented using NAND gates as in Fig. 4.4.

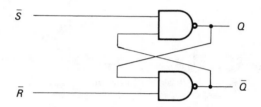

Fig. 4.4 NAND implementation of an *SR*/FF

From Fig. 4.4, $\bar{Q} = \overline{\bar{R} \cdot Q}$

$$Q = \overline{\bar{S} \cdot \overline{Q\bar{R}}} = S + \bar{R}Q$$

Similarly the *SR*/FF can be implemented using NOR gates, as in Fig. 4.5.

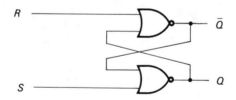

Fig. 4.5 NOR implementation of an *SR*/FF

From Fig. 4.5, $\bar{Q} = \overline{S + Q}$

$$
\begin{aligned}
Q &= \overline{R + \overline{S + Q}} \\
&= \bar{R}(S + Q) \\
&= \bar{R}S + \bar{R}Q \\
&= \bar{R}S + \bar{R}Q + RS \qquad \text{since } SR = 0 \\
&= S + \bar{R}Q
\end{aligned}
$$

Up to now, the *S* and *R* inputs have been applied directly. In a clocked *SR*/FF, the logic levels on *S* and *R* are applied only when the device is clocked. This is shown in Fig. 4.6. The block diagram representation is given in Fig. 4.7.

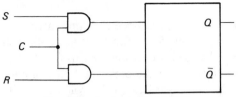

Fig. 4.6   A clocked *SR*/FF

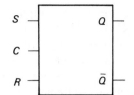

**Fig. 4.7   A block diagram
of a clocked *SR*/FF**

Master–Slave *SR*/FF

Fig. 4.8 shows the construction of a master–slave *SR*/FF. To illustrate its operation consider the following condition:

$$C = 0, \quad Q_1 = 0, \quad Q_2 = 0$$

If *S* is made 1 and *R* is made 0, and the clock *C* is kept 0, the flip-flop does not change state since the outputs of gates 1 and gate 2 are low. If *C* is made 1, the output of gate 1 goes high and the output of gate 2 stays low. The output of the inverter is low, thus inhibiting gates 3 and 4.

The output of gate 1 is applied to the set input of the master flip-flop. This will set the output Q_1 to the logic level 1.

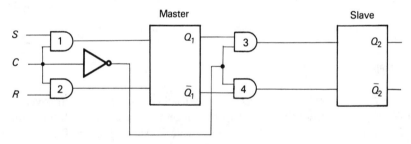

Fig. 4.8   Master–slave *SR*/FF

The output of the slave flip-flop remains at logic level 0 and the outputs $Q_1 = 1$, $Q_2 = 0$ continue as long as *C* is at 1.

If the clock goes low (*C* changes from 1 to 0), gates 1 and 2 are disabled, gates 3 and 4 are enabled. The output $Q_1(= 1)$ is now applied to the set input of the slave flip-flop causing its output Q_2 to change from 0 to 1.

In brief, the master changes state when the clock changes from 0 to 1. The slave changes state when the clock changes from 1 to 0. This is illustrated in Fig. 4.9.

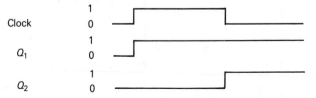

Fig. 4.9   Waveform of a master–slave flip-flop

Asynchronous Inputs

The S and R inputs are called *synchronous inputs* since they only affect the state of the flip-flop when the clock pulse is present. Some flip-flops have other inputs called *preset* (or *set*) and *clear* that can be used for setting the flip-flop to 1 or resetting it to 0 by applying the appropriate signal to the preset and clear inputs. These inputs can change the state of the flip-flop regardless of the synchronous inputs or the clock.

If the signal required is logic 0 (active low) a circle on the input connection is shown, as in Fig. 4.10. If the signal required is logic 1 (active high) no circle is shown.

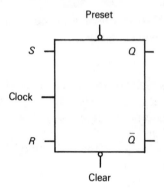

Fig. 4.10   An *SR*/FF with preset and clear inputs

4.2.2 JK Flip-Flop

The *JK* flip-flop is the most commonly used type of flip-flop. It is very similar to the *SR* flip-flop except that a *JK* flip-flop accepts a logic 1 on both the *J* and *K* inputs. The truth table for a *JK*/FF shown in Table 4.3 shows that the flip-flop changes state when both inputs are at 1. Table 4.1 shows that the state of the *SR*/FF is indeterminate when both inputs are at 1.

Input		Output		Excitation requirement			
		Present state	Next state				
J	*K*	*Q*	Q_+	*Q*	Q_+	*J*	*K*
0	0	0	0 ⎫				
0	1	0	0 ⎭	0	0	0	*X*
1	0	0	1 ⎫				
1	1	0	1 ⎭	0	1	1	*X*
0	0	1	1 ⎫				
1	0	1	1 ⎭	1	1	*X*	0
0	1	1	0 ⎫				
1	1	1	0 ⎭	1	0	*X*	1

Table 4.3   Truth table for *JK*/FF

Apart from the toggle operation, the table indicates that an input $J = 1$ sets the flip-flop to 1 while an input $K = 1$ resets the flip-flop to 1. In this sense J and K are equivalent to S and R. In fact a JK flip-flop can be made from an SR flip-flop as shown in Fig. 4.11.

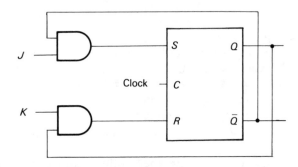

Fig. 4.11   Logic diagram of a JK/FF

Fig. 4.12 is that of a master–slave JK/FF. The diagram indicates that the device is that of a master slave SR/FF with feedback from Q and $\bar{Q}$ to K and J respectively.

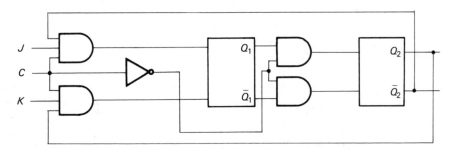

Fig. 4.12   Logic diagram of master–slave JK/FF

The block diagram for a JK/FF is given in Fig. 4.13.

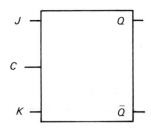

Fig. 4.13   Block diagram for a JK/FF

In clocked *JK* flip-flops the signal levels on *J* and *K* have no effect until the device is clocked. If the flip-flop changes when the clock rises from 0 to 1 it is called *positive-edge triggered flip-flop*. If the flip-flop changes when the clock signal changes from 1 to 0 it is called *negative-edge triggered flip-flop*. This is sometimes indicated by a small circle at the clock terminal.

Asynchronous Inputs
The *J* and *K* inputs in a clocked flip-flop can affect the output only when the device is clocked. They are called synchronous inputs (as in *SR/FF*).

Some flip-flops have asynchronous inputs that are independent of the synchronous inputs and the clock. These are normally called preset (or set) and clear.

A signal on the preset input sets the output to 1 while a signal on the clear input sets the output to 0, regardless of the conditions at the synchronous inputs or the clock.

If the signal required to set or reset the flip-flop is logic 0 (active low) a circle is shown as in Fig. 4.14. If the signal required to set or clear the flip-flop is logic 1 (active high), the circle is omitted.

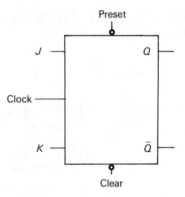

Fig. 4.14 A *JK* flip-flop with preset and clear inputs

Next-state Equation
It should be clear that all types of *JK* flip-flops obey the truth table given in Table 4.3. From this table an expression can be derived for the next state Q_+ in terms of the present state Q and the inputs *J* and *K*, as shown in Fig. 4.15.

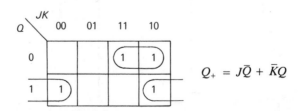

$$Q_+ = J\bar{Q} + \bar{K}Q$$

Fig. 4.15   Next-state equation for a *JK* flip-flop

The change function

Another way of looking at the flip-flop is in terms of its change function X_Q, which we shall find very useful later on. X_Q is 1 if the output Q changes and is 0 if the output Q does not change. Consider the truth table given in Table 4.4.

A K-map is now plotted for X_Q, as in Fig. 4.16.

J	K	Q	Q_+	X_Q
0	0	0	0	0
0	1	0	0	0
1	0	0	1	1
1	1	0	1	1
0	0	1	1	0
0	1	1	0	1
1	0	1	1	0
1	1	1	0	1

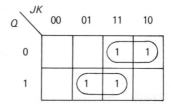

Table 4.4   The change function for a *JK*/FF

Fig. 4.16   K-map for the change function

From the K-map we find the following expression for X_Q:

$$X_Q = J\bar{Q} + KQ \tag{4.1}$$

The thing to notice at this stage is that the coefficient of $\bar{Q}$ is J and the coefficient of Q is K. The use of this will become apparent when designing sequential circuits and counters.

4.2.3   D Flip-Flop

The D flip-flop shown in Fig. 4.17 is different from the *SR* and *JK* flip-flops in that it only has one input. The flip-flop transfers the logic value at the input D to the output when a clock pulse is applied. It can therefore be used as a delay device or for storing a single bit which may be 1 or 0.

The operation of the flip-flop can be summarized as shown in Table 4.5. It can be seen that the next state is the same as D, i.e. $Q_+ = D$.

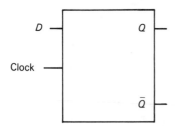

Input D	Output Q_+
0	0
1	1

Fig. 4.17   Block diagram for a *D* flip-flop

Table 4.5   Truth table for a *D*/FF

As with *JK* and *SR* flip-flops, some *D* flip-flops have preset and clear inputs. A *D* flip-flop can be made using either a *JK* or *SR* flip-flop with one inverter as shown in Fig. 4.18.

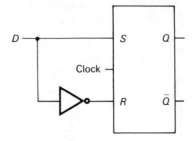

Fig. 4.18   An *SR*/FF connected as a *D*/flip-flop

4.2.4 T-type Flip-Flop

The trigger *T*-type flip-flop (*T*/FF) is a less popular type which can be obtained by connecting both *J* and *K* to logic 1 as shown in Fig. 4.19.

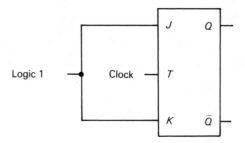

Fig. 4.19   *T*-type flip-flop

The state of the device changes each time an input pulse is received. The equation for the next state is the EXclusive-OR of *Q* and *T* as can be seen from the state table given in Table 4.6.

$$Q_+ = \bar{Q}T + Q\bar{T}$$

Input T	Present state Q	Next state Q_+
0	0	0
1	0	1
0	1	1
1	1	0

Table 4.6   Truth table for a *T*/FF

4.3   REGISTERS

Memory and shift registers are constructed with flip-flops and used for storing binary data. One flip-flop is used for each bit. A memory register holding the number 1001 is shown in a block diagram form in Fig. 4.20.

| 1 | 0 | 0 | 1 |

Fig. 4.20   A four-bit memory register block diagram

Different types of flip-flop are used for this purpose. If *D*-type flip-flops are used then a two-bit memory register can be constructed as shown in Fig. 4.21.

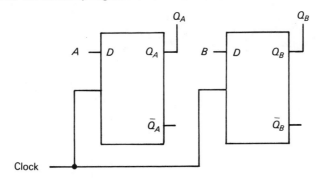

Fig. 4.21   A two-bit memory regis̤ci circuit

The input is applied by *A* and *B* and transfered to the outputs Q_A and Q_B when a clock pulse is applied. We note here that the data are loaded in parallel and are available at the output in parallel. This is known as *parallel in* and *parallel out*. We can also observe that the complement of the output is also available if required.

The data stored in Fig. 4.21 cannot be moved from one storage cell to another. If, however, the output Q_A is connected to input *B* the data can be shifted from left to right and the arrangement is called a shift register, as shown in Fig. 4.22.

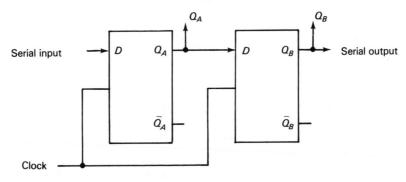

Fig. 4.22   A two-bit shift register

Let Q_A and Q_B be set to 0 initially. The first bit is applied to the input of flip-flop A; when the device is clocked this first bit is transfered to the output Q_A. This is now applied to the input of flip-flop B and the second bit of the input data applied to the input of flip-flop A. When the second clock pulse arrives, the first bit moves to the output of flip-flop B (i.e. Q_B) while the second bit of the input data moves to the output Q_A.

The data can now be read in parallel form Q_B and Q_A simultaneously. If, however, a serial output is desired, the first bit is taken from Q_B and another clock pulse is applied to move the second bit to Q_B. It is clear that more flip-flops can be added to obtain a larger register.

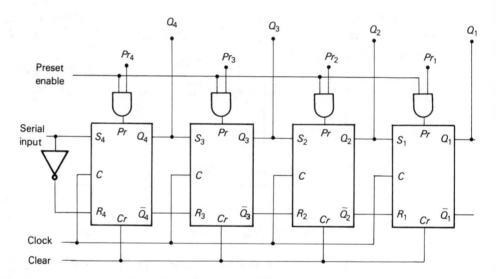

Fig. 4.23   A four-bit shift register using *SR* flip-flops

Fig. 4.23 shows a different shift register using *SR* master – slave flip-flops (*JK* master–slave flip-flops can also be used), with the additional facilities of preset and clear. The circuit can be used in different ways:

(i)   Serial Input

The shift register can be cleared using the clear input. This sets all outputs, namely Q_1 to Q_4, to 0. The least significant bit (lsb) of the data is now applied to the serial input. When the device is clocked, the data are transfered to Q_4. All other outputs remain at 0. Since the output of each stage is connected to the input of the next stage, the new output Q_4 is now applied to S_3 while the second lsb of the data is applied to S_4. After the second clock pulse, these two bits are transfered to Q_3 and Q_4 respectively, and Q_2 and Q_1 remain at 0. After four clock pulses, the four flip-flops contain four bits of data; each bit can be 0 or 1.

The data can be read from Q_1, Q_2, Q_3 and Q_4 in parallel; hence serial-in/ parallel-out register or serial-to-parallel conversion. If further clock pulses are applied the data can be taken serially from Q_1, hence serial-in/serial-out. We note that the spacing in time between the binary data can be changed if required.

(ii) Parallel Input
If the input data are available in parallel, the bits can be applied to the preset inputs with lsb applied to Pr_1 and the most significant bit (msb) applied to Pr_4. When the chip is enabled, by the preset enable, the data are written in the shift register. Obviously the same data can be taken again in parallel from the output Q_1 to Q_4 or taken from Q_1 in a serial form by applying further clock pulses to shift the data serially along the register. The parallel-to-serial convention is useful in transmitting data over a single line instead of using one line for each bit as in parallel transmission.

4.3.1 Bi-directional Shift Register

It is necessary in many applications to shift the data in a register in one direction or another by a specified number of positions. A shift right moves all bits stored in the register to the right, while a shift left moves all bits to the left.

The basic idea can be seen from Fig. 4.24, which illustrates the shift operation for a single stage.

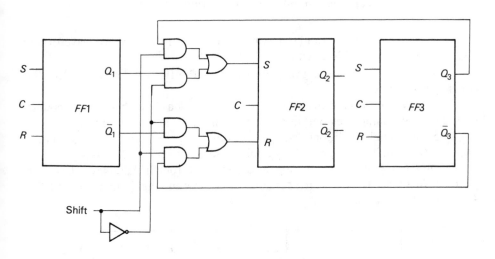

Fig. 4.24   A block diagram for a bidirectional shift register

From Fig. 4.24 it can be seen that a logic 0 on the shift input will enable the output of FF1 to go to the input of FF2 – hence a shift right. A logic 1 on the other hand will enable the output of FF3 to go to the input of FF2 – hence a shift left. It is also possible to use other types of flip-flop or use separate lines for the shift left and shift right functions.

4.4   COUNTERS

4.4.1 Ring Counters

If the output of the least significant stage of a shift register is connected to the input of the most significant stage, a ring counter is obtained. Fig. 4.25 shows a three-stage ring counter. If one stage is at logic 1 while the others are at logic 0, the 1 will circulate around the register and can be seen at any stage once every three pulses.

This is called a *mod-3* or *divide-by-three* ring counter. For *n*-stages we have a mod-*n* ring counter.

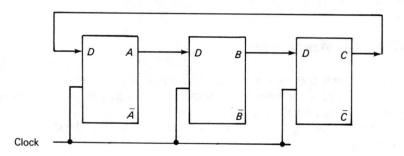

Fig. 4.25   Mod-3 shift register ring counter

The circuit behaviour can be shown with the aid of the idealized waveform of Fig. 4.26 or the truth table given in Table 4.7.

We note that a mod-*n* ring counter requires *n* flip-flops and no other decoding circuitry is required.

If the five-bit shift register in Appendix 4.1 is used, a divide-by-five ring counter is obtained by feeding back output Q_E (pin 10) to the serial input (pin 9).

For the initial conditions, we may set output $Q_A = 1$ and all other outputs $Q_B, \ldots, Q_E = 0$. This is achieved by applying logic 1 to A (pin 2) and logic 0 to B, $\ldots, E$ (pins 3, 4, 6 and 7) and making the preset enable (pin 8) high. Next the clear input (pin 16) is made high, the preset enable is made low and the circuit is clocked. The 1 on Q_A will circulate as usual.

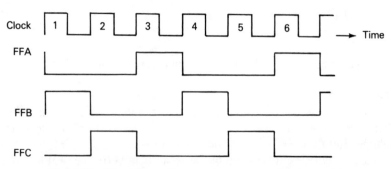

Fig. 4.26   Waveform diagram for the circuit of Fig. 4.25

Count	Output A	Output B	Output C
0	1	0	0
1	0	1	0
2	0	0	1
3	1	0	0
4	0	1	0
5	0	0	1
6	1	0	0
7	0	1	0
.			
.			
.			

Table 4.7   Truth table for the circuit of Fig. 4.25

All flip-flops can be cleared by applying logic 0 to the clear and preset enable. Presetting and clearing are independent of the level of the clock input.

4.4.2 Twisted Ring Counters

If the complemented output of the least significant stage of a shift register is fed back to the input of the most significant stage as shown in Fig. 4.27, we have a twisted ring counter.

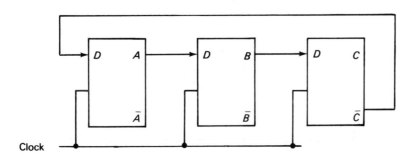

Fig. 4.27   Mod-6 twisted ring counter

If all the outputs are initally at logic 0, then $\bar{C}(\bar{Q}_c)$ is at logic 1. This is fed back to the D input of the A stage.

When the device is clocked, A becomes 1, B and C stay at 0. After the second clock pulse A and B become 1, C stays at 0. After the third clock pulse, A, B and C become 1. Then $\bar{C} = 0$. After the fourth clock pulse A becomes 0, B and C stay at 1, and so on, as shown in the truth table given in Table 4.8.

The truth table indicates that this twisted ring counter has six distinguishable states (mod-6). In general, a mod-n twisted ring counter requires $n/2$ flip-flops.

Count	Output sequence		
	A	B	C
0	0	0	0
1	1	0	0
2	1	1	0
3	1	1	1
4	0	1	1
5	0	0	1
0	0	0	0
1	1	0	0

Table 4.8   Truth table for a mod-6 twisted ring counter

The truth table also shows that additional logic is required if we are to read the output of the circuit. In fact, six two-input decoding gates are required, or 3–6 decoders. This is shown in Fig. 4.28. Any of these gates can be used for a divide-by-six.

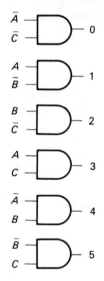

Fig. 4.28   Decoding logic for the twisted ring counter of Fig. 4.27

4.4.3 Asynchronous Counters

Counting circuits are a class of sequential circuits that require no external input and may be clocked (synchronous) or unclocked (asynchronous). Let us consider first the asynchronous type.

Binary Asynchronous Counter
Consider the four-bit binary counter in Fig. 4.29. The reset inputs reset all flip-flops to 0; the count is initially at 0 0 0 0. This is a four-bit binary counter, which could count from 0000 (decimal 0) to 1111 (decimal 15) as shown in Table 4.9.

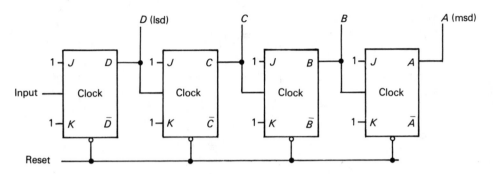

Fig. 4.29   A four-bit binary counter

Count	Binary output			
	A	B	C	D
0	0	0	0	0
1	0	0	0	1
2	0	0	1	0
3	0	0	1	1
4	0	1	0	0
5	0	1	0	1
6	0	1	1	0
7	0	1	1	1
8	1	0	0	0
9	1	0	0	1
10	1	0	1	0
11	1	0	1	1
12	1	1	0	0
13	1	1	0	1
14	1	1	1	0
15	1	1	1	1

Table 4.9   Truth table for the counter in Fig. 4.29

All flip-flops are connected as toggle flip-flops with their J and K inputs connected to 1 and change one after the other. For example when the first input arrives, FFD will change to 1, all others are still at 0. The second input changes D back to 0. The 1 to 0 transition on D changes C to 1, resulting in a 1 on C while all others are at 0. The third input changes D back to 1, FFC remains at 1 and all others are 0. The least significant position (FFD) will change every time there is an input. The second flip-flop (FFC) will change for every other input. FFB changes every four inputs. FFA changes every eight inputs. In fact each flip-flop changes when its input changes from 1 to 0 (negative edge). After a count of 15 FFD will return to 0. This 1 to 0 change on FFD will change FFC to 0, which changes FFB to 0. The change on FFB will in turn change A to 0, with a return to the initial count of 0000.

Reverse Binary Asynchronous Counter

A reverse asynchronous counter (down counter) can be achieved by simply con-

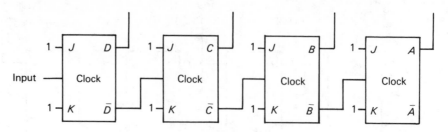

Fig. 4.30 Reverse binary counter

necting the complemented output of each stage to the clock input of the next, as shown in Fig. 4.30. This will function as an up counter if the outputs are taken from $\bar{D}$, $\bar{C}$, $\bar{B}$ and $\bar{A}$.

Up/Down Asynchronous Binary Counter

The circuits in Fig. 4.30 and Fig. 4.29 can be combined in a circuit that counts up or down depending on the signal level applied to the control input.

In Fig. 4.31 if the control input is high (logic 1), gates 1, 2 and 3 are enabled gates 4, 5 and 6 are disabled. This allows the changes at true outputs to go through; hence we have an up counter as in Fig. 4.29. Similarly a logic 0 at the control will enable gates 4, 5 and 6 and allow the complement output to go through, resulting in a down counter as in Fig. 4.30.

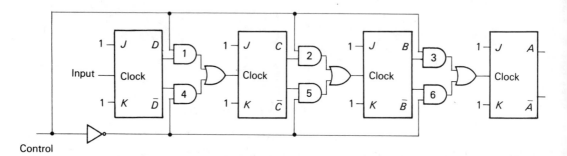

Fig. 4.31 Up/down counter

Decimal Asynchronous Counter (Mod-10)

A decimal counter is required to count from 0 to 9 and reset back to 0.

Looking at Table 4.9, we can see that the counter in Fig. 4.32(a) is required to count from 0000 to 1001 as shown in Table 4.10. The counter starts to count as in Table 4.9.

In counting 1010, FFA and FFC are both 1 for the first time. This condition can be used to reset all the flip-flops to 0. If logic 0 resets the flip-flops, a NAND gate can be used as shown in Fig. 4.32(a).

		Binary output		
Count	A	B	C	D
0	0	0	0	0
1	0	0	0	1
2	0	0	1	0
3	0	0	1	1
4	0	1	0	0
5	0	1	0	1
6	0	1	1	0
7	0	1	1	1
8	1	0	0	0
9	1	0	0	1
0	0	0	0	0

Table 4.10   Truth table for a decimal counter

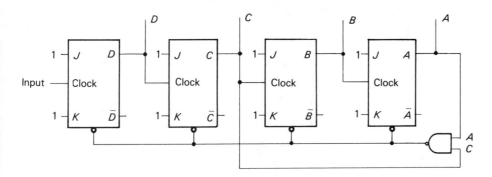

Fig. 4.32(a)   Decimal counter

The decimal, also called decade or binary coded decimal (BCD), counter is frequently used in practice. It is useful to look at another version as shown in Fig. 4.32(*b*).

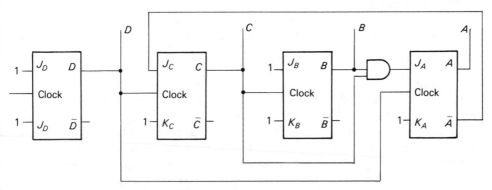

Fig. 4.32(b)   Alternative decimal counter

The counter starts like the counters in Fig. 4.32(a) and Fig. 4.29. During counts 0 to 7, J_C is 1 since $\bar{A}$ is 1, and hence flip-flops B, C and D toggle for every 1 to 0 transition at their inputs (clock), assuming negative-edge triggered flip-flops.

At count 7 (0111) J_A is 1; therefore a 1 to 0 transition at D will change A to 1 while B, C and D change to 0 as usual, resulting in count 8 (1000). Since A is now 1, $\bar{A}$ is 0. $\bar{A}$ is connected to J_C and will hold C at 0; B is also held at 0. The next input changes D to 1 (1001). J_A is now 0 and flip-flop A will be reset to 0 at the next 1 to 0 transition at D. The counter returns to its initial state of 0000.

Self-Stopping Counter

In Fig. 4.32(a) the output of the NAND gates was used to reset all the flip-flops and hence go back to zero. It is possible to use this signal to actually stop the counter. This is shown in Fig. 4.33.

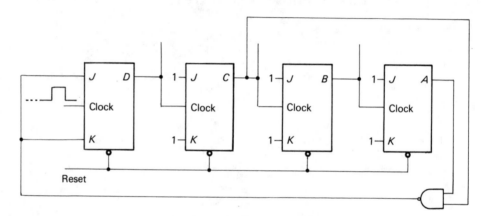

Fig. 4.33   Self-stopping counter

While the circuit counts from 0 to 9 the output of the NAND gate is 1. The circuit behaves as in Fig. 4.32(a). When both A and C become 1 at count 10 (1010) the output of the NAND gates becomes zero. Flip-flop D cannot change after that unless the counter is reset externally.

Similarly one could stop at other counts. For example A and B are fed to the NAND gate to stop at count 1100, B and C to stop at 0110, A B C to stop at 1110, and so on.

Example (4.1)

Using the 50 Hz 240 V mains supply, design a 12-hour digital clock

SOLUTION: In digital clocks a stable clock signal is required to drive the clock. This is normally obtained using a crystal oscillator. In this example we are required to use the mains supply. Also in digital clocks batteries are used. In this case we should try to use the mains again to supply our circuit. We shall assume TTL chips are available. These require a 5 V supply.

The design of d-c power supplies is outside the scope of this book, but a simple power supply can be constructed using a step-down transformer, a rectifier and a regulator. A 5 V 1 A supply is sufficient.

The signal taken from the secondary of the transformer is a 50 Hz sinusoidal wave. This can be converted to a square wave pulse using a Schmitt trigger. The 50 pulses per second (pps) square pulse can be changed to a 1 pps using a mod-50 (divide-by-50) counter (Fig. 4.34).

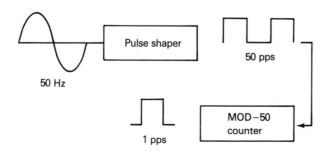

Fig. 4.34 Generation of the clock pulse

The 1 pps signal is passed through a mod-10 (BCD) and a mod-6 (divide by six) counter. These two counters will count from 0 to 59 (60 s).

The mod-10 counter counts from 0 to 9, which is 0000 to 1001, and resets to zero. The most significant digit (msd) is used to trigger the mod-6 counter. When the msd changes from 1 to 0 (once for every 10 input pulses) the mod-6 counter is incremented by 1. The mod-6 counter counts from 000 to 101 and resets to zero, in other words the two counters act as tens and units for the seconds section. The msd of the mod-6 counter changes to zero once every sixty seconds and hence can be used as a 1 pulse per minute (1 ppm) signal to trigger the minutes section.

The minutes section is identical to the seconds section and counts from 0 to 59 before reseting. In other words the msd of the mod-6 counter changes from 1 to 0 once every 60 minutes. The 1 pulse-per-hour signal is used to trigger the last section, which counts the hours.

The hours section is required to count up to twelve. The units section is a mod-10 (BCD) counter but the tens unit needs only two states 0 or 1. Hence one flip-flop can be used. The BCD counter could be the circuit already described with slight modification imposed by the fact that the hours must return to 1 o'clock after 12 o'clock and not to 0. The complete arrangement is shown in Fig. 4.35.

The 1 to 0 transition on A will change H to 1. Now we have 10 (1 0000) then 11 (1 0001) and 12 (1 0010).

Next the circuit will try to go to (1 0011), at which point C D and H are all 1s. This state is used to clear all the flip-flops except D so that the correct state of 1 o'clock (0 0001) is arrived at after 12. This can be achieved simply by using a NAND gate with CDH as inputs, as shown in the hours section of the clock.

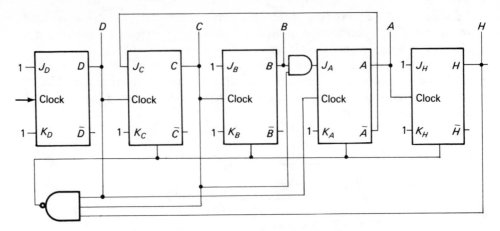

Fig. 4.35 The hours section of the digital clock

H	A	B	C	D	
0	0	0	0	0	
0	0	0	0	1	
0	0	0	1	0	
0	0	0	1	1	
0	0	1	0	0	0–9
0	0	1	0	1	
0	0	1	1	0	
0	0	1	1	1	
0	1	0	0	0	
0	1	0	0	1	
1	0	0	0	0	
1	0	0	0	1	
1	0	0	1	0	

The outputs of the counters are fed to decoders (BCD to a seven-segment decoder driver), before being displayed on a suitable device (seven segment display).

It should be emphasized that most of the counters required should be available in standard IC form. Further a 24-hour clock can be designed in a simlar manner. In fact the seconds and minutes sections are the same. The hours section requires a mod-24 instead of a mod-12 counter.

4.5   DESIGN OF SYNCHRONOUS COUNTERS

An asynchronous circuit has no clock pulse to control its operation. The circuit operates at its own speed moving from one state to another. Since the circuit does not need to wait for the clock pulse, one expects such a circuit to be fast, but also one expects some problems if the inputs change before the circuit has a chance to settle.

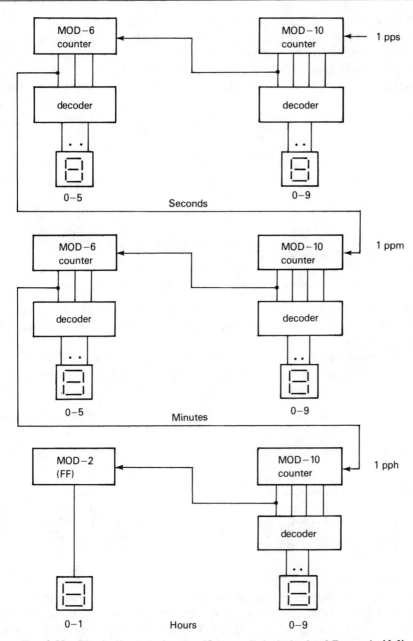

Fig. 4.36   Block diagram for the 12-hour digital clock of Example (4.1)

These kind of problems will be discussed when we deal with asynchronous circuits. The rest of this chapter, however, will be devoted to synchronous (clocked) circuits which only operate when clocked. The clock pulses are equally spaced square pulses. The time between the pulses is long enough for the circuit to settle.

A Mod-8 Binary Counter

To illustrate the procedure, consider the design of a mod-8 binary counter. A mod-8 binary counter requires three flip-flops; it counts from 000 to 111 and then repeats. The count sequence is shown in Table 4.11. From the table it can be seen that the next state to any count is the succeeding count. For example the next state to 000 is 001.

Present state			Next state		
A	B	C	A_+	B_+	C_+
0	0	0	0	0	1
0	0	1	0	1	0
0	1	0	0	1	1
0	1	1	1	0	0
1	0	0	1	0	1
1	0	1	1	1	0
1	1	0	1	1	1
1	1	1	0	0	0

Table 4.11 Table for a mod-8 binary counter

The second step after deciding the number of flip-flops required is to select the type to be used. To illustrate the design procedure the circuit will be designed twice, using (i) SR/FFs and (ii) JK/FFs.

At this stage it is necessary to remember the behaviour of the device to be used. For this purpose the excitation requirements for both flip-flops are reproduced here:

Excitation requirements for SR/FF:

Present state		Next state	Inputs	
Q		Q_+	S	R
0	to	0	0	X
0	to	1	1	0
1	to	0	0	1
1	to	1	X	0

Excitation requirement for JK/FF:

Q		Q_+	Inputs	
			J	K
0	to	0	0	X
0	to	1	1	X
1	to	1	X	0
1	to	0	X	1

(i) Using SR/FFs

By comparing the present and next states of FFA (A and A_+), we can see that A changes from 0 to 1 only once (count 3 to 4). For that to happen the set input must be 1. If both A and A_+ are 1, we have a DON'T CARE condition represented by X.

Therefore $S_A = \bar{A} B C$

Similarly A changes from 1 to 0 only once (count 7 to 0). This requires a reset R = 1. If both A and A_+ are 0, we have a DON'T CARE condition.

Therefore $R_A = A B C$

Entering this on K-maps we have:

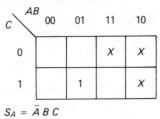

$S_A = \bar{A} B C$

$R_A = A B C$

Similarly for FFB and FFC.

It is better, however, to draw a complete table for the set and reset condition for all the flip-flops as shown in Table 4.12.

A	B	C	A$_+$	B$_+$	C$_+$	S$_A$	R$_A$	S$_B$	R$_B$	S$_C$	R$_C$
0	0	0	0	0	1	0	X	0	X	1	0
0	0	1	0	1	0	0	X	1	0	0	1
0	1	0	0	1	1	0	X	X	0	1	0
0	1	1	1	0	0	1	0	0	1	0	1
1	0	0	1	0	1	X	0	0	X	1	0
1	0	1	1	1	0	X	0	1	0	0	1
1	1	0	1	1	1	X	0	X	0	1	0
1	1	1	0	0	0	0	1	0	1	0	1

Table 4.12 **The state of the *SR* flip-flops for a mod-8 binary counter**

From the table, maps are drawn for S and R of all the flip-flops:

$S_A = \bar{A} B C$

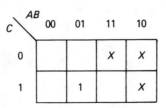

$R_A = A B C$

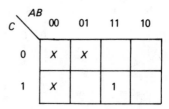

$S_B = \bar{B} C$

$R_B = B C$

$S_C = \bar{C}$

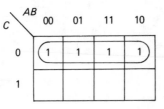

$R_C = C$

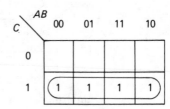

The complete circuit diagram is given in Fig. 4.37.

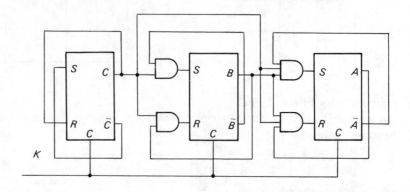

Fig. 4.37 *SR* **implementation of a Mod−8 binary counter**

(ii) Using *JK*/FFs

(a) DESIGN PROCEDURE USING THE TRUTH TABLE: A table is drawn for the present state, next state and the transition for the *JK* flip-flops as in Table 4.13.

A	B	C	A_+	B_+	C_+	J_A	K_A	J_B	K_B	J_C	K_C
0	0	0	0	0	1	0	X	0	X	1	X
0	0	1	0	1	0	0	X	1	X	X	1
0	1	0	0	1	1	0	X	X	0	1	X
0	1	1	1	0	0	1	X	X	1	X	1
1	0	0	1	0	1	X	0	0	X	1	X
1	0	1	1	1	0	X	0	1	X	X	1
1	1	0	1	1	1	X	0	X	0	1	X
1	1	1	0	0	0	X	1	X	1	X	1

Table 4.13   The states of *JK* flip-flops for a mod-8 binary counter

As for *SR*/FF, maps are drawn for all the *J* and *K* inputs as follows:

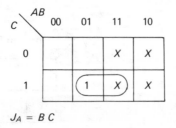

$J_A = B\,C$

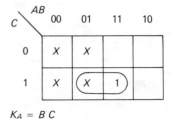

$K_A = B\,C$

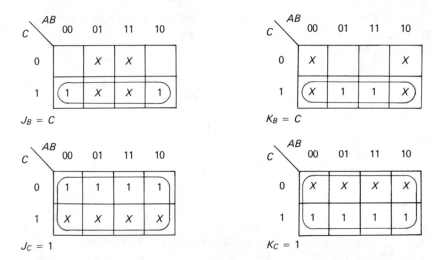

$J_B = C$ $K_B = C$

$J_C = 1$ $K_C = 1$

Now the J and K inputs of all flip-flops are known, the circuit can be constructed as in Fig. 4.38.

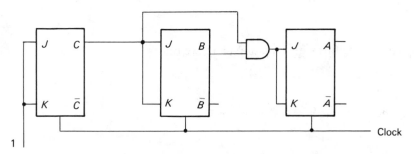

Fig. 4.38 JK implementation of a Mod−8 binary counter

(*b*) DESIGN PROCEDURE USING THE CHANGE FUNCTION: To illustrate the use of the change function mentioned earlier we have another look at the state table for the mod-8 counter. Present and next states are compared for each flip-flop and a 1 is placed if the corresponding flip-flop changes state. This is shown in Table 4.14.

Present state A B C			Next state A_+ B_+ C_+			Change X_A X_B X_C		
0	0	0	0	0	1	0	0	1
0	0	1	0	1	0	0	1	1
0	1	0	0	1	1	0	0	1
0	1	1	1	0	0	1	1	1
1	0	0	1	0	1	0	0	1
1	0	1	1	1	0	0	1	1
1	1	0	1	1	1	0	0	1
1	1	1	0	0	0	1	1	1

Table 4.14   State change for the flip-flops of a mod-8 counter

The next step is to draw three maps for X_A, X_B and X_C instead of the six maps drawn earlier for J and K of each flip-flop. There are no DON'T CARE conditions in this case. First, the map for X_A:

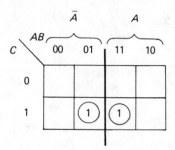

Looking back at the change function equation for a JK/FF it can be seen that the J_A input is the coefficient of $\bar{A}$ and the K_A input is the coefficient of A. To obtain an expression similar to that of X_Q in Equation (4.1) we split the X_A map into two sections for A and $\bar{A}$ as shown. The ones in each section are looped separately, resulting in:

$$X_A = \bar{A}BC + ABC$$

By comparing this with Equation (4.1), we find:

$$J_A = BC$$
$$K_A = BC$$

Similarly the following map is drawn for X_B:

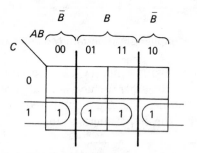

This time the map is split into B and $\bar{B}$ sections:

$$X_B = \bar{B}C + BC$$

Hence:

$$J_B = C$$
$$K_B = C$$

A third map can be drawn for X_C, but in this case one can easily see that X_C is always 1, which results in $J_C = K_C = 1$. The result is the same as that already obtained, leading to a circuit diagram as shown in Fig. 4.38.

Example (4.2)

Design an autonomous synchronous circuit to count the first nine states of the 8421 binary code. Investigate what happens if the count should go into any of the CAN'T HAPPEN states.

SOLUTION: Since the circuit counts from 0 to 8, four bistables are required; these are labeled A, B, C and D.

(i) *Design using the change function method.* The preset state, next state and change function for the four bistables are shown in Table 4.15.

Count	Present state A B C D	Next state A_+ B_+ C_+ D_+	Change X_A X_B X_C X_D
0	0 0 0 0	0 0 0 1	0 0 0 1
1	0 0 0 1	0 0 1 0	0 0 1 1
2	0 0 1 0	0 0 1 1	0 0 0 1
3	0 0 1 1	0 1 0 0	0 1 1 1
4	0 1 0 0	0 1 0 1	0 0 0 1
5	0 1 0 1	0 1 1 0	0 0 1 1
6	0 1 1 0	0 1 1 1	0 0 0 1
7	0 1 1 1	1 0 0 0	1 1 1 1
8	1 0 0 0	0 0 0 0	1 0 0 0
9	1 0 0 1	X X X X	
10	1 0 1 0	X X X X	
11	1 0 1 1	X X X X	
12	1 1 0 0	X X X X	
13	1 1 0 1	X X X X	
14	1 1 1 0	X X X X	
15	1 1 1 1	X X X X	

Table 4.15   State table and change function for Example (4.2)

From Table 4.15, four maps are drawn for the four change functions:

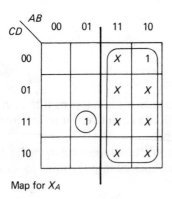

Map for X_A

The map is divided into $\bar{A}$ and A regions.

$$X_A = \bar{A}(BCD) + A$$
$$J_A = BCD$$
$$K_A = 1$$

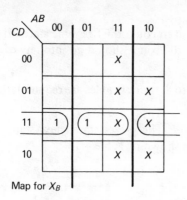

The map is divided into $\bar{B}$ and B regions.

$$X_B = \bar{B}(CD) + B(CD)$$
$$J_B = CD$$
$$K_B = CD$$

Map for X_B

The map is divided into $\bar{C}$ and C regions.

$$X_C = \bar{C}D + CD$$
$$J_C = D$$
$$K_C = D$$

Map for X_C

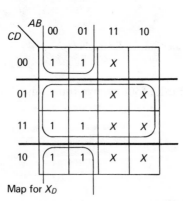

The map is divided into $\bar{D}$ and D regions.

$$X_D = \bar{D}(\bar{A}) + D(1)$$
$$J_D = \bar{A}$$
$$K_D = 1$$

Map for X_D

The design is now complete, but we must consider what happens if the circuit starts, say in state 12 or any disallowed state. Will the circuit return to the correct sequence, or will it stay circulating in a minor loop?

To do this, we consider the way the CAN'T HAPPEN states were looped. Some were looped as 1s, others were left as 0s. These are entered into the table to see the next state resulting from this choice:

	Present state A B C D	Change function X_A X_B X_C X_D	Next state A_+ B_+ C_+ D_+
9	1 0 0 1	1 0 1 1	0 0 1 0
10	1 0 1 0	1 0 0 0	0 0 1 0
11	1 0 1 1	1 1 1 1	0 1 0 0
12	1 1 0 0	1 0 0 0	0 1 0 0
13	1 1 0 1	1 0 1 1	0 1 1 0
14	1 1 1 0	1 0 0 0	0 1 1 0
15	1 1 1 1	1 1 1 1	0 0 0 0

The above table shows that the next states for the CAN'T HAPPEN are as in Fig. 4.39.

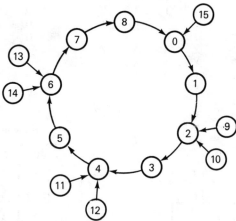

Fig. 4.39 State transition for Example (4.2)

From Fig. 4.39 it can be seen that the circuit always returns to the correct sequence.

The circuit diagram is then as shown in Fig. 4.40.

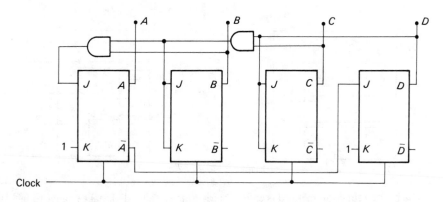

Fig. 4.40 Mod−9 binary counter for Example (4.2)

The same circuit can be designed by the method used for the Mod-8 binary counter. That is to draw two maps for each flip-flop from the state table and the *JK* table.* The answer is the same as shown in (ii) below.

(ii) *Alternative design procedure for Example (4.2)*

	Present state A B C D	Next state A_+ B_+ C_+ D_+
0	0 0 0 0	0 0 0 1
1	0 0 0 1	0 0 1 0
2	0 0 1 0	0 0 1 1
3	0 0 1 1	0 1 0 0
4	0 1 0 0	0 1 0 1
5	0 1 0 1	0 1 1 0
6	0 1 1 0	0 1 1 1
7	0 1 1 1	1 0 0 0
8	1 0 0 0	0 0 0 0
9	1 0 0 1	X X X X
10	1 0 1 0	X X X X
11	1 0 1 1	X X X X
12	1 1 0 0	X X X X
13	1 1 0 1	X X X X
14	1 1 1 0	X X X X
15	1 1 1 1	X X X X

J	K	Q	Q_+
0	0	0	0
0	1	0	0
0	0	1	1
1	0	1	1
1	0	0	1
1	1	0	1
0	1	1	0
1	1	1	0

State table of a *JK*/FF

$J_A = BCD$

CD \ AB	00	01	11	10
00	0	0	X	X
01	0	0	X	X
11	0	(1)	X	X
10	0	0	X	X

Maps for FF/A $K_A = 1$

CD \ AB	00	01	11	10
00	X	X	X	1
01	X	X	X	X
11	X	X	X	X
10	X	X	X	X

$J_B = CD$

CD \ AB	00	01	11	10
00	0	X	X	0
01	0	X	X	X
11	(1	X	X	X)
10	0	X	X	X

Maps for FF/B $K_B = CD$

CD \ AB	00	01	11	10
00	X	0	X	X
01	X	0	X	X
11	(X	1	X	X)
10	X	0	X	X

*The change function method is easier, but it is up to the individual designer to use the method he prefers.

K-map for FF/C — $J_C = D$:

CD \ AB	00	01	11	10
00	0	0	X	0
01	1	1	X	X
11	X	X	X	X
10	X	X	X	X

$J_C = D$

K-map for FF/C — $K_C = D$:

CD \ AB	00	01	11	10
00	X	X	X	X
01	X	X	X	X
11	1	1	X	X
10	0	0	X	X

$K_C = D$

Maps for FF/C

K-map for FF/D — $J_D = \bar{A}$:

CD \ AB	00	01	11	10
00	1	1	X	0
01	X	X	X	X
11	X	X	X	X
10	1	1	X	X

$J_D = \bar{A}$

K-map for FF/D — $K_D = 1$:

CD \ AB	00	01	11	10
00	X	X	X	X
01	1	1	X	X
11	1	1	X	X
10	X	X	X	X

$K_D = 1$

Maps for FF/D

As before the way the loops are taken decides what happens if we start from a disallowed state, e.g. $15 = 1\,1\,1\,1$

This is taken as follows:

FF A $J_A = 1$ $K_A = 1$ Therefore A changes to 0
   B $J_B = 1$ $K_B = 1$ Therefore B changes to 0
   C $J_C = 1$ $K_C = 1$ Therefore C changes to 0
   D $J_C = 0$ $K_D = 1$ Therefore D changes to 0

Therefore the next state for 1111 is 0000, i.e. the next state for 15 is 0.
If in state 1100

$J_A = 0$ $K_A = 1$, then A changes to 0
$J_B = 0$ $K_B = 0$, then B stays as 1
$J_C = 0$ $K_C = 0$, then C stays as 0
$J_D = 0$ $K_D = 1$, then D stays as 0

The next state for 1100 is 0100. Therefore the next state for 12 is 4, and so on as before.

Example (4.3)

Design a synchronous sequential generator, using three *JK* flip-flops and any necessary logic gates, to count the sequence, 0, 2, 4, 6, when the control line *D* is at logic 0, and count the sequence 6, 4, 2, 0 when the control line is at logic 1: should the circuit fall into a disallowed state, it should always return to the 0 state.

SOLUTION: The circuit behaviour is described in Table 4.16.

Count	Present state A B C	Control D	Next state A_+ B_+ C_+	Change function X_A X_B X_C
0	0 0 0	0	0 1 0	0 1 0
2	0 1 0	0	1 0 0	1 1 0
4	1 0 0	0	1 1 0	0 1 0
6	1 1 0	0	0 0 0	1 1 0
6	1 1 0	1	1 0 0	0 1 0
4	1 0 0	1	0 1 0	1 1 0
2	0 1 0	1	0 0 0	0 1 0
0	0 0 0	1	1 1 0	1 1 0
1	0 0 1	0	0 0 0	0 0 1
1	0 0 1	1	0 0 0	0 0 1
3	0 1 1	0	0 0 0	0 1 1
3	0 1 1	1	0 0 0	0 1 1
5	1 0 1	0	0 0 0	1 0 1
5	1 0 1	1	0 0 0	1 0 1
7	1 1 1	0	0 0 0	1 1 1
7	1 1 1	1	0 0 0	1 1 1

Table 4.16 State table and change function for Example (4.3)

Next, three K-maps are drawn for the three change functions:

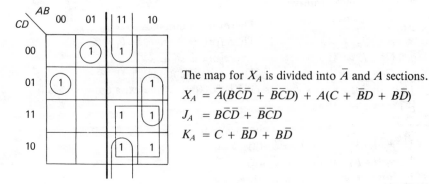

The map for X_A is divided into $\bar{A}$ and A sections.

$X_A = \bar{A}(B\bar{C}\bar{D} + \bar{B}\bar{C}D) + A(C + \bar{B}D + B\bar{D})$

$J_A = B\bar{C}\bar{D} + \bar{B}\bar{C}D$

$K_A = C + \bar{B}D + B\bar{D}$

K-map for X_A

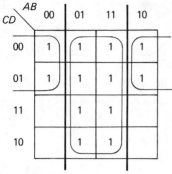

K-map for X_B

The map for X_B is divided into $\bar{B}$ and B sections.

$X_B = \bar{B}(\bar{C}) + B(1)$

$J_B = \bar{C}$

$K_B = 1$

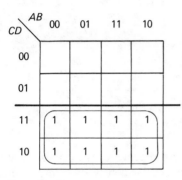

K-map for X_C

The map is divided into $\bar{C}$ and C sections.

$X_C = \bar{C}(0) + C(1)$

$J_C = 0$

$K_C = 1$

Note that FF/C is required to stay at logic 0 all the time. This is ensured by making $J_C = 0$, $K_C = 1$. The final circuit is given in Fig. 4.41.

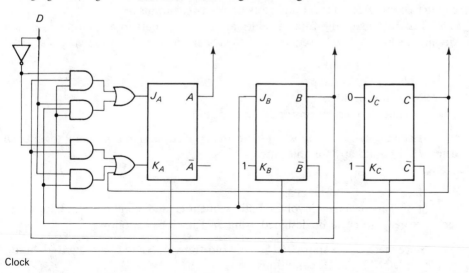

Fig. 4.41   Circuit diagram for the sequence generator of Example (4.3)

Design Examples involving Combinational and Sequential Circuits

Example (4.4)

Design a synchronous sequence generator to count the sequence 0, 1, 2, 3, 4, 5 and repeat.

A long chain of lights is used to produce the effect of a moving band of lights along the chain, by turning on every other three lights. The illuminated bands move by one position for each clock pulse. Design a logic system using the sequence generator and any suitable logic to achieve this.

Example (4.5)

Using the counter of Example (4.4) and any combinational logic, design a logic system to control ten lights in the following manner:

Initially all lights are out.
After the first clock pulse lights 1 and 10 come on
After the second clock pulse lights 2 and 9 come on
After the third clock pulse lights 3 and 8 come on
and so on until all lights are on.
After another clock pulse all lights go out and the sequence is repeated.

Example (4.6)

A binary counter which counts from 0 to 4 repeatedly is available. Design a minimal logic system to produce a binary sequence which is five times the count sequence.

Example (4.7)

A clock pulse of 10 s period is used to clock a four-bit binary counter. The counter counts from 0 to 11 and rests on the twelfth pulse.

The outputs of the four bistables are used as inputs to a combinational logic circuit which controls the sequence of the traffic lights in the following way:

Green is on for 40 s.
Amber is on for 20 s.
Red is on for 40 s.
Red and Amber are on for 20 s., and so on.

Assuming a light is on when logic 1 is applied, design a minimal logic circuit using JK/FFs and any suitable gates.

SOLUTIONS TO EXAMPLES:

(4.4) The problem involves both sequential and combinational logic. The sequence generator will be designed using JK flip-flops as follows:

Since there are six states 0 to 5, three flip-flops A, B and C are required. States 6 and 7 are DON'T CARE conditions.

The circuit diagram for this is given in Fig. 4.42.

	Present state A B C	Next state A_+ B_+ C_+	Change functions X_A X_B X_C
0	0 0 0	0 0 1	0 0 1
1	0 0 1	0 1 0	0 1 1
2	0 1 0	0 1 1	0 0 1
3	0 1 1	1 0 0	1 1 1
4	1 0 0	1 0 1	0 0 1
5	1 0 1	0 0 0	1 0 1
6	1 1 0	X X X	X X X
7	1 1 1	X X X	X X X

Table 4.17   State and change function table for Example (4.4)

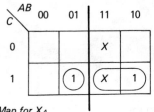

$$X_A = \bar{A}BC + BC$$
$$J_A = BC$$
$$K_A = C$$

Map for X_A

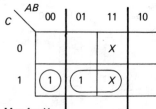

$$X_B = \bar{B}\bar{A}C + BC$$
$$J_B = \bar{A}C$$
$$K_B = C$$

Map for X_B

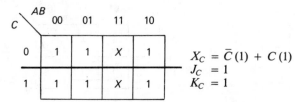

$$X_C = \bar{C}(1) + C(1)$$
$$J_C = 1$$
$$K_C = 1$$

Map for X_C

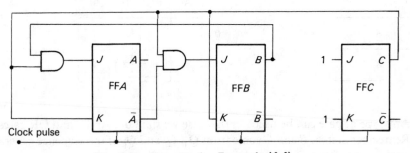

Fig. 4.42   Counter for Example (4.4)

Next we develop the combinational logic to control the lights:

a	b	c	d	e	f	a	b	c	d	e	f
1	2	3	4	5	6	7	8	9	10	11	12
*	*	*	*	*	*	*	*	*	*	*	..

Since there are three lights on and three lights off which is repeated all along, we consider any six *a, b, c, d, e, f*

$Q_A\ Q_B\ Q_C$ correspond to the states of the bistables *A, B* and *C*:

	A B C	a b c d e f
0	0 0 0	1 1 1 0 0 0
1	0 0 1	0 1 1 1 0 0
2	0 1 0	0 0 1 1 1 0
3	0 1 1	0 0 0 1 1 1
4	1 0 0	1 0 0 0 1 1
5	1 0 1	1 1 0 0 0 1

and repeat

Six K-maps are drawn for the six variables *a,b,c,d,e,f*:

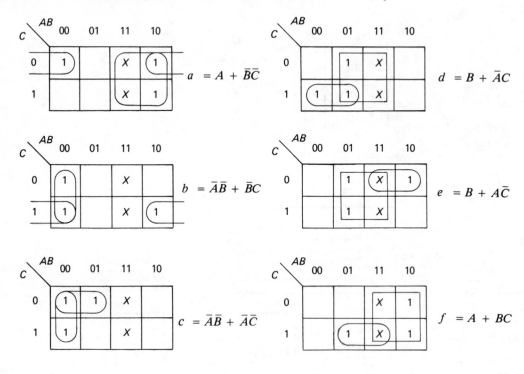

$$a = A + \bar{B}\bar{C}$$

$$d = B + \bar{A}C$$

$$b = \bar{A}\bar{B} + \bar{B}C$$

$$e = B + A\bar{C}$$

$$c = \bar{A}\bar{B} + \bar{A}C$$

$$f = A + BC$$

Any type of gate can be used. In this case we shall use AND, and OR gates.

Remember that *A, B* and *C* come from Q_A, Q_B and Q_C of the sequence generator. $\bar{A}$, $\bar{B}$, and $\bar{C}$ come from $\bar{Q}_A$, $\bar{Q}_B$ and $\bar{Q}_C$.

The outputs are used as follows:

  a   drives lamps   1, 7, 13, etc.
  b   drives lamps   2, 8, 14, etc.
  c   drives lamps   3, 9, 15, etc.
  d   drives lamps   4, 10, 16, etc.
  e   drives lamps   5, 11, 17, etc.
  f   drives lamps   6, 12, 18, etc.

the complete circuit is given in Fig. 4.43.

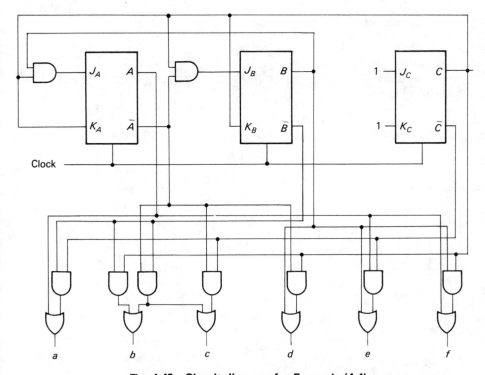

Fig. 4.43   Circuit diagram for Example (4.4)

With minor modification to the combinational logic other variations can be designed. For example, one can turn two lights on and four off, five on and one off, etc.

(4.5) The outputs of the counter are A, B and C

let   *a*   drive lights   1 and 10
      b   drive lights   2 and 9
      c   drive lights   3 and 8
      d   drive lights   4 and 7
      e   drive lights   5 and 6

A, B and C are taken from the outputs of the three flip-flops Q_A, Q_B and Q_C respectively. Then the circuit performance can be described as in the following table:

	A B C	a b c d e
0	0 0 0	0 0 0 0 0
1	0 0 1	1 0 0 0 0
2	0 1 0	1 1 0 0 0
3	0 1 1	1 1 1 0 0
4	1 0 0	1 1 1 1 0
5	1 0 1	1 1 1 1 1

and repeat

The K-maps are:

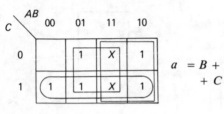

$$a = B + A + C$$

K-map for a

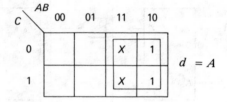

$$d = A$$

K-map for d

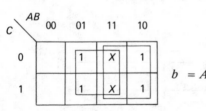

$$b = A + B$$

K-map for b

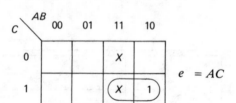

$$e = AC$$

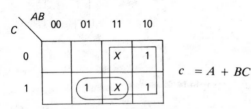

$$c = A + BC$$

K-map for c

The circuit is given in Fig. 4.44.

(4.6) The outputs of the counter and binary sequence are as follows:

	Counter output A B C		Required sequence a b c d e
0	0 0 0	0	0 0 0 0 0
1	0 0 1	5	0 0 1 0 1
2	0 1 0	10	0 1 0 1 0
3	0 1 1	15	0 1 1 1 1
4	1 0 0	20	1 0 1 0 0

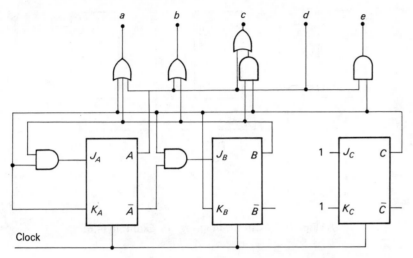

Clock

Fig. 4.44   Circuit diagram for Example (4.5)

The K-maps are:

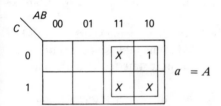

K-map for a

$$a = A$$

K-map for c

$$c = A + C$$

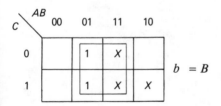

K-map for b

$$b = B$$

K-map for d

$$d = B.$$

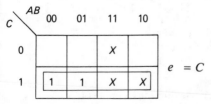

K-map for e

$$e = C$$

The circuit is given in Fig. 4.45. The connections for the counter are not shown.

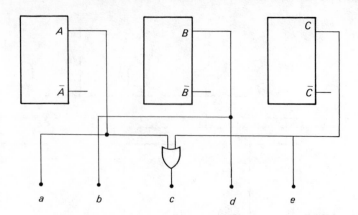

Fig. 4.45   Circuit diagram for Example (4.6)

The results can in fact be predicted directly from the truth table.

(**4.7**) If the counter is to count twelve states, it requires four bistables; *JK* bistables will be used.

Now we are more familiar with the design procedure, some short cut will be made, for example the next state will not be included in the table since it is obvious that the next state for every row is the row below. The next state for 11 in this case is zero.

If the four bistables are *A, B, C,* and *D,* then the count sequence and the corresponding change function are as in Table 4.18.

	A	*B*	*C*	*D*	X_A	X_B	X_C	X_D
0	0	0	0	0	0	0	0	1
1	0	0	0	1	0	0	1	1
2	0	0	1	0	0	0	0	1
3	0	0	1	1	0	1	1	1
4	0	1	0	0	0	0	0	1
5	0	1	0	1	0	0	1	1
6	0	1	1	0	0	0	0	1
7	0	1	1	1	1	1	1	1
8	1	0	0	0	0	0	0	1
9	1	0	0	1	0	0	1	1
10	1	0	1	0	0	0	0	1
11	1	0	1	1	1	0	1	1

Table 4.18   Count sequence and change function for Example (4.7)

We draw two K-maps, for X_A and X_B:

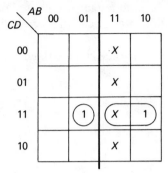

Map for X_A

$$X_A = \bar{A}BCD + ACD$$
$$\text{Therefore } J_A = BCD$$
$$K_A = CD$$

Note that unused states are DON'T CARE.

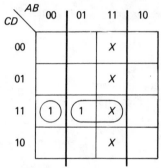

Map for X_B

$$X_B = \bar{B}\bar{A}CD + BCD$$
$$J_B = \bar{A}CD$$
$$K_B = CD$$

We could also draw maps for X_C and X_D, but it is not necessary since by inspection X_C is identical to D.

Therefore $J_C = K_C = D$

Also from the table, X_D is always 1.

Therefore $J_D = K_D = 1$

The complete circuit is given in Fig. 4.46.

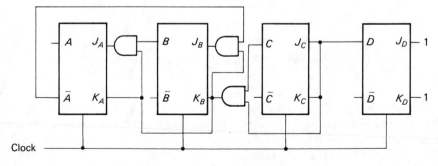

Fig. 4.46   Counter for Example (4.7)

It is useful to see what happens if the counter should fall into an unused state. Would it go back to the correct sequence, or would it stay circulating between the unused states?

This can be seen from the following table, in which the change functions are entered as they were used in the K-maps:

	A B C D	X_A X_B X_C X_D	A_+ B_+ C_+ D_+
12	1 1 0 0	0 0 0 1	1 1 0 1
13	1 1 0 1	0 0 1 1	1 1 1 0
14	1 1 1 0	0 0 0 1	1 1 1 1
15	1 1 1 1	1 1 1 1	0 0 0 0

From the table, we can see that if the counter should fall into state 12, its next state is 13 which takes it to 14, 15 and back to 0 to start its correct sequence.

Had the counter been unable to return to the correct sequence, the DON'T CARE state must be changed even though this might require more hardware.

Next we consider the combinational logic required to control the lights. For this we compile Table 4.19.

	A	B	C	D	Z_g	Z_a	Z_r
0	0	0	0	0	1	0	0
1	0	0	0	1	1	0	0
2	0	0	1	0	1	0	0
3	0	0	1	1	1	0	0
4	0	1	0	0	0	1	0
5	0	1	0	1	0	1	0
6	0	1	1	0	0	0	1
7	0	1	1	1	0	0	1
8	1	0	0	0	0	0	1
9	1	0	0	1	0	0	1
10	1	0	1	0	0	1	1
11	1	0	1	1	0	1	1

Table 4.19   Truth table for the light sequence

Let the variables Z_g, Z_a and Z_r refer to the logic functions controlling the green, amber and red lights respectively. Since the period for each pulse is 10 s, the green will stay on for four pulse periods followed by amber which stays on for two periods, and so on, as shown in the truth table.

The K-maps are:

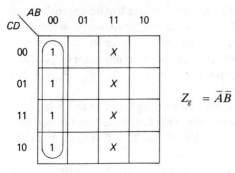

K-map for Z_g

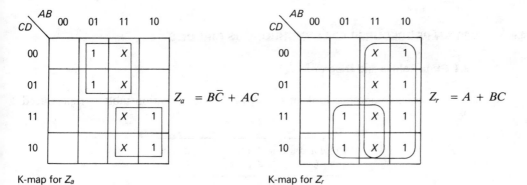

K-map for Z_a K-map for Z_r

The complete circuit can now be drawn as in Fig. 4.47.

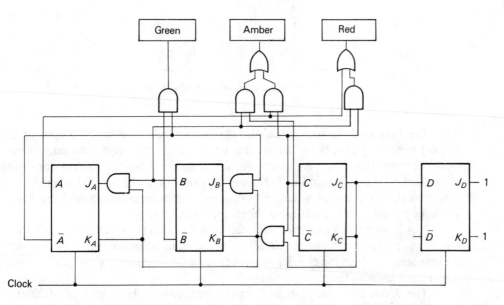

Fig. 4.47   Circuit diagram for the traffic controller of Example (4.7)

Microprocessor implementation of the traffic controller. The traffic controller can also be implemented using a microprocessor. This would be particularly attractive if we wanted to control a large number of lights. The program is given in Appendix 4.2, and it is for a 6800 microprocessor. The first part of the program makes P_{A7} of the programmable interface adapter (PIA) an input port. It also makes P_{A0}, P_{A1} and P_{A2} output ports. This is known as the initialization.

The three outputs control the green, amber and red lights. The clock input in case is taken from a pulse generator. The microprocessor keeps the appropriate outputs high for the number of pulses specified (40, 20, 40 and 20 in this case) and then repeats.

This is not necessarily the best implementation but it is easy to follow.

4.6   DESIGN PROCEDURES FOR SYNCHRONOUS CIRCUITS

4.6.1 State Tables and Diagrams

Let us have another look at the *JK*/FF. The transition table is reproduced below:

Present state Q	Inputs J	K	Next state Q_+
0	0	0	0
0	0	1	0
0	1	0	1
0	1	1	1
1	0	0	1
1	0	1	0
1	1	0	1
1	1	1	0

Transition table of a *JK* flip-flop

The first thing to note is that the next state (which is also the output of the device) depends not only on the present input but on the previous inputs. As an example when *J* and *K* are both zero, the next state could be zero (low) or one (high) depending on the present state. This represents the history of the device which, in a way, remembers previous inputs. Thus, depending on previous inputs, the device could be in one of two states (also called internal states).

If *A* is used to represent the low state and *B* is used to represent the high state, we may say that the state of the circuit when the input is applied is the present state, and the state to which the circuit goes is the next state. The next state might, or might not, be the same as the present state.

The circuit behaviour can be described by a state table, as in Table 4.20(a). It is also possible to write the output with the next state as in Table 4.20(b).

Present state	Next state				Output			
$JK =$	00	01	10	11	$JK =$ 00	01	10	11
A	A	A	B	B	0	0	1	1
B	B	A	B	A	1	0	1	0

Table 4.20(a)   State table of a *JK* flip-flop

Present state	Next-state/Output			
	$JK =$ 00	01	10	11
A	$A/0$	$A/0$	$B/1$	$B/1$
B	$B/1$	$A/0$	$B/1$	$A/0$

Table 4.20(b)   Alternative form of Table 4.20(a)

From the state table, it can be seen that if the present state is *B* and the inputs at *JK* are 11 the circuit will go to state *A* and produce an output 0.

Another way of representing the behaviour of the circuit is by means of a state diagram. In this diagram the states are represented by circles. The arrows represent the transitions. The labels on the arrows (arcs) specify the inputs and resulting output. For example 00/1 indicates that both inputs *J* and *K* are 0 and that the output is 1.

This is called a *Mealy model*, in which the output depends on both the present state and the external inputs. In a *Moore model*, the output is a function of the state only and is independent of the external inputs. In this case, the output is identical to the state, and can be written inside the circle. The external inputs only are shown on the arrows as in Fig. 4.48.

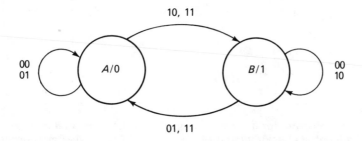

Fig. 4.48   State diagram for a *JK* flip-flop using a Moore model

Sequential systems can be described by means of statements, state diagrams or state tables. Consider the following statement:

A synchronous sequential checking circuit is required to give an output of logic 1 when three or more consecutive logic 1s are detected in a serial flow of binary data.

This requirement can be described by means of a state diagram or a state table.

In both cases the circuit starts at a state which we may call state 1 and produces an output 0. If a 1 is received the circuit moves to a new state which we call state 2. The output in state 2 is still 0. In state 2 the circuit looks for two logic 1s to produce an output of 1, while in state 1 it looks for three logic 1s and this explains why a new state is required. In a similar manner we can see that the circuit moves to a new state every time an input of 1 is received until a total of three 1s is received. By then the circuit is in state 4 and an output of 1 is produced.

If the circuit receives a logic 0 at any time it will go back to the original state (state 1), and start looking again for three logic 1 inputs.

Another point to observe is that the output table is omitted since, from the specification, the output is 1 only when the circuit is in state 4. But this is not always true of course as will be seen in future examples.

The state diagram and state table are shown below in Fig. 4.49 and Table 4.21 respectively.

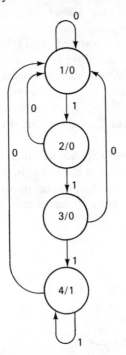

Present state	Next state	
	$D = 1$	$D = 0$
1	2	1
2	3	1
3	4	1
4	4	1

Fig. 4.49   State diagram for a synchronous   **Table 4.21   State table for a synchronous**
sequential checking circuit   **sequential checking circuit**

Note that the state table and the state diagram contain the same information. The state diagram is useful for the designer to see the general behaviour of the system under various input conditions. Deriving the state table represents a first step towards the design of the circuit.

In the block diagram representation of a sequential circuit, the feedback paths which represent the state of the circuit are called the state variables or secondary variables. A system with n-states requires $\lceil \log_2 n \rceil$ state variables. In other words, s

state variables can represent up to 2^s states. Similarly if the circuit has p primary (external) input variables, there may be up to 2^p different input combinations.

In the present design example there are four states requiring two state variables and there are two possible inputs 0 and 1 requiring one input variable D.

The states of the system which are defined by the state variables are called the *internal states*. The states which are defined by the state variables and primary inputs are called the *total states* and there are 2^{p+s} of them. As the external inputs change, the circuit moves from one total state to another.

The output variables may be the same as the state variables as in counters. Generally there is an output with every total state, with the outputs being defined by the state variables and the inputs. It should also be clear now why counters are called *autonomous sequential circuits*. They are sequential circuits without external inputs and simply move from one state to another governed by the speed of the components used as in asynchronous circuits or by the timing clock as in synchronous circuits. The type of circuits we are dealing with are also called *deterministic sequential systems*, in which the next state $S(t+1)$ is uniquely determined by the present state $S(t)$, and the present input $x(t)$, i.e.

$$S(t+1) = v[S(t), x(t)]$$

where v is the state transition function.

If the output $Z(t)$ is a function only of the present state, the machine is called a Moore machine, and

$$Z(t) = \omega[S(t)]$$

where ω is defined as the output function.

In a Mealy machine, the output is a function of the present states as well as the inputs, and

$$Z(t) = \omega[S(t), x(t)]$$

A general way of describing a sequential machine is by means of a state table, as in Table 4.22. The present state S_i and input x_j uniquely determine the next state S_k. The sequential machine is therefore a mapping of pairs $S_i x_j$ into S. The state table shows a unique output for every input and present state. In a Moore machine the outputs in any row will be the same, since they are independent of primary inputs. This effectively means a one-column output table.

Present state	Next state, output Z			
	x_1	x_2	$\ldots$	x_m
S_1	S_{11}, Z_{11}	S_{12}, Z_{12}	$\ldots$	S_{1m}, Z_{1m}
S_2	.			
.	.			
.	.			
S_n	S_{n1}, Z_{n1}	S_{n2}, Z_{n2}	$\ldots$	S_{nm}, Z_{nm}

Table 4.22 A state table for any sequential machine

State Assignment

Having obtained a state table with the states of the system labelled with numerical symbols (states 1, 2, ..., etc.), alphabetical symbols (*A, B, ...*, etc.) or in fact any other symbols, we proceed with what is known as the *state assignment*.

The state assignment refers to the process of giving the states a binary code. This is in fact the assignment of the states of the memory device (*JK* flip-flops, say), to the states of the circuit. The memory device being a two-state device has an output which may be 0 or 1. The number of bits of the binary code is equal to the number of secondary variables (feedback paths). For a state table with *n* states (*n*-rows), $\lceil \log_2 n \rceil = s$ state variables are required. In our present example two state variables are required. In other words a two-bit code is used for the state assignment and two flip-flops are required for the implementation.

A state table with 5, 6, 7 or 8 states requires three state variables. A state table with 9–16 states requires four state variables, and so on.

Returning to Table 4.21, the states can now be given a binary assignment. The two state variables are called *A* and *B*, each could be 0 or 1.

In the state assignment in Table 4.23 state 1 is given the code 00, state 2 is given the code 01, and so on. Obviously this is an arbitrary choice and one could easily assign 01 to state 1 and 00 to state 2. In fact any assignment will do, although in Chapters 8 and 9 we shall see that some state assignments result in simpler expressions and hence more economical circuits. There is no easy way of knowing which is the best assignment, but techniques that result in good assignments will be discussed.

Now having completed the state assignment, we are in a position to design and implement the synchronous sequential checking circuit of Fig. 4.49 and Table 4.21.

Present state		Next state	
A	*B*	*D* = 1	*D* = 0
0	0	0 1	0 0
0	1	1 1	0 0
1	1	1 0	0 0
1	0	1 0	0 0

Table 4.23   State assignment for Table 4.21

4.6.2 Design Using JK Flip-Flops

We decide at this stage what to use to implement the circuit. Suppose *JK* flip-flops and NOR gates are available.

Method 1

The behaviour of a *JK*/FF is given at the beginning of Section 4.6. Two tables are drawn for each flip-flop, one for the *J* input and one for the *K* input. *A* and *B* are the state variables assigned to the two flip flops *A* and *B*. *D* is the variable assigned to the input. The tables are:

A	B	D = 1	D = 0
0	0	0	0
0	1	1	0
1	1	X	X
1	0	X	X

J_A table

D = 1	D = 0
X	X
X	X
0	1
0	1

K_A table

For the 1st row (00) under column $D = 1$ the present and next states for FFA are 0 as can be seen from the state assignment table. This means FFA does not change state.

From the JK table, the first two rows show that if Q and Q_+ are both a logic 0, J must be 0, but K can be 0 or 1. Therefore in the J_A table we put 0 for J_A and in the K_A table we put X (DON'T CARE) for K_A.

Still under $D = 1$ column, we now consider the second row (01).

From the assignment in Table 4.23, the next state for $A = 0$ under $D = 1$ is $A = 1$. In other words FFA must be set to 1.

Returning to the JK/FF table, rows 3 and 4 show that if $Q = 0$ and $Q_+ = 1$, J must be 1 but K can be 0 or 1.

In the J_A table we put 1 and in the K_A table we put X at the same positions, namely row 01 column 1, and so on.

From the J_A table we obtain

$$J_A = DB$$
$$\quad = \overline{\overline{D} + \overline{B}} \quad \text{for NOR implementation}$$

From the K_A table we obtain

$$K_A = \overline{D}$$

Similarly for FFB, we have:

$$J_B = D\overline{A}$$
$$\quad = \overline{\overline{D} + A}$$

$$K_B = \overline{D} + A$$
$$\quad = \overline{\overline{D} + A}$$
$$\quad = \overline{J_B}$$

A	B	D = 1	D = 0
0	0	1	0
0	1	X	X
1	1	X	X
1	0	0	0

J_B table

A	B	D = 1	D = 0
0	0	X	X
0	1	0	1
1	1	1	1
1	0	X	X

K_B table

Method 2

The change function equation for a *JK*/FF as obtained earlier is given by:

$$X = J\bar{Q} + KQ$$

Since we only require two flip-flops, two maps are requires, one for X_A and one for X_B.

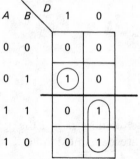

Change function map for FF*A*

Logic 1 is entered in every position where the flip-flop is to change state and logic 0 if the FF does not change state; consider the position under $D = 1$, row 1.

From the state assignment table $A = 0$ under present state and $A = 0$ under next state conditions. FF*A* does not change hence a zero is entered.

Now consider the position under $D = 1$, row 01. At the same position in the state assignment table $A = 0$ under present state and $A = 1$ under next state conditions. FF*A* changes state; hence a 1 is entered, and so on.

Since we are dealing with FF*A* the change function map is split into $\bar{A}$ and A sections in order to arrive at an equation similar to the change function equation of any *JK*/FF:

$$X = J\bar{Q} + KQ$$

From the map we obtain

$$X_A = \bar{A}DB + A\bar{D}$$

The coefficient of $\bar{A}$ must be *J*. The coefficient of *A* must be *K*. Therefore:

$$J_A = BD = \overline{\bar{B} + \bar{D}}$$
$$K_A = \bar{D}$$

Similarly for the FF*B*

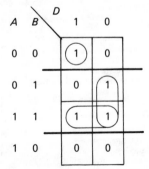

Change function map for FF*B*

The map is divided into B and $\bar{B}$ before making the loops. From the map we have

$$X_B = \bar{B}D\bar{A} + B(\bar{D}+A)$$

Therefore

$$J_B = D\bar{A}$$
$$ = \overline{\bar{D}+A}$$

$$K_B = \overline{\bar{D}+A}$$
$$ = \overline{\bar{D}+A}$$
$$ = \bar{J}_B$$

The answer in both cases is the same and results in the circuit of Fig. 4.50. The output is logic 1 only if the circuit is in state 4. Therefore

The output $Z = A\bar{B} = \overline{\bar{A}+B}$

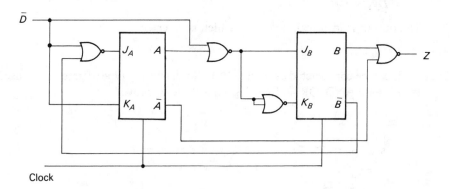

Fig. 4.50   **Diagram of a synchronous sequential checking circuit using** JK **flip-flops**

4.6.3   Design Using D Flip-Flops

With a D flip-flop, two state variables are required, as in the last case, to distinguish between the four states, hence two flip-flops are also required. Let us call the state variables y_1 and y_2 instead of A and B. We also need to remember that the next state of a D flip-flop is the same as its present state, but delayed. In the following Y is used to indicate the next state. After the next state is fed back it becomes a present state for the next operation. Remember the flip-flop has one input not two as in the JK/FF.

Let us start from the state table in Table 4.21, to which reference should be made. Binary codes are given to the four states as shown in the state assignment in Table 4.24.

Note that the same state assignment is used as for the JK/FF. From Table 4.24, expressions for Y_1 and Y_2 are derived terms of y_1, y_2 and the input variable D:

Present state		Next state			
		$D = 1$		$D = 0$	
y_1	y_2	Y_1	Y_2	Y_1	Y_2
0	0	0	1	0	0
0	1	1	1	0	0
1	1	1	0	0	0
1	0	1	0	0	0

Table 4.24   State assignment table

$$Y_1 = D\bar{y}_1 y_2 + Dy_1 y_2 + Dy_1 \bar{y}_2$$
$$ = Dy_2 + Dy_1$$

$$Y_2 = D\bar{y}_1 \bar{y}_2 + D\bar{y}_1 y_2$$
$$ = D\bar{y}_1$$

The output is identical to state 4, which is given by:

$$Z = y_1 \bar{y}_2$$

The circuit can be constructed as in Fig. 4.51. Naturally any type of gate can be used. In this case AND/OR and NOT gates will be employed

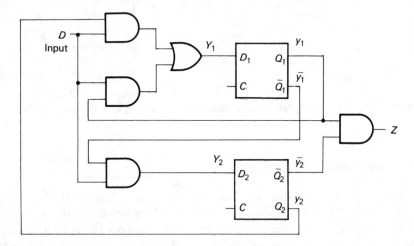

Fig. 4.51   Diagram of a synchronous sequential checking circuit, using D flip-flops

The circuit is checked by placing a 1 on D and applying three clock pulses (0 to 1) to get an output of 1 on Z. Further clock pulses have no effect.

Example (4.8)
A circuit is required to produce an output of 1 when the sequence 1010 is detected, and output of 0 at all other times.

SOLUTION: The requirement may be represented by means of a state diagram (Fig. 4.52) and state table (Table 4.25).

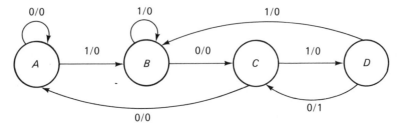

Fig. 4.52   State diagram for Example (4.8)

Present state	Next state/Output	
	$D = 1$	$D = 0$
A	$B/0$	$A/0$
B	$B/0$	$C/0$
C	$D/0$	$A/0$
D	$B/0$	$C/1$

Table 4.25   State table for Example (4.8)

Present state		Next state				Output Z	
		$D = 1$		$D = 0$		$D = 1$	$D = 0$
Q_1	Q_2	Q_{1+}	Q_{2+}	Q_{1+}	Q_{2+}		
0	0	0	1	0	0	0	0
0	1	0	1	1	1	0	0
1	1	1	0	0	0	0	0
1	0	0	1	1	1	0	1

Table 4.26   State assignment and output table for Example (4.8)

The input data arrive serially one at a time and can be represented by one variable D. JK flip-flops will be used, and their outputs are usually called Q and $\bar{Q}$. In this example we call the two present states Q_1 and Q_2 and their next states Q_{1+} and Q_{2+}, where Q refers to the true output. Obviously we could have named the flip-flops A and B or Y_1 and Y_2 as we did in previous examples.

We shall use the change function method in this case, and so we need two maps, one of X_1 and one of X_2.

Change function tables can be drawn, but by now the reader should be familiar with the method and can find X by comparing the present and next states for each flip-flop under each input:

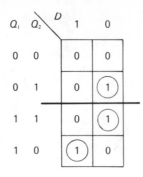

Map for X_1 Map for X_2

$$X_1 = \bar{Q}_1(\bar{D}Q_2) + Q_1(\bar{D}Q_2 + D\bar{Q}_2) \qquad X_2 = \bar{Q}_2(D+Q_1) + Q_2(Q_1)$$

This results in

$$J_1 = \bar{D}Q_2$$
$$K_1 = \bar{D}Q_2 + D\bar{Q}_2$$
$$J_2 = D+Q_1$$
$$K_2 = Q_1$$

The output $Z = \bar{D}Q_1\bar{Q}_2$
The circuit diagram is shown in Fig. 4.53.

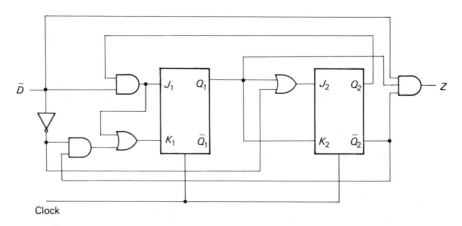

Clock

Fig. 4.53   Implementation of Example (4.8) using *JK* flip-flops

4.6.4 Design Using ROM Devices

To illustrate the use of ROMs for synchronous sequential circuits we have another look at Example (4.8). The procedure is not much different from that already explained for combinational logic. The external inputs and present states are used as address lines. The next states and outputs are stored inside the ROM. The next states are fed back to achieve the sequential action. The delay round the feedback loop is obtained by using latches (clocked D flip-flops may be used). The enable line of the ROM can be used to clock the system.

The starting point is the state assignment and output table. The table is rearranged as a ROM transition Table. We shall assume that the ROM has four outputs though we need only three. If the number of inputs of the ROM is greater than the three required, we use the three least significant (normally labeled $A_2 A_1 A_0$). This means that the ROM will be partially used. A block diagram of the ROM and latches (L) appears in Fig. 4.54.

Address			Output			
A_2	A_1	A_0	D_3	D_2	D_1	D_0
Q_1	Q_2	D		Q_{1+}	Q_{2+}	Z
0	0	0	0	0	0	0
0	0	1	0	0	1	0
0	1	0	0	1	1	0
0	1	1	0	0	1	0
1	0	0	0	1	1	1
1	0	1	0	0	1	0
1	1	0	0	0	0	0
1	1	1	0	1	0	0

Table 4.27 ROM table for Example (4.8)

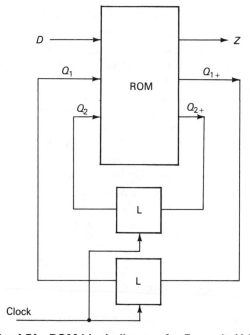

Fig. 4.54 ROM block diagram for Example (4.8)

4.7   STATE REDUCTION

Consider the state table in Table 4.28.

Present state	Next state		Output	
	$I = 0$	$I = 1$	$I = 0$	$I = 1$
1	4	5	1	0
2	4	6	0	0
3	2	5	1	0
4	2	6	0	0
5	6	3	1	0
6	3	2	0	0

Table 4.28   State table for a sequential system

Following the previous discussion, the next step would be to give the states a binary code (state assignment) and proceed with the design. For this three state variables are required (three flip-flops) and any of the binary codes 000 to 111 can be given to any state with no repetition. Two codes will be unused (DON'T CARE). This should result in a correct solution.

A second look at the table, however, shows that states 2 and 4 have the same output under all input conditions. This may raise the questions: 'Could state 2 be equivalent to state 4?'; 'If the two states are equivalent, can they be replaced by one state?'.

It should be clear that two states having the same output at a particular time does not necessarily imply that they are equivalent unless the next states are also the same, since two states with the same output can lead to next states that are different, which in turn results in different outputs at a later stage. It is a fact, however, that by starting from a logical statement, one may over-specify and use more states than is necessary. Systematic procedures are available for the reduction of the number of states and computing equivalent states. This will be explained with the aid of the sequential system in Table 4.28, but first we define the term equivalent.

Two states are *equivalent* if and only if the output sequence is the same for every input sequence regardless of which of the two states is the initial state. If states A and B are equivalent ($A = B$), and states B and C are equivalent ($B = C$), then using the transitive property, states A and C are equivalent ($A = C$). This results in a larger equivalent group ABC. In other words, the equivalent subsets of states are disjoint. The reader is reminded that a binary relation in a set is an equivalence relation if it is reflexive, transitive and symmetric.

4.7.1 Tabular Method for State Reduction

The states in Table 4.28 are divided into two groups according to their output. States within a group must have the same output under all input conditions. This results in the following groups:

$$135, 246 \tag{4.2}$$

Next the successors of these two groups are examined. Group 135 has 246 as successors under input $I = 0$ and 553 or simply 53 under input $I = 1$. Since both successors are members of the groups in (4.2), states 135 stay in a single group.

States 246 have 662(62) as successors under $I = 1$. Both 6 and 2 are members of a single group in (4.2). Under $I = 0$ the successors are 423. Since 42 and 3 are in two different groups in (4.2), group 246 is split into two groups. These are 24 and 6, because the successors of 24 are in one group in (4.2) while the successor of 6 is in another group. Now we have three groups:

$$135, 24, 6 \tag{4.3}$$

The procedure is repeated again. 135 have 35 as successors under $I = 1$, 135 have 246 as successors under $I = 0$. Since 24 and 6 are in two groups in (4.3) 135 is split into 13 and 5. 24 have 24 and 6 as successors, both in the same groups in (4.3). Hence we have the following groups:

$$13, 24, 5, 6 \tag{4.4}$$

This implies that states 1 and 3 and states 2 and 4 can be combined resulting in a state table shown as Table 4.29. The states are relabeled to avoid confusion.

Present state	Next state		Output	
	$I = 0$	$I = 1$	$I = 0$	$I = 1$
13 (A)	B	C	1	0
24 (B)	B	D	0	0
5 (C)	D	A	1	0
6 (D)	A	B	0	0

Table 4.29   A reduced form of Table 4.28

This new table with four states performs the same function as the original table with six states. The reduced table requires two state variables (two flip-flops). The original table requires three state variables (three flip-flops).

Though any state assignment can be used, it is useful to know that a good assignment could result in a more economical circuit, as will be shown in later chapters. This 'good' assignment might not be possible if the state table were reduced. In other words, state reduction might spoil our chance of finding an optimal state assignment ⟨44⟩.

4.7.2 Implication Chart

Determining the equivalent states can be achieved by means of the implication chart. The chart is constructed so that all state pairs can be considered for possible equivalence.

Consider the chart for the sequential system given in Table 4.28.

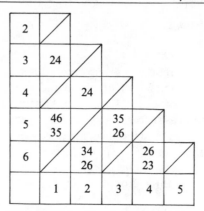

Table 4.30(a) First implication chart

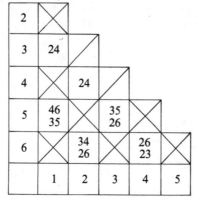

Table 4.30(b) Second implication chart

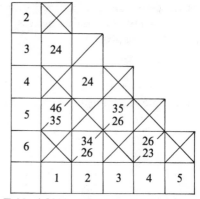

Table 4.30(c) Third implication chart

From the state table given in Table 4.28, the non-equivalent states are identified. States 1 and 2 are non-equivalent because they have different outputs. The corresponding box on the chart is crossed by a diagonal line. Similarly the following state pairs are crossed: (1 4), (1 6), (2 3), (2 5), (3 4), (3 6), (4 5) and (5 6).

States that are equivalent (have the same outputs and the same next states) are ticked. There is none in this case.

The remaining state pairs may be equivalent, if their next states are equivalent. These next state pairs are now entered in the appropriate boxes. For example state pair (4 6) can only be equivalent if (2 6) and (2 3) are equivalent. These two pairs are entered in box (4 6), and so on, as in Table 4.30(a). Next the crossed boxes are examined one at a time. The fact that state pairs (5 6), (4 5), (3 6), (2 5), (1 6), (1 4) and (1 2) are not equivalent does not affect the rest of the state pairs. These are therefore crossed again to show that they have been examined, as in Table 4.30(b).

State pair box (3 4) is also crossed, and so any box containing pair (3 4) must also be crossed. For example state pair (2 6) depend for their equivalence on pair (3 4) being equivalent, but since pair (3 4) is not equivalent, state pair (2 6) also is not equivalent. Box (3 4) is crossed again. Box (2 6) is crossed for the first time, as in Table 4.30(c).

Crossing box (2 6) means that state pairs (4 6) and (3 5) are not equivalent and must be crossed. This in turn implies that state pair (1 5) cannot be equivalent and must be crossed. This leaves state pairs (2 4) and (1 3) as equivalent, which agrees with our previous finding. Normally, it is not necessary to redraw the chart and all crosses can be put on the first chart.

Once the equivalent state pairs are obtained, the transitive relationship can be used to obtain larger groups. For example if states A and B are equivalent, states B and C are equivalent. Then a larger equivalent group ABC is obtained.

As a check we must confirm that the groups form what is known as a closed partition, which simply means that the next states implied by any groups are contained in a single group. In this example the next states for group (1 3) are members of one group (2 4).

4.7.3 Incompletely Specified State Tables

In some situations, it is immateral what the output is when the system is in a particular state. The output can be represented by a dash. It is also possible that some input conditions are not allowed to happen and the output and next states can be considered as DON'T CARE conditions and replaced by dashes. These are called *incompletely specified state tables*, as shown in Table 4.31.

Incompletely specified tables can be made into completely specified tables by replacing all the dashes in the next state table by Xs and adding an extra state X with unspecified outputs, as shown in Table 4.32.

Present state	Next-state/Output	
	$I = 0$	$I = 1$
A	$B/0$	$A/0$
B	$B/0$	$C/0$
C	$D/1$	$E/0$
D	$B/0$	$C/-$
E	$-/0$	$A/1$

Table 4.31   A sequential circuit with unspecified transitions

Present state	Next-state/Output	
	$I = 0$	$I = 1$
A	$B/0$	$A/0$
B	$B/0$	$C/0$
C	$D/1$	$E/0$
D	$B/0$	$C/-$
E	$X/0$	$A/1$
X	$X/-$	$X/-$

Table 4.32   Specified version of Table 4.31

State reduction applies also to incompletely specified state tables. The combined states are said to be *compatible*. The term 'equivalent' is not used here since this implies that both outputs and next states are defined for all inputs. Two states are compatible if and only if the output sequence, whenever specified, is the same for every input sequence regardless of which of the two states is the initial state. A set of states A, B and C is compatible, if and only if all members of the set are compatible. In other words, the transitive property which is used for completely specified tables for generating larger equivalent groups does not necessarily hold for incompletely specified tables. The consequence of the last constraint is that the subsets are non-disjoint and may overlap. A compatible group of states that is not a subset of a large compatible group is called a *maximal compatible*. A compatible can consist of one state, and the complete set of maximal compatibles must cover all the original states. Obviously large compatible states result in a smaller state table.

The unspecified output can be specified as 1 or 0 since it is a DON'T CARE condition. If the state reduction procedure is adopted, one finds that the size of the final table may depend on the way the output is specified. In Table 4.32, for example, if the output for state D, with input $I = 1$, is taken as 0, states B and D become compatible. States B and D are incompatible if the same output is chosen as 1.

4.7.4 A Programmable Procedure for Computing Maximum Compatibles

At this stage it is useful to look at the procedure for finding the maximal exclusive groups described in Chapter 7. The procedure for computing the maximal compatibles is the same and can be programmed and run on a digital computer. Consider as a simple illustration the example of Table 4.28, for which the compatible pairs are (1 3) and (2 4).

Looking at the procedure in Section 7.3, we draw a table as shown in Table 4.33 and all non-compatible pairs are crossed. The E-list (see Section 7.3.3) used for the maximal exclusive groups is called the maximal compatibles list (*MC*) for our present purposes.

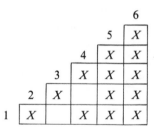

Table 4.33   Non-compatible states

We have:

$S_1 =$		MC	$= 1$
$S_2 =$		MC	$= 1, 2$
$S_3 = 1$		MC	$= (1\ 3), 2$
$S_4 = 2$		MC	$= (1\ 3), (2\ 4)$
$S_5 =$		MC	$= (1\ 3), (2\ 4)$
$S_6 =$		MC	$= (1\ 3), (2\ 4)$

The final answer shows that the pairs obtained earlier, namely (1 3) and (2 4), are in fact the maximal compatibles.

Example (4.9)
Consider the state and output table given in Table 4.34 and try to minimise the number of states if possible.

Present state	Next-state/Output	
	$I = 0$	$I = 1$
1	6/1	4/1
2	7/1	8/1
3	4/1	1/0
4	1/0	8/1
5	3/1	4/1
6	4/1	5/0
7	8/1	2/0
8	2/0	4/1
9	6/0	1/0

Table 4.34   State table of a sequential system

SOLUTION: The implication chart for this state machine is given in Table 4.35. The implication chart can be used for completely (incompletely) specified tables to find equivalent (compatible) pairs.

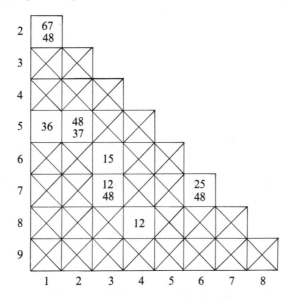

Table 4.35   Implication chart for Example (4.9)

From the implication table we see that the following are the set of compatibles (equivalent states):

(1 2), (1 5), (2 5), (3 7), (4 8), (6 7), (3 6), (9)

State pairs (1 2), (1 5) and (2 5) result in one large group (1 2 5), similarly (3 7), (3 6) and (6 7) result in (3 6 7). Note that this is true whether the transitive property is used or not. The sub-groups can therefore be considered equivalent or compatible. The procedure for computing maximal compatible can be used to the same effect.

Using the procedure outlined earlier, we can compute the set of maximal compatibles (MC). First, the set of compatible pairs is displayed on the chart in Table 4.36. As usual all non-compatible pairs are crossed out.

	2	3	4	5	6	7	8	9
1		X	X		X	X	X	X
2		X	X		X	X	X	X
3			X	X			X	X
4				X	X	X		X
5					X	X	X	X
6							X	X
7							X	X
8								X

Table 4.36   Conflicting states for Example (4.9)

From Table 4.36 we obtain:

$S_1 =$		MC	$= 1$
$S_2 = 1$		MC	$= (1\ 2)$
$S_3 =$		MC	$= (1\ 2),\ 3$
$S_4 =$		MC	$= (1\ 2),\ 3,\ 4$
$S_5 = 12$		MC	$= (1\ 2\ 5),\ 3,\ 4$
$S_6 = 3$		MC	$= (1\ 2\ 5),\ (3\ 6),\ 4$
$S_7 = 36$		MC	$= (1\ 2\ 5),\ (3\ 6\ 7),\ 4$
$S_8 = 4$		MC	$= (1\ 2\ 5),\ (3\ 6\ 7),\ (4\ 8)$
$S_9 =$		MC	$= (1\ 2\ 5),\ (3\ 6\ 7),\ (4\ 8),\ 9$

The final set of MC gives the maximal compatible groups which agrees with our previous finding using the implication chart. Either method may be used for small examples. The last procedure, however is programmable, which makes it more suitable for large problems. The program described in Section 7.3 may be used for this purpose.

The result indicates that the original state table with nine states can now be replaced by four states. This is shown in Table 4.37(a).
If the states are relabeled, we obtain Table 4.37(b).

Present state		New-state/Output	
		$I = 0$	$I = 1$
(1)	125	367/1	48/1
(2)	367	48/1	125/0
(3)	48	125/0	48/1
(4)	9	367/0	125/0

Table 4.37(a)   Minimized state table for Example (4.9)

Present state	New-state/Output	
	$I = 0$	$I = 1$
1	2/1	3/1
2	3/1	1/0
3	1/0	3/1
4	2/0	1/0

Table 4.37(b)   The state table for Example (4.9) minimized and relabeled

4.8   CONCLUSION

The reader may conclude that simple sequential circuits may be designed intuitively. For more complex circuits, it is safer to follow the systematic procedure outlined in Chapter 4. The procedure involves a logic diagram, a state table, state assignment and a set of logic equations. It is true that the choice of the state assignment code may influence the logic complexity, but as yet there is no way of predicting the optimum state assignment. Some useful techniques are discussed in Chapters 8 and 9 for finding 'good' state assignments.

The problem a designer may face is that large state tables are difficult to handle. In fact difficulties can arise if the number of primary inputs is greater than five or six. Fortunately there is what is known as the algorithmic state machine (ASM) method which offers a good design technique especially for multi-input system controllers. The technique is not described here but the interested reader is referred to the appropriate references ⟨19, 35⟩.

This subject is discussed further in Chapter 11.

Exercises

Q(4.1)   (*a*)   Show how the five-bit shift register given in Appendix 4.1 can be used as a twisted ring counter. Derive suitable logic to provide a divide-by-ten counter.

(*b*)   Consider the first three flip-flops of the register in Q[4.1(*a*)], and assume that Q_C is initially set to 1 while other flip-flops are set to 0. Predict the quasi-random sequence generated if the outputs Q_B and Q_C are fed back to the serial input through an EX-OR gate.

Q(4.2)   The digital clock in Example (4.1) (page 146) requires a signal for setting or correcting the minutes and hours sections. A manual switch is used for each section. These switches pass 1 pps clock pulses for setting when pressed and the normal pulse when released. Design bounce-free switches and derive the necessary logic to achieve this.

Q(4.3)   Design a 12-state binary counter capable of counting from 0000 up to 1011. Hence derive and optimize the necessary logic to give an output when the binary count is divisible by decimals two or three.

Q(4.4)   Using *JK* flip-flops and any logic gates, design a synchronous error-checking circuit to detect the presence of the sequence 101 in a serial flow of binary data.

Q(4.5)   A motor is to run when the start button is pressed and the following conditions are satisfied:

(*a*)   A safety door is closed.
(*b*)   The level of the liquid in the tank is above a certain level.

The motor is to stop and a lamp is to glow if any of the following conditions occur:

(*a*)   The stop button is pressed.
(*b*)   The safety door is opened.
(*c*)   The liquid level is below the satisfactory level.

The lamp stops glowing when the motor starts.
Design a suitable circuit using logic gates and a flip-flop.

APPENDIX 4.1   A SAMPLE OF AVAILABLE ICs

73 Dual JK negative edge-triggered Flip-Flop

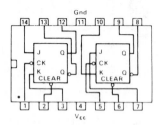

70 J-K flip-flop

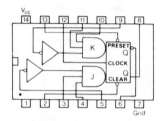

72 J-K master-slave flip-flop

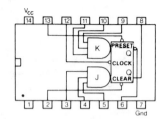

74 Dual D-type edge-triggered Flip-Flop

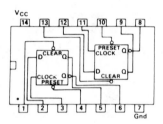

75 4-bit D Latch

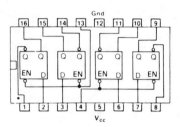

76 Dual JK Flip-Flop with set and clear

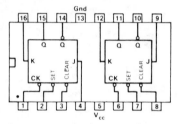

90 Decade counter

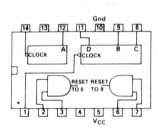

92 Divide-by-twelve counter

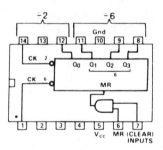

93 4-bit binary counter

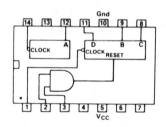

96 5-bit shift register

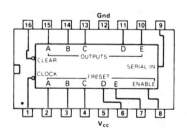

164 Serial-in parallel-out shift register

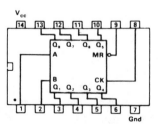

165 8-bit parallel to serial converter

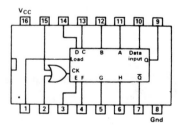

APPENDIX 4.2   SOFTWARE IMPLEMENTATION OF EXAMPLE (4.7) USING THE 6800 MICROPROCESSOR

```
0010   7F   F511
0013   86   7F
0015   B7   F510
0018   86   04
001A   B7   F511
```

```
001D   86   01
001F   B7   F510
0022   5F
0023   B6   F510
0026   2A   FB
0028   B6   F510
002B   2B   FB
002D   5C
002E   C1   28      ④⓪
0030   26   F1
```

```
0032   86   02
0034   B7   F510
0037   5F
0038   B6   F510
003B   2A   FB
003D   B6   F510
0040   2B   FB
0042   5C
0043   C1   14      ②⓪
0045   26   F1
```

```
0047   86   04
0049   B7   F510
004C   5F
004D   B6   F510
0050   2A   FB
0052   B6   F510
0055   2B   FB
0057   5C
0058   C1   28      ④⓪
005A   26   F1
```

```
005C   86   06
005E   B7   F510
0061   5F
0062   B6   F510
0065   2A   FB
0067   B6   F510
006A   2B   FB
006C   5C
006D   C1   14
006F   26   F1
0071   20   AA
*
```

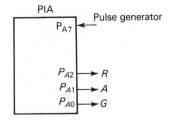

PIA
P_{A7} ← Pulse generator

P_{A2} → R
P_{A1} → A
P_{A0} → G

Light sequence

P_{A2}	P_{A1}	P_{A0}	
0	0	1	40 S
0	1	0	20 S
1	0	0	40 S
1	1	0	20 S

5

Asynchronous Sequential Circuits

5.1   INTRODUCTION

Like synchronous sequential circuits, the outputs and next states of asynchronous sequential circuits are functions of both the primary (external) inputs and the present internal states, as shown in the general model of Fig. 5.1. The difference is the absence of the timing (clock) pulse. The absence of the clock pulse may be considered an advantage since it increases the overall speed of the circuit, but it makes the design more difficult and increases the probability of malfunction. To understand the difference, consider a sequential circuit with a static input. A clocked (synchronous) circuit will sample the input every clock pulse and will, therefore, see the input as a repetitive series of pulses. The unclocked (asynchronous) circuit will only see one input level.

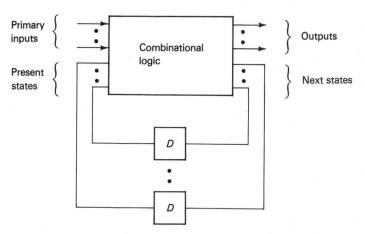

Fig. 5.1   General model for asynchronous circuits

The total state of the circuit is defined by the primary inputs and internal (secondary) states. The next-state secondary variables are usually represented by capital letters, Y, say. These are delayed and fed back to represent the present states. The delay is not always shown, but is implied. The expressions for the next states in terms of the external inputs and present states are called the *excitation functions*.

If the next state, of the asynchronous circuit, is equal to the present state, the circuit is said to be stable. When the input changes, the excitation variables change and the circuit moves to an unstable state. Sometime later the secondary variables assume their new values, the next state becomes equal to the present state, and the circuit enters a new stable state. The circuit moves from one stable state to another as the input changes.

In fundamental mode circuits, once an input changes, no further change is allowed until the circuit enters a stable state. Otherwise the new destination becomes uncertain. The time between consecutive input changes should be greater than the largest feedback delay. For multi-inputs then, only one input is allowed to change at any time since it is very difficult in practice to ensure simultaneous change in inputs. If these conditions are not satisfied races could occur.

There is another class of asynchronous circuits called *pulse mode* which accepts pulse inputs. Because the circuit is not clocked, the input pulses may arrive at random but not more than one pulse can arrive at any time. For the circuit to operate correctly, the duration of the pulse must be long enough to change the state of the memory device but short enough for it to be no longer there after the state change. Obviously if the last restriction is violated, one input could cause more than one change. The circuit is stable when there are no inputs; hence the circuit can be treated as a special class of synchronous circuit and will not be considered further here.

5.2   CIRCUIT ANALYSIS

Consider the circuit of Fig. 5.2, in which only one of the two inputs A and B can change at any one time.

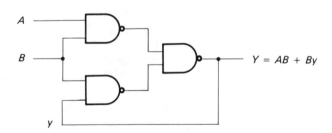

Fig. 5.2   Asynchronous sequential circuit

y is the present-state variable.
A and B are the primary inputs.
Y is the next-state variable.

Note that the circuit has a feedback input y as well as the primary inputs. The output may be derived separately, or may be the same as the next state as in this case.

5.2.1 The Excitation Map

If the feedback path is disconnected, we may write:

$$Y = AB + By$$

If a K-map is now drawn with the primary inputs used as column variables as shown in Table 5.1, we have what is known as the *excitation map*.

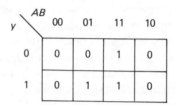

Table 5.1   Excitation map for Y

On this map, the states where $Y = y$ are called *stable states* and those where $Y \neq y$ are called *unstable states*. To see the reason, consider the position on the map under column 10, row 0 ($ABy = 100$). Here $y = 0$, $Y = 0$.

If the feedback is reconnected, the circuit will remain in this state, which is a stable state and is usually identified by a circle, as shown in Table 5.2 below.

Now consider position $ABy = 110$ where $y = 0$, $Y = 1$. If the feedback is reconnected, one of the variables must change. This state is then unstable and the circuit must go to a stable state where $Y = y$. Since A and B are fixed, the change must be in the vertical direction.

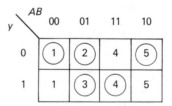

Table 5.2   Flow table

5.2.2 Flow Table

The stable states are now given numbers as shown in Table 5.2. In this case five stable states can be recognized.

The stable states are circled and the unstable states are given the same number as the adjacent stable states in the same column. If, while the circuit is in stable state ④, the input changes from 11 to 10, the circuit moves to the unstable (transient) state 5 and then settles at stable state ⑤.

5.2.3 **The Output**

The output in this introductory example is assumed to be the same as the next state. Generally speaking, however, the output may be derived separately and could be a function of the state variables as well as some or all the input variables, as will be shown later.

5.2.4 **The State Diagram**

In the state diagram, circles are used to represent the stable states. The states are joined by arrows (arcs) to indicate transitions. For example, when the circuit is in state ① and input B changes from 0 to 1, the circuit moves to state ②. The input causing this is 01 as shown in Fig. 5.3. Normally the states producing a logic 1 output are shaded on the diagram.

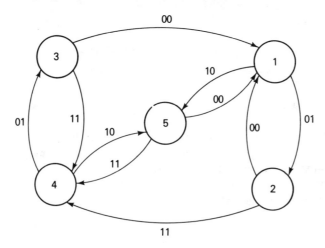

Fig. 5.3   The state diagram

5.2.5 **Primitive Flow Table**

In the primitive flow table, one column is reserved for each possible input combination. Since there are two primary inputs, we must have $2^2 = 4$ columns. Similarly one row is reserved for each state. The stable states occupy the column leading to those states. On the side of each stable state are the unstable states through which the circuit passes when an input changes.

In Table 5.3, we have two unstable states next to each stable state, both at a distance-one variable (one variable changing). Since the circuit is limited to one input change, we have one CAN'T HAPPEN state in each row. The CAN'T HAPPEN state is represented by a dash.

Suppose, for example, that the circuit is in stable state ① while the inputs A and B are both zero. Then if B changes we have a new input combination $AB = 01$.

From the state diagram of Fig. 5.3, we know that the circuit goes to state ②. Therefore transient state 2 is entered in the first row, the second column of the primitive flow table. Similarly a change in variable A while in state ① results in inputs $AB = 10$, which takes the circuit to state ⑤. Therefore transient state 5 is entered in the first row, fourth column. Since A and B cannot change simultaneously, the first rwo, third column is a CAN'T HAPPEN state represented by a dash.

The primitive flow table contains the same information as the state diagram, but in a tabular form.

	00	01	11	10
1	①	2	—	5
2	1	②	4	—
3	1	③	4	—
4	—	3	④	5
5	1	—	4	⑤

Table 5.3   Primitive flow table

5.2.6 Merger Diagram

Consider the first two rows in Table 5.3. Since in the same columns, the transient and stable states are the same, the two rows can be merged. The CAN'T HAPPEN states can be merged with any other state.

To achieve a merged table, a merger diagram is drawn as shown in Fig. 5.4. If any two states can be merged, they are joined by a straight line, and they are said to be *mergeable*.

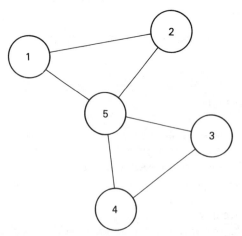

Fig. 5.4   Merger diagram

In Table 5.3 each state is compared with all succeeding states. State 1 can be merged with state 2 or state 5. State 2 can be merged with state 5. State 3 can be merged with state 4 or state 5. State 4 can be merged with state 5. This is shown in the merger diagram (Fig. 5.4). It will be shown later that states can be merged even if their outputs are different. Usually these are joined by dotted lines.

The two closed triangles indicate that states 1, 2 and 5 can be merged with each other since each pair is mergeable. Similarly states 3, 4 and 5 can be merged together. State 5 can be in either group. If we choose to merge 3, 4 and 5 then 1 and 2 are merged on their own.

Four states are mergeable only if each state can be merged with every other one. On the merger diagram (Fig. 5.5) this is shown as in (*d*).

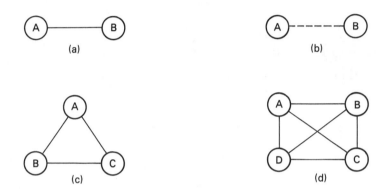

(a)   (b)

(c)   (d)

Fig. 5.5   Typical Merger diagrams

(*a*) Two mergeable states having the same output
(*b*) Two mergeable states having different outputs
(*c*) Three mergeable states
(*d*) Four mergeable states

The primitive flow table can now be rewritten in a more compact (merged) form, as in Table 5.4.

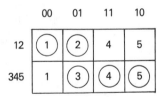

Table 5.4   Merged flow table

In the merged table, a stable and a transient state produce a stable state.

5.2.7 The State Assignment

If we arbitrarily assign logic 0 to the first row, then in the first row in Table 5.4 all the stable states are assigned 0 while all the unstable states are assigned 1. Similarly

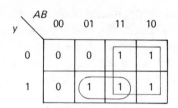

Table 5.5   Excitation map

the second row, assigned logic 1, results in all the stable states in that row being assigned 1 while all the unstable states are assigned 0, as in Table 5.5.

By looping the 1s in Table 5.5, we obtain an expression for the next-state variable:

$$Y = A + By$$
$$= \overline{\overline{A} \cdot \overline{By}}$$

This can be implemented using NAND gates, as shown in Fig. 5.6.

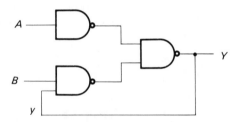

Fig. 5.6   Alternative implementation of the circuit of Fig. 5.2

It is obvious by now that we have completed a full circle. We started from a circuit diagram, analyzed it and ended up with a different circuit. The reason, in fact, is our choice of merged states. If we merge states 125 and 34, we would obtain the flow table shown in Table 5.6, the entries of which are identical to those of Table 5.2.

Again, if we assign logic 0 to the first row and logic 1 to the second row, we obtain Table 5.7, the entries of which are identical to those of Table 5.1.

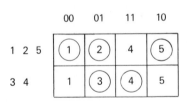

	00	01	11	10
1 2 5	1	2	4	5
3 4	1	3	4	5

Table 5.6   Alternative merged table

y \ AB	00	01	11	10
0	0	0	1	0
1	0	1	1	0

Table 5.7   Alternative excitation map

If the 1s are looped as shown we obtain:

$$Y = AB + By$$
$$= \overline{\overline{AB} \cdot \overline{By}}$$

The NAND implementation for this is the same as that shown in Fig. 5.2.

In brief, different ways of merging states (if a choice is possible) and different assignments (as was shown in Chapter 4) could result in different circuits. The circuits should fulfill the specified specifications.

5.3   DESIGN OF SINGLE STATE VARIABLE SYSTEMS

In Section 5.2 we started from a designed circuit, analyzed it and redesigned it. In most practical situations, a circuit specification or perhaps a waveform diagram is given.

The engineer is required to design a working circuit. The best way to illustrate the design procedure is by means of an example. We shall assume also that *SR* flip-flops and discrete gates are to be used.

Example (5.1)

Design an asynchronous sequential circuit whose output becomes logic 1 if either of the two inputs rises to logic 1 when the other input is already at logic 1. The output is at logic 0 if input *A* drops to logic 0 regardless of the value of input *B*. The inputs *A* and *B* cannot change simultaneously.

5.3.1 Waveform Diagram

The behaviour of the circuit is shown by means of a waveform, as in Fig. 5.7.

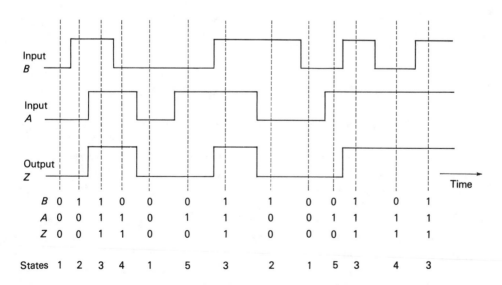

B	0	1	1	0	0	0	1	1	0	0	1	0	1
A	0	0	1	1	0	1	1	0	0	1	1	1	1
Z	0	0	1	1	0	0	1	0	0	0	1	1	1

States 1   2   3   4   1   5   3   2   1   5   3   4   3

Fig. 5.7   Waveform diagram for Example (5.1)

The waveform portrays all possible combinations of the two binary inputs A and B. As specified, the output Z rises to 1 only when one input rises to 1 provided the other input is at logic 1 as in state ③. Once $Z = 1$ it drops to 0 only if input level A drops to zero.

Every new possible input/output combination is allocated a stable state and is given a new number. For example if when $A = B = 0$, $Z = 0$. This is given number ① when input B rises to 1; we have a new situation which we call ② even though Z is still at logic 0, and so on.

5.3.2 Primitive Flow Table

From the waveform diagram, or from the problem specifications we can develop a flow table. The problem has five states; the table will have five rows.

The number of inputs is two; the table will have $2^2 = 4$ columns, as shown in Table 5.8.

Inputs

B A

	0 0	0 1	1 1	1 0
1	①	5	–	2
2		–		②
3	–		③	
4		④		–
5		⑤		–

Table 5.8   Partial primitive flow table
– signifies CAN'T HAPPEN

First, the stable states are entered one stable state per row in the appropriate columns as dictated by the waveform.

Next the CAN'T HAPPEN states are marked. In row 1 for example, the circuit is at stable state ① and the inputs are 0 0. The circuit cannot move to column 3 since this means both inputs changing simultaneously. This position is marked by a dash (—). Similarly other rows are marked as shown in Table 5.8.

From Fig. 5.7 we can see that when the circuit is in state ① it could go to state 2 if B changes or to state 5 if A changes. These are the next transient states entered on the primitive flow table.

If in state ③ say, the circuit could go to state 4 if B changes or state 2 if A changes, and so on until all the states are entered in the table. The complete primitive state table is given in Table 5.9.

Note that the table contains one stable state per row.

BA

	00	01	11	10
1	①	5	—	2
2	1	—	3	②
3	—	4	③	2
4	1	④	3	—
5	1	⑤	3	—

Table 5.9 Primitive flow table for Example (5.1)

It should be emphasized at this stage that, when the circuit is in a transient state, it moves within the same column until a stable state is found. Further the input changes only when the system is in a stable state, and even then only one input is allowed to change.

5.3.3 State Diagram

It is not always necessary to draw the state diagram, but to do so usually helps to visualize the circuit behaviour. It effectively contains the same information as the flow table. As mentioned in Section 5.2, the states resulting in a high output are shaded. From Fig. 5.7, the output is high when the circuit is in states 3 or 4. States ③ and ④ are shaded in Fig. 5.8.

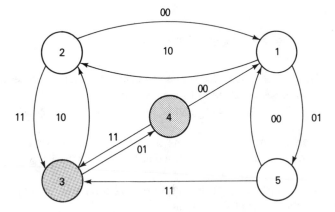

Fig. 5.8 Total state diagram for Example (5.1)

5.3.4   Equivalent States

Some over-specified tables may contain equivalent states as in a synchronous circuit. Some of these can be detected by comparing all pairs of stable states. States are equivalent if they are in the same column, they have the same output, and their unstable successor states under all possible inputs are the same, equivalent, or DON'T CARE.

Implication charts, as described in Chapter 4, can be used to detect equivalent states for complex systems.

5.3.5 State Merging

The number of internal states remains equal to the number of stable states even after eliminating equivalent states. In order to reduce the number of state variables, we must try to combine rows by the merging process.

Merging states means that they can be assigned the same combination of state variables. The stable states are from different columns. The outputs are not necessarily the same, but merging of states with the same output usually leads to simpler circuits.

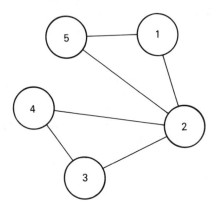

Fig. 5.9   Merger diagram for Example (5.1)

In Fig. 5.9 the five states are drawn, and following the merger procedure outlined in Section 5.2 it can be seen that states (1) and (5) can be merged, and that states (1) and (5) are linked. Similarly states (2) and (4), (2) and (5), (3) and (4) are linked, and so on.

The two triangles in Fig. 5.9 indicate that states 1 2 5, 3 4 or 1 5, 2 3 4 can be merged. In other words state 2 can be merged either with 3 and 4 or with states 1 and 5.

Figs 5.7 and 5.8 indicate that the output occurs when the circuit is in states (3) and (4). It seems reasonable to merge these states.

This leads to Table 5.10.

5.3.6 The State Assignment

Table 5.10 contains two rows, and requires one state variable y. If in the first row y is assigned the binary code 0 and in the second row y is assigned the binary code 1 we have the state assignment table shown in Table 5.11.

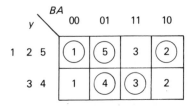
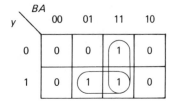

Table 5.10   Merged table for Example (5.1)   **Table 5.11   State assignment**

By definition, the stable states are identical to the present states. Hence 0s and 1s are entered in the table.
From Table 5.11, we obtain

$$Y = BA + Ay$$

5.3.7 The Output Table

In Fig. 5.7, the output Z is at logic 1 when the circuit is in states ③ and ④. These are shown shaded in the total state diagram (Fig. 5.8).
An output table can now be drawn as in Table 5.12

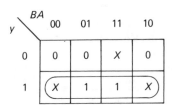

Table 5.12   Output table for Example (5.1)

In the output table 1s are inserted in states ③ and ④, where the output is known to be high, and 0s are inserted in states ①, ② and ⑤, where the output is known to be low. In the transient states Xs are inserted to indicate DON'T CARE. The DON'T CARES can be used to simplify the logic, depending on whether the output is required to change as quickly as possible or as slowly as possible.
Consider the circuit in state ③ when the inputs are 11 and the output is 1. If the inputs change to 10, the circuit moves to transient state 2 before going to stable state ②. Since in state ②, the output is 0, the output variable must now change from 1 to 0. If in transient state 2 the output is made 1 (same as old state ③), the output will

change as slowly as possible. If it is made 0 (same as the new state ②), it will change as quickly as possible. In this example the choice leading to the simplest logic is used.

Note that if all the Xs are made 0 we get

$$Z = Ay$$

If the Xs in the bottom row are made 1 we get

$$Z = y$$

The circuit can be implemented as in Fig. 5.10.

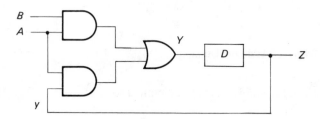

Fig. 5.10   Gate implementation of Example (5.1)

5.3.8 Flip-Flop Implementation

If it is required to implement the circuit using flip-flops, the input equations for the flip-flop must be derived.

Assume set-reset flip-flops (SR/FF) are to be used. In this case the present and next state variables y and Y are assigned to the present and next states of the flip-flop, which are called Q and Q_+ in Chapter 4.

It would be useful at this stage to reproduce the excitation table of the SR/FF given in Table (4.1(b)). Knowing the present and next state conditions of the flip-flop as dictated by the assignment in Table 5.11, the excitation table enables us to determine the required inputs for S and R.

Present state Q	Next state Q_+	Inputs S R	
0	0	0	X
0	1	1	0
1	0	0	1
1	1	X	0

Excitation table for the SR/FF [based on Table 4.1(b)]

To determine the input requirements for the flip-flop two maps are drawn for S and R as in Table 5.13.

To determine the content of the two tables, examine Table 5.11 one square at a time ($a - h$):

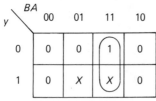

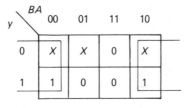

Map for *S*   Map for *R*

Table 5.13   *SR* maps for Example (5.1)

(a)   Consider square $BAy = 000$ in Table 5.11. In this position, the present state y $= 0$ (first row). The next state $Y = 0$ (first row, first column). Since both the present and next states are 0, S must be 0 and R is DON'T CARE; as shown in the excitation map of the SR/FF. Hence in square $BAy = 000$ in Table 5.13, 0 is inserted for S and X is inserted for R.

(b)   Square $BAy = 010$.
Present state $y = 0$ (first row)
Next state $Y = 0$ (first row, second column)
Hence in square $BAy = 010$ of Table 5.13 we enter
$S = 0$ $R = X$

(c)   Square $BAy = 110$.
Present state $y = 0$
Next state $Y = 1$
For the flip-flop to change from 0 to 1 the excitation map for the SR/FF shows that
$S = 1, R = 0$
Hence in square $BAy = 110$ of Table 5.13 we enter
$S = 1$, $R = 0$

(d)   Square $BAy = 100$
Same as (a) and (b)
$S = 0, R = X$

(e)   Square $BAy = 001$.
Present state $y = 1$
Next state $Y = 0$
The flip-flop is reset from 1 to 0
The excitation map of the flip-flop shows that $S = 0, R = 1$
Hence in square $BAy = 001$ of Table 5.13 we enter
$S = 0, R = 1$

(f)   Square $BAy = 011$
Present state $y = 1$
Next state $Y = 1$
The excitation map of the flip-flop shows that when the present and next states are 1, $S = X$ and $R = 0$.
Hence in square $BAy = 011$ of Table 5.13 we enter
$S = X, R = 0$

(g) Square $BAy = 111$
     Same as (f)
     $S = X, R = 0$

(h) Square $BAy = 101$
     Same as (e)
     $S = 0, R = 1$

Now Table 5.13 is completed, the 1s and Xs are looped in the normal way, giving

$$S = BA$$
$$R = \bar{A}$$

From Table 5.12 it was found that

$$Z = y$$

5.3.9 Circuit Diagram

The circuit diagram for Example (5.1) can now be implemented using an SR/FF and gates as in Fig. 5.11(a).

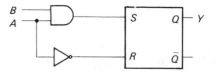

Fig. 5.11(a)   Implementation of Example (5.1) using an *SR* flip-flop

The use of an SR/FF has the advantage that it eliminates static hazards by including all the prime implicant terms. To see this the circuit diagram in Fig. 5.11(a) is redrawn in Fig. 5.11(b) with the SR flip-flop replaced by its NOR gate circuit equivalent, which was described in Chapter 4.

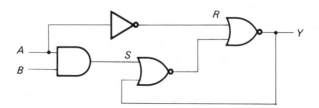

Fig. 5.11(b)   Alternative implementation of Example (5.1)

For the SR/FF, the next state equation is given by

$$Y = S + \bar{R}y$$

For this circuit

$$Y = BA + Ay$$

Entering this on a K-map, we can see that no static hazard is present.

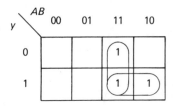

5.3.10 Alternative SR Map

For the experienced designer, it is more convenient to use one map for both S and R to replace the two maps in Table 5.13. In this case capital S and R are used for the set and reset conditions while small s and r are used for the DON'T CARE conditions. This results in Table 5.14.

	BA			
y	00	01	11	10
0	r	r	S	r
1	R	s	s	R

Table 5.14   Alternative *SR* map for Example (5.1)

5.4   HAZARD CONDITIONS

The absence of a clock pulse in asynchronous systems means that signals travelling along paths with different delays can lead the circuit to the wrong state. In other words the final destination depends on which signal wins the race. If the race takes the circuit to a different next state from the intended state, the race is called *critical*. Races that cannot cause a false transition are called *non-critical*. In the following the main types of race hazards are briefly described.

5.4.1 Static and Dynamic Hazards

Static hazards are a property of the combinational logic, and occur due to the existence of two paths with different delays. Their effect becomes serious if the momentarily incorrect signal is fed back to the input when a permanent error could occur. Whether a static hazard causes a malfunction or not depends on the various delays. Static hazards can be eliminated logically, as was described in Chapter 3.

If the output is required to remain 1, the static hazard is in the 1s. If the output is required to remain 0, the static hazard is in the 0s. If the output is required to change, the circuit may go temporarily to the new value, then go back to the initial value before making the final permanent transition; that is $0 \rightarrow 1 \rightarrow 0 \rightarrow 1$, instead of $0 \rightarrow 1$. This is called dynamic hazard, and occurs when paths of at least three different delays are present.

A helpful rule, widely used, is that a second-order sum-of-product circuit that is free of all static hazards in the 1s will be free of all static and dynamic hazards.

5.4.2 Essential Hazards (29, 46, 107, 111)

Essential hazards are peculiar to sequential circuits, and are caused by different path lengths of a signal to two different feedback loop responses (at least two feedback paths). This can cause false transitions to an incorrect state. The hazard is caused by the circuit specifications and is not affected by the particular form of realization. Unlike static hazards, essential hazards cannot be corrected logically.

Unger showed that an essential hazard occurs in fundamental mode circuits whenever three consecutive input changes take the circuit to a different stable state from that caused by the first change.

Circuits having no steady-state hazards are called *S-proper* regardless of whether there are transient hazards. Unger described a procedure for *S-proper* delay-free circuit realizations of single-output change functions without essential hazards assuming single-input change operation, though transient hazards may be present. He also proved that flow tables with essential hazards cannot be realized without delay elements if steady-state hazards (entering the wrong state) are to be eliminated. Further Unger proved that any function can be realized in a trouble-free manner by a circuit containing just one delay element. If essential hazards are a problem, delays should be used to ensure that the required signal wins the race.

Eichelberger used ternary algebra for detecting hazards in combinational and sequential switching circuits resulting from the simultaneous changing of two or more input signals, and Unger also studied the case of unrestricted input change. In Hill and Peterson the case of multiple input change is briefly summarized in the following two paragraphs:

If there is only one stable state in a column, there can be no malfunction. Sooner or later all changes propagate sufficiently to bring the circuit into the proper column, and if there is only one stable state in that column, the circuit must terminate there.

If the multiple change is into a column with two or more stable states, reliable operations will be ensured if the following two rules are satisfied.

(i)   There must be enough delay in each feedback loop to allow all input changes to have propagated through at least the first level of gating before any secondary changes have propagated around a loop.

(ii)   If a change in a subset of the inputs involved in a multiple change can cause the circuit to leave a row in the state table, this behaviour must be consistent with the intended behaviour associated with the multiple change.

Designers prefer using a single chip to implement logic functions since the variation in delay times along various paths is small compared with the multi-chip case. Even then it is still necessary to insert delays in the feedback loop so that secondaries do not change until the input change has propagated to all parts of the circuits. This slowing of the operation is not always desirable, but with clocks absent from asynchronous circuits, the use of delays is often the only way to cure essential hazards. The use of delays is particularly necessary when modules are used in cascade, a situation where the variation in delay time along the various paths is more pronounced. If a single ROM, say, is used, the risk of an essential hazard is minimized, but even then there is no guarantee that the next states at the output of the device will appear simultaneously. Delays, sometimes of different magnitudes, must therefore be inserted in the feedback loop.

5.5   MULTIPLE STATE VARIABLE SYSTEMS

In Example (5.1) the system had only two rows (internal states) and hence one state variable was used to distinguish between the states. The variable was assigned the binary digit 0 for the first row and binary digit 1 for the second row. For n rows, s state variables are necessary where $2^s \geqslant n$.

In synchronous circuits (Chapter 4) it was mentioned that any state assignment should give a working circuit, though different state assignments could result in different complexities.

For asynchronous circuits with more than one state variable, care must be taken to ensure that not more than one state variable changes at any time. Otherwise critical races may arise, as explained in Section 5.4. The problem of optimum design is considered in Chapter 10. For the time being, our priority is to design a working system free of critical races.

To illustrate this point consider a system with seven or eight rows. In synchronous circuits, three state variables are required, as was explained in Chapter 4. In asynchronous circuits three state variables might not be enough to avoid race conditions and ensure that all possible transitions from transient to stable states can be achieved with only one state variable changing. The extra DON'T CARE states available with the additional state variables give the designer more flexibility.

Consider part of a flow table as shown with two inputs and two state variables:

Inputs

AB

$y_1\ y_2$	00	01	11	10
1	3			
2	②			
3	③			
4	3			

Let us assume the following state assignment:

$y_1\,y_2$	00	01	- - -
10	01		
00	00		
01	01		
11	01		

Let the system be in state $y_1y_2 = 10$, with the final destination 01 and the inputs AB = 00. Both y_1 and y_2 must change. If y_1 changes first, the system moves to stable state 00 and stays there, since 00 under input 00 is a stable state. If y_2 changes first, the system moves to transient state 11 and then to 01 when y_1 changes.

The final destination depends on which variable changes first. This state assignment is not acceptable since the final destination is thus unpredictable.

Now consider the following state assignment:

$y_1\,y_2$	00	01	- -
11	01		
10	10		
01	01		
00	01		

If again we start with $y_1y_2 = 10$ and the final destination is 01 both y_1 and y_2 must change.

If y_1 changes first, the system moves to 00, then to 01, when y_2 changes.

If y_2 changes first, the system moves to 11, and then to 01 when y_1 changes.

The final destination is not dependent on the order in which y_1 and y_2 change. This is then a valid assignment.

It is possible to arrive at a valid assignment by ensuring that the assignment for rows containing an unstable state j and the assignment for rows containing a stable state ⓙ differ in one variable only (adjacent).

In the above two state variable table, the state assignment code for rows containing 3 (rows 1 and 4) must be adjacent to the state assignment code for the row containing state ③ (row 3).

Generally speaking it is always possible to achieve a valid assignment by ensuring that every row is adjacent to all other rows. This very safe approach, however, could use more secondary variables than necessary. It is always worth while to attempt a race-free assignment using the minimum number of variables.

Example (5.2)

An asynchronous sequential circuit has two binary inputs A and B as shown in the time sequence. The two inputs can never change together.

In Fig. 5.12 input A is a train of sampling pulses which are used to sample the data input B. The output of the circuit must assume the value of the data B when the sampling pulse A changes and hold that value until A changes again.

Design the circuit using suitable logic.

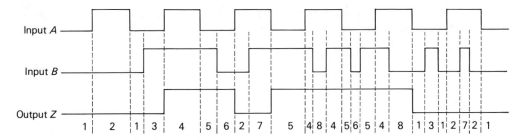

Fig. 5.12 Waveform diagram for Example (5.2)

SOLUTION: Let Z be the output variable. The waveforms in Fig. 5.12 show the possible input/output combinations which indicate a logical output at states 4, 5, 6 and 8. The same information can be translated into a total state diagram (Fig. 5.13) or a primitive flow table (Table 5.15) as follows:

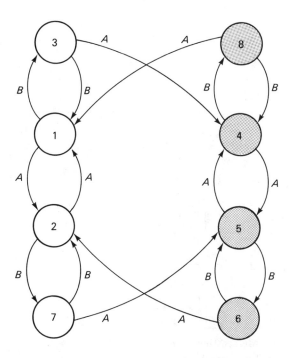

Fig. 5.13   Total state diagram for Example 5.2

	State			
Present state	Input AB			
	00	01	11	10
1	①	3	—	2
2	1	—	7	②
3	1	③	4	—
4	—	5	④	8
5	6	⑤	4	—
6	⑥	5	—	2
7	—	5	⑦	2
8	1	—	4	⑧

Table 5.15   Primitive flow table

At this stage one would try to merge some of the states in order to reduce the number of state variables required for the state assignment.

The merger diagram (Fig. 5.14) gives us a choice of merger. State 1 can be merged with either 2 or 3. State 8 can be merged with 3 or 4. State 6 can be merged with 5 or 7.

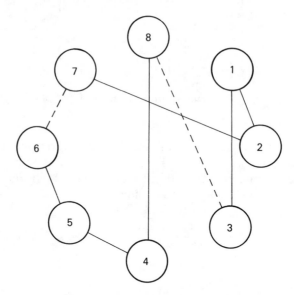

Fig. 5.14   Merger diagram for Example (5.2)

If states 12, 38, 45 and 67 are merged, the primitive flow table can be reduced to a four-state table.

Note that states are merged even though they have different outputs.

	00	01	11	10
12	①	3	7	②
38	1	③	4	⑧
45	6	⑤	④	8
67	⑥	5	⑦	2

**Table 5.16   Merged flow table
for Example (5.2)**

A binary code is given to the four rows of the merged table as in Table 5.17, according to the state assignment rules discussed in Section 5.5.

		Y_1Y_2			
		AB			
y_1	y_2	00	01	11	10
0	0	00	01	10	00
0	1	00	01	11	01
1	1	10	11	11	01
1	0	10	11	10	00

Table 5.17   Assigned flow table for Example (5.2)

5.5.1 Gate Implementation of Example (5.2)

Expressions for the next state variables Y_1 and Y_2 can be obtained in terms of the input variables A and B and the present state variables y_1 and y_2, e.g.

$$Y_1 = \bar{A}\bar{B}\,(y_1y_2 + y_1\bar{y}_2) +$$
$$\bar{A}B\,(y_1y_2 + y_1\bar{y}_2) +$$
$$AB\,(\bar{y}_1\bar{y}_2 + \bar{y}_1y_2 + y_1y_2 + y_1\bar{y}_2)$$

This expression can be simplified, using a K-map or otherwise. In this case, however, the state assignment is chosen in such a way that the state assignment table is in K-map form, but the map is used twice for Y_1 and Y_2, which are entered side by side. Hence we can write the expressions for Y_1 and Y_2 directly in a minimized form.

Hence $Y_1 = AB + \bar{A}y_1 + By_1$

and $Y_2 = \bar{A}B + Ay_2 + By_2$

The output is high in states 4, 5, 6, and 8 as can be seen from the waveform or state diagram. This is shown in Table 5.18.

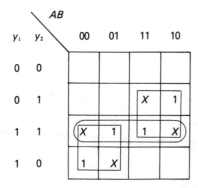

Table 5.18   Output map for Example (5.2)

From the output map

$$Z = y_1y_2 + \bar{A}y_1 + Ay_2$$

The circuit can be implemented using suitable logic as shown in Fig. 5.15. Note that the delays D are not always shown.

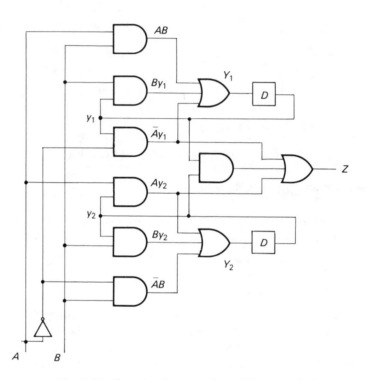

Fig. 5.15   Gate implementation of Example (5.2)

5.5.2 Flip-Flop Implementation of Example (5.2)

Since we have two state variables, two flip-flops are required. From the state assignment table, the present and next states for each flip-flop are compared. The resulting set and reset conditions are entered on the SR maps. For convenience, one map is drawn for each flip-flop (Tables 5.19 and 5.20). Alternatively one might draw individual maps for S and R for both flip-flops, a total of four maps.
The output Z is the same as before. The circuit is shown in Fig. 5.16.

5.5.3 ROM Implementation of Example (5.2)

Asynchronous circuits like other logic circuits can be implemented using memory devices such as a ROM.

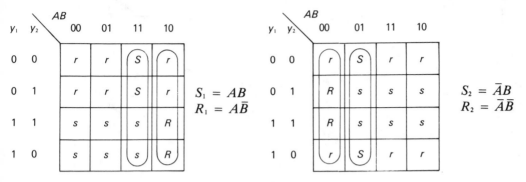

$S_1 = AB$
$R_1 = A\bar{B}$

$S_2 = \bar{A}B$
$R_2 = \bar{A}\bar{B}$

Table 5.19 *SR* map for flip-flop 1   Table 5.20 *SR* map for flip-flop 2

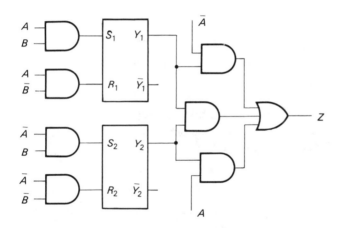

Fig. 5.16 **Flip-flop implementation of example (5.2)**

The starting point for a ROM implementation is the assigned flow table and output table which are rearranged to form the transition table given in Table 5.21. In a ROM the present state and input variables are used as address lines to find the corresponding next states and outputs which are stored in the ROM.

It is obvious that the need for minimization of combinational logic and the derivations of input equations for the flip-flops, etc., are no longer necessary. The state reduction and row merging can still be useful in reducing the number of words, which implies that smaller ROMs can be used. The number of inputs of the ROM must be at least equal to the number of bits required to code the input and the state variables of the circuits. Also the number of outputs of the ROM (word length) must be at least equal to the number of bits required for the output functions and state code.

Address				Output			
I_3	I_2	I_1	I_0	D_3	D_2	D_1	D_0
A	B	y_1	y_2	Y_1	Y_2	Z	
0	0	0	0	0	0	0	0
0	0	0	1	0	0	0	0
0	0	1	0	1	0	1	0
0	0	1	1	1	0	1	0
0	1	0	0	0	1	0	0
0	1	0	1	0	1	0	0
0	1	1	0	1	1	1	0
0	1	1	1	1	1	1	0
1	0	0	0	0	0	0	0
1	0	0	1	0	1	1	0
1	0	1	0	0	0	0	0
1	0	1	1	0	1	1	0
1	1	0	0	1	0	0	0
1	1	0	1	1	1	1	0
1	1	1	0	1	0	0	0
1	1	1	1	1	1	1	0

Table 5.21   Transition table for Example (5.2)

For Example (5.2) a ROM with four inputs (16 words) and four outputs (four-bit word length) is used as shown in the block diagram of Fig. 5.17. One of the outputs is not used.

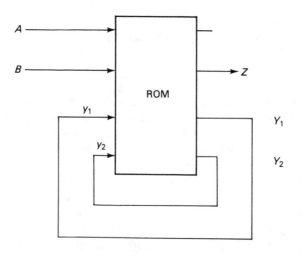

Fig. 5.17   ROM implementation of Example (5.2)

5.6   PRIMARY PLANE METHOD

The last few sections clearly show that the aim of obtaining an optimum circuit, by a systematic method, is still beyond our reach. Different mergers and different state assignments result in different circuits and there is no way of knowing which will result in the optimum circuit except by exhaustive search.

Vingron ⟨113⟩ recently described a new design procedure which employs a sequential input output diagram, a coherency tree and reduced Karnaugh maps, but no merging or encoding. This method is still under development but early indications suggest that it might be more suitable for computerized design than the traditional Huffman method. The problem of optimization, however, is still unsolved.

Another approach which is described in an internal report at Heriot-Watt University is called the *primary plane method*. This approach attempts to overcome one of the problems with the normal design method, namely that the number of columns in the primitive table doubles for every additional primary input. If the number of primary inputs is large the table becomes difficult to handle. This method assumes that the circuit is combinational, and secondary states are introduced as necessary to separate the plane into secondary planes.

Consider again example (5.1). A primary plane is drawn as shown in Fig. 5.18, with the total state diagram showing all transitions.

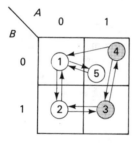

Fig. 5.18   Primary plane map

The map shows that the circuit can be in states 4 or 5 when the primary inputs *A* and *B* are 1 and 0 respectively. A secondary variable *y* is introduced to distinguish between the states. This results in two maps as in Fig. 5.19.

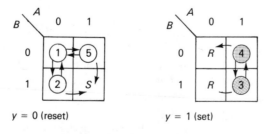

$y = 0$ (reset) $y = 1$ (set)

Fig. 5.19   Secondary plane maps

If an *SR* flip-flop is used as the memory device, a 4 to 1 transition requires the flip-flop to reset since 4 and 1 are in two different planes. Similarly for the 3 to 2 transition. *R* (for reset) is entered in these two positions. The 2 to 3 transition requires the flip-flop to be set. *S* (for set) is entered.

From Fig. 5.19, expressions for *S* and *R* can be derived:

$$S = AB$$
$$R = \bar{A}$$

The output is high in the shaded states 3 and 4. This gives an expression for the output as $Z = Ay$. If the output is assumed to be high during the transient states labeled *R*, the output expression simplifies to $Z = y$. The variable *y* is assigned to the present state of the *SR* flip-flop.

The answer agrees with the previous solution. This however depends on the way the states are split between the secondary maps. Normally more than one solution is possible, and the question of optimization is unresolved. Example (5.2) requires two secondary variables and results in four secondary plane maps.

The design algorithm is summarized as follows:

(*a*)   Draw a total state diagram on the primary plane map showing all transitions.

(*b*)   Primary states with more than one total state require at least as many secondary planes as the maximum number of total states contained in any primary state.

(*c*)   Separate the total states onto the secondary planes such that any two different total states occupying the same primary state are separated onto different secondary planes. A total state with an immediate predecessor and immediate successor with the same primary state is placed on a different secondary plane from the predecessor. Further a total state with more than one immediate predecessor on different secondary planes is duplicated on these planes, but only one of these is a stable state.

(*d*)   Secondary variables are assigned, and expressions derived for bistables and output.

The method is useful at least when there is a large input redundancy. The obvious limitation, like all Karnaugh maps, is that the map size becomes impractical for large numbers of variables. It seems that the Huffman normal design techniques, plus the designer's own intuition and experience, will continue to be in demand for the foreseeable future. Further aspects of the state assignment problem and design of asynchronous circuits are discussed in Chapter 10.

Exercises

Q (5.1)   Using NAND gates, design a hazard-free asynchronous sequential circuit for the system with the state diagram given in Fig. Q5.1.

Q (5.2)   Analyze the asynchronous sequential circuit given in Fig. Q5.2, given that its inputs *A* and *B* cannot change simultaneously. Produce a flow table, a total state diagram and an output map.

Q (5.3)   A factory produces blocks of wood of two lengths 2 m and 4 m. The blocks are placed on a conveyor belt passing under two photocells A and B spaced 3 m apart as shown in Fig. Q5.3. The blocks are separated by more than 4 m. A trap

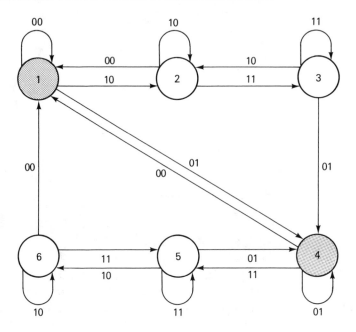

Fig. Q 5.1   State diagram

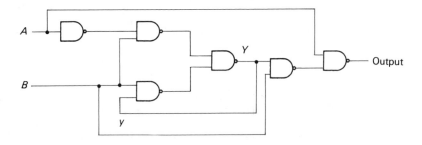

Fig. Q 5.2   Asynchronous sequential circuit

door is placed immediately to the right of photocell B. To separate the two types of block it is required that the trap door should open when the small blocks pass under B and close when the blocks have passed beyond B. The door opens when the output Z is at logic 1. Draw the waveform diagram and design an asynchronous sequential circuit using SR/FF and gates to achieve this.

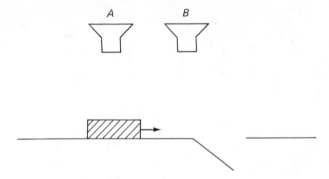

Fig. Q 5.3   A conveyor belt arrangement

Q (5.4)   A Katz network, as given in Fig. Q5.4 employs feedback and is thus an asynchronous sequential circuit. Assuming the binary inputs ABC have the weighting 124, derive expressions and state tables for Y_1, Y_2, Y_3 and Q_+ in terms of A,B,C and Q. Show that the circuit can be reduced to a combinational one with the three outputs depending only on the inputs A,B and C.

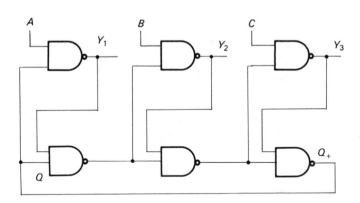

Fig. Q 5.4   Katz network

6

Arithmetic Logic Circuits

6.1   INTRODUCTION

Mankind has always had a need for counting, whether it is the animals people hunted, the cows they have owned or the pounds they save in the bank.

One way of counting, probably used in early civilization, is to assign a symbol for every item. Thus 111 would represent three, and 11111 would represent five, and so on. A more advanced method, introduced by the Romans, gave symbols for various numbers. These are the well-known Roman numerals I, II, III, IV, etc., which helped to simplify counting and visual recognition. This method was used throughout Europe until the Arabs introduced the more familiar Arabic numerals 0, 1, ..., 9. The Arabic numerals provide a much easier system since there are only ten symbols to remember. The introduction of the number zero and giving a numeral a value according to its position further simplified matters. Arabic numerals were quickly adopted, and were found very suitable for addition, subtraction and other arithmetical operations. With Roman numerals, such operations could be done only by experts. Of course, other number systems existed as well. In ancient Iraq, a system based on 60 was used over 4000 years ago by the Sumerians and Babylonians. This is still used in digital watches.

Nowadays, binary, octal and hexadecimal number systems, as shown in Table 6.1 are widely used in the fields of digital computers and microprocessors. Because binary digits (0s and 1s) relate easily to physical realization in one form or another, the binary number system is the most suitable for use in machines like digital computers. It is not surprising, therefore, that the arithmetic logic circuits developed in this chapter employ binary numbers. The binary output of these circuits can easily be converted to decimal for the benefit of the human operator.

6.2   BINARY CODES

6.2.1 Binary Coded Decimal Codes (BCD)

Binary digits can be used to represent not only numbers but also letters and symbols. Combinations of binary digits are called *binary codes*. Let us first consider the coding of the decimal numbers 0 to 9. To code numbers up to decimal 9, four binary digits are required. It is also known that four binary digits can be combined in

Decimal		Binary					Octal		Hexadecimal	
10^1	10^0	2^4	2^3	2^2	2^1	2^0	8^1	8^0	16^1	16^0
0	0	0	0	0	0	0	0	0	0	0
0	1	0	0	0	0	1	0	1	0	1
0	2	0	0	0	1	0	0	2	0	2
0	3	0	0	0	1	1	0	3	0	3
0	4	0	0	1	0	0	0	4	0	4
0	5	0	0	1	0	1	0	5	0	5
0	6	0	0	1	1	0	0	6	0	6
0	7	0	0	1	1	1	0	7	0	7
0	8	0	1	0	0	0	1	0	0	8
0	9	0	1	0	0	1	1	1	0	9
1	0	0	1	0	1	0	1	2	0	A
1	1	0	1	0	1	1	1	3	0	B
1	2	0	1	1	0	0	1	4	0	C
1	3	0	1	1	0	1	1	5	0	D
1	4	0	1	1	1	0	1	6	0	E
1	5	0	1	1	1	1	1	7	0	F
1	6	1	0	0	0	0	2	0	1	0
1	7	1	0	0	0	1	2	1	1	1
1	8	1	0	0	1	0	2	2	1	2
1	9	1	0	0	1	1	2	3	1	3
2	0	1	0	1	0	0	2	4	1	4
.			.					.		.
.			.					.		.

Table 6.1   Comparison of number systems

many different ways, which suggest that it is possible to devise different codes. In general, any binary code used to represent the decimal digits is called BCD.

The 8421 Code

This is the natural binary code where the decimal numbers are represented by their binary equivalent. The name 8421 indicates the binary weight of the four binary digits (bits).

$$2^3 \ 2^2 \ 2^1 \ 2^0 = 8421$$

The main advantage of this code is the ease of conversion to and from decimal.

The 8421 code uses four binary digits. Table 6.1 shows that four bits can represent sixteen numbers. In BCD only ten are used, as shown in Table 6.2. The rest are not valid. In BCD each decimal number (zero to nine) is represented by its four-bit code. For example decimal 15 can be represented by 1111 in binary. In binary coded decimal both decimal digits are replaced by their binary code. Decimal 1 is replaced by 0001 and decimal 5 is replaced by 0101 resulting in 0001 0101, which is the BCD code for decimal 15.

Complementary Codes

The 2421 is a common complementary code. This is similar to the 8421 code in

Decimal	Weighted codes				Unweighted	
	8241	2421	5211	7421	Excess 3	Reflected
0	0000	0000	0000	0000	0011	0000
1	0001	0001	0001	0001	0100	0001
2	0010	0010	0100	0010	0101	0011
3	0011	0011	0110	0011	0110	0010
4	0100	0100	0111	0100	0111	0110
5	0101	1011	1000	0101	1000	0111
6	0110	1100	1001	0110	1001	0101
7	0111	1101	1011	1000	1010	0100
8	1000	1110	1110	1001	1011	1100
9	1001	1111	1111	1010	1100	1000

Table 6.2   Binary codes

the sense that the positions of the digits carry a known weight except that the most significant digit carries a weight of 2 instead of 8.

To overcome ambiguity, the most significant digit is always 0 for the first five codes and 1 for the last five codes. The advantage of this code is that it is self-complementing, a property which is useful in arithmetic operations.

The complements will be discussed later, but to illustrate the meaning consider any number say 263:

$$263_{10} = (0010\ 1100\ 0011)_{2421}$$

The nine complement of a decimal number N is defined as $9 - N$, so that the nine complement of 263 is 736.

The 2421 code for 736 is:

$$736_{10} = (1101\ 0011\ 1100)_{2421}$$

The 2421 code for 736 is obtained by complementing (inverting) all the bits of the 2421 code for 263. This property is true for codes in which the sum of the weights is equal to 9 and which are symmetrical about the centre. This is true for the 2421 and 5211 code shown in Table 6.2.

7421 Code
Another useful code is the 7421 code, in which the weight of the most significant digit is 7. This has the useful property that it has a minimum number of 1s, which could be useful for economy in power consumption.

Weighted and Non-Weighted Codes
In all codes mentioned above the position of the digits carries a weight. Many codes are possible provided their weights have a sum that is not less than 9 and not more than 15. Also one weight must be 1 and a second weight must be 1 or 2.

Other codes are possible in which the digits do not carry a weight but have useful properties for certain applications.

The Excess-3 Code

This unweighted code is obtained by adding decimal 3 (0011) to the 8421 code. It is self-complementing as can be seen by comparing the codes for any digit with the codes for nine minus that digit.

EXAMPLES
Code for 5 = 1000
Code for 4 = 0111

Code for 3 = 0110
Code for 6 = 1001

The Reflected Code

This is also unweighted but has the useful property that only one digit changes in going from one number to another. This is useful for shaft position encoding as can be seen in Fig. 6.1. The light and dark positions represent 0 and 1 respectively.

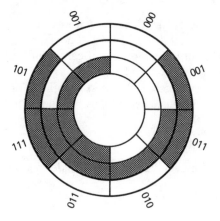

Fig. 6.1   Shaft encoding

6.2.2 Error-Detecting Codes

Binary codes, or binary data, are sensitive to error which may be due to noise or failure of equipment. Suppose the binary code (0110) for decimal six using the 8421 code is transmitted. If, because of noise, a digit is picked up, then at the receiving end 1110 or 0111 is detected. Obviously 1110 is non-valid and indicates an error; 0111 however will be accepted as decimal 7 and the error will not be detected.

One way of overcoming this problem is by using a code in which an error in one position results in a non-valid code. A typical example is the 2-out-of-5 code, which employs five digits to represent the numbers 0 to 9. In each code two digits are 1 and three are 0. A single error will result in either a single 1 or three 1s which can be detected.

Another way of detecting single errors is by using a parity bit. In this case an extra bit is added so as to make the total number of 1s either even or odd. In even

Decimal	8421 code
0	1 0000
1	0 0001
2	0 0010
3	1 0011
4	0 0100
5	1 0101
6	1 0!10
7	0 0111
8	0 1000
9	1 1001

Table 6.3   8421 code with odd parity

parity the total number of 1s including the parity bit is made even. In odd parity the total number of 1s including the parity is made odd. A typical application is computer punched tape. Employing odd parity, the 8421 code will become as in Table 6.3. Here the most significant digit is the parity bit.

Using the odd parity code above we can detect only single errors or an odd number of errors which result in the total number of 1s becoming even. We cannot detect an even number of errors, or the position of the error. Further, the error cannot be corrected.

6.2.3 Error-Detecting and -Correcting Codes

Even and odd parity are a form of error-detecting techniques. These result from the addition of one parity bit. Now we consider the possibility of adding more than one bit.

Consider the four coded characters A, B, C and D:

A   00000
B   11100
C   00111
D   11011

Note that each code differs from all other codes in at least 3 of the 5 bits, i.e. the *minimum distance* of the code is 3, the minimum distance between codes being the minimum number of bits in which any two characters differ. This minimum distance has been achieved at the price of adding three redundant bits.

Suppose D was transmitted as shown but received as 11000. In spite of the two errors, message D can still be distinguished from A, B and C. In brief, errors of 2 or less can be detected in a minimum-distance-3 code.

Errors in 3 or more bits cannot always be detected in a minimum distance 3 code. If the first three digits in B are in error, B will be mistaken for A. If the probability of an error in two-bits is assumed small, it is possible to employ a minimum-distance-3 code to correct single errors. Suppose 11000 is received; we can conclude that the correct message is 11100.

Thus a circuit can be provided to detect and correct single errors. Note that the code will differ in one bit from the transmitted character and in at least two bits from all other codes.

In general a minimum-distance-3 code can be used to detect and correct all single errors without retransmission. Alternatively it can be used to detect two-bit errors, and ask for the message to be retransmitted.

Hamming Code

A commonly used error-correcting code is the Hamming code. In this code the bit positions numbered as powers of 2 are reserved for parity check. The others are information bits.

In the seven-bit code shown, P_1, P_2 and P_4 indicate parity bits and X_3, X_5, X_6 and X_7 are the data to be transmitted. The two are combined as follows:

$$
\begin{array}{ccccccc}
1 & 2 & 3 & 4 & 5 & 6 & 7 \\
P_1 & P_2 & X_3 & P_4 & X_5 & X_6 & X_7
\end{array}
$$

P_1 is chosen to establish even parity over bits 1, 3, 5 and 7.
P_2 is chosen to establish even parity over bits 2, 3, 6 and 7.
P_4 is chosen to establish even parity over bits 4, 5, 6 and 7.

Assume the binary message to be transmitted is 1010. Let $X_3X_5X_6X_7$ be 1010. For even parity:

$$P_1 \oplus X_3 \oplus X_5 \oplus X_7 = 0$$
$$P_1 = 1 \oplus (0 \oplus 0) = 1 \oplus 0 = 1$$

Similarly:

$$P_2 = X_3 \oplus X_6 \oplus X_7 = 1 \oplus (1 \oplus 0) = 0$$
$$\text{and} \quad P_4 = X_5 \oplus X_6 \oplus X_7 = 0 \oplus (1 \oplus 0) = 1$$

If the parity bits are added to the message the code becomes 1011010.

Assuming a single error, the following test is carried out at the receiving end:

$$C_1 = P_1 \oplus X_3 \oplus X_5 \oplus X_7$$
$$C_2 = P_2 \oplus X_3 \oplus X_6 \oplus X_7$$
$$C_4 = P_4 \oplus X_5 \oplus X_6 \oplus X_7$$

If $C_1 = C_2 = C_4 = 0$, no error has occurred.

If $C_1 = 1$ but $C_2 = C_4 = 0$, we conclude that one of the bits 1, 3, 5 and 7 is in error.

But 4, 5, 6 and 7 and 2, 3, 6 and 7 are correct. Therefore bit 1 must be in error.

If $C_1 = C_2 = C_4 = 1$, bit 7 is in error.

Let us assume that the message 1010 is transmitted using the Hamming code as 1011010. At the receiving end, the message 1011110 is received.

The error-detecting circuit will compute C_1, C_2 and C_3 as follows:

$$C_1 = P_1 \oplus X_3 \oplus X_5 \oplus X_7$$
$$= 1 \oplus 1 \oplus 1 \oplus 0 = 1$$

$$C_2 = P_2 \oplus X_3 \oplus X_6 \oplus X_7$$
$$= 0 \oplus 1 \oplus 1 \oplus 0$$
$$= 0$$
$$C_4 = P_4 \oplus X_5 \oplus X_6 \oplus X_7$$
$$= 1 \oplus 1 \oplus 1 \oplus 0 = 1$$

Since only X_5 is involved in C_1 and C_4, position X_5 must be in error. This can be complemented to obtain the correct code.

If on the other hand 1011010 is received, then

$$C_1 = 1 \oplus 1 \oplus 0 \oplus 0 = 0$$
$$C_2 = 0 \oplus 1 \oplus 1 \oplus 0 = 0$$
$$C_4 = 1 \oplus 0 \oplus 1 \oplus 0 = 0$$

This indicates that the message is correct.

Example (6.1)

Design a minimal logic circuit to convert the binary numbers 0 to 9 to excess-3 code using NAND gates.

SOLUTION:

Code	Binary				Excess-3			
	A	B	C	D	W	X	Y	Z
0	0	0	0	0	0	0	1	1
1	0	0	0	1	0	1	0	0
2	0	0	1	0	0	1	0	1
3	0	0	1	1	0	1	1	0
4	0	1	0	0	0	1	1	1
5	0	1	0	1	1	0	0	0
6	0	1	1	0	1	0	0	1
7	0	1	1	1	1	0	1	0
8	1	0	0	0	1	0	1	1
9	1	0	0	1	1	1	0	0

Code conversion table

From the code conversion table four K-maps are drawn for each of the four outputs *W, X, Y* and *Z*.

From the maps the following expressions are derived:

W	$= A + BD + BC$	Using AND/OR gates
W	$= \overline{\bar{A} \cdot \overline{BD} \cdot \overline{BC}}$	Using NAND gates
X	$= B\bar{C}\bar{D} + \bar{B}D + \bar{B}C$	
X	$= \overline{\overline{B\bar{C}\bar{D}} \cdot \overline{\bar{B}D} \cdot \overline{\bar{B}C}}$	
Y	$= \bar{C}\bar{D} + CD$	
	$= \overline{\overline{\bar{C}\bar{D}} \cdot \overline{CD}}$	
Z	$= \bar{D}$	

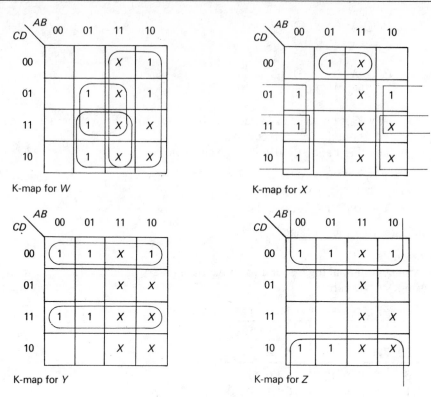

K-map for W

K-map for X

K-map for Y

K-map for Z

The implementation using NAND gates appears in Fig. 6.2.

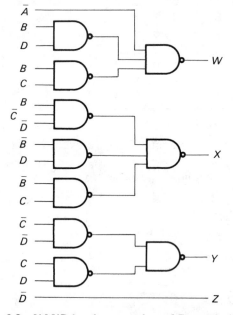

Fig. 6.2 NAND implementation of Example (6.1)

Example (6.2)

Four data bits are to be transitted. Design a parity bit generator to give an output of 1 if the number of logic 1s in the message is: (i) odd; (ii) even.

SOLUTION:

(i)   If an output is required for an odd number of 1s the problem can be represented by the following flow table:

Data bits	Output
A B C D	f
0 0 0 0	0
0 0 0 1	1
0 0 1 0	1
0 0 1 1	0
0 1 0 0	1
0 1 0 1	0
0 1 1 0	0
0 1 1 1	1
1 0 0 0	1
1 0 0 1	0
1 0 1 0	0
1 0 1 1	1
1 1 0 0	0
1 1 0 1	1
1 1 1 0	1
1 1 1 1	0

The function is entered on a K-map:

CD \ AB	00	01	11	10
00		1		1
01	1		1	
11		1		1
10	1		1	

The map shows that f can't be minimized; in fact this sort of map represents an EX-OR operation:

$$f = A \oplus B \oplus C \oplus D$$

The EX-OR expression can also be derived from the truth table as follows:

$$f = \bar{A}\bar{B}\bar{C}D + \bar{A}\bar{B}C\bar{D} + \bar{A}B\bar{C}\bar{D} + \bar{A}BCD +$$
$$A\bar{B}\bar{C}\bar{D} + A\bar{B}CD + AB\bar{C}D + ABC\bar{D}$$
$$= \bar{A}\bar{B}(\bar{C}D + C\bar{D}) + AB(\bar{C}D + C\bar{D}) +$$
$$\bar{C}\bar{D}(A\bar{B} + \bar{A}B) + CD(A\bar{B} + \bar{A}B)$$
$$= (C \oplus D)(\bar{A}\bar{B} + AB) + (A \oplus B)(\bar{C}\bar{D} + CD)$$
$$= (C \oplus D)(\overline{A \oplus B}) + (A \oplus B)(\overline{C \oplus D})$$
$$= (C \oplus D) \oplus (A \oplus B)$$

It can be seen, from the EX-OR truth table, given below, for any two variables A and B, that the EX-OR relationship is $A\bar{B} + \bar{A}B$. The complement of the EX-OR relationship is $\bar{A}\bar{B} + AB$.

A	B	EX-OR
0	0	0
0	1	1
1	0	1
1	1	0

EX-OR truth table

This can be implemented as in Fig. 6.3(a).

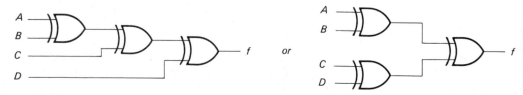

Fig. 6.3(a)   Parity generator for (i)

(ii)   If the output is 1 for an even number of 1s in the message, another gate is required as in Fig. 6.3(b).

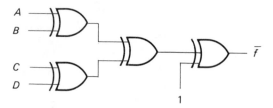

Fig. 6.3(b)   Parity generator for (ii)

6.2.4 Alphanumeric Codes

Not only numbers are given binary codes but also alphabets and other symbols such as $+$, $-$, (, $=$, clear, etc. Their codes are called *alphanumeric*.

Naturally, if we represent 10 decimal digits, 26 letters (for the English language) and other symbols and control statements, at least 6 binary digits are required.

The American Standard Code for Information Interchange (ASCII) given in Table 6.4 is widely used, although other codes are available. The ASCII code employs seven bits, giving 128 possible characters. A parity bit is usually added to detect single errors, but this is not shown in Table 6.4.

Character	Code	Character	Code
A	100 0001	a	110 0001
B	100 0010	b	110 0010
C	100 0011	c	110 0011
D	100 0100	d	110 0100
E	100 0101	e	110 0101
F	100 0110	f	110 0110
G	100 0111	g	110 0111
H	100 1000	h	110 1000
I	100 1001	i	110 1001
J	100 1010	j	110 1010
K	100 1011	k	110 1011
L	100 1100	l	110 1100
M	100 1101	m	110 1101
N	100 1110	n	110 1110
O	100 1111	o	110 1111
P	101 0000	p	111 0000
Q	101 0001	q	111 0001
R	101 0010	r	111 0010
S	101 0011	s	111 0011
T	101 0100	t	111 0100
U	101 0101	u	111 0101
V	101 0110	v	111 0110
W	101 0111	w	111 0111
X	101 1000	x	111 1000
Y	101 1001	y	111 1001
Z	101 1010	z	111 1010
0	011 0000	{	111 1011
1	011 0001	}	111 1101
2	011 0010	Blank	010 0000
3	011 0011	!	010 0001
4	011 0100	"	010 0010
5	011 0101	#	010 0011
6	011 0110	%	010 0101
7	011 0111	&	010 0110
8	011 1000	/	010 1111
9	011 1001	(	010 1000
;	011 1011	)	010 1001
⟨	011 1100	*	010 1010
=	011 1101	+	010 1011
⟩	011 1110	−	010 1101
?	011 1111	.	010 1110

Table 6.4   Partial listing of the ASCII code

6.3   ARITHMETIC OPERATIONS

Binary arithmetic operations are the simplest to implement physically and most widely associated with digital computers. To start with, however, it is useful to appreciate that any number system can be used.

Consider adding two numbers 19_{10} and 26_{10} using the four-number systems previously mentioned:

(i)   Decimal

1	← Carries	A carry of 1 to the next higher position is generated when the sum in any column exceeds 9.
1 9		
+2 6		
4 5		

(ii)   Binary

1 0 0 1 0	← Carries	A carry of 1 to the next higher position is generated when the sum in any column exceeds 1.
1 0 0 1 1		
+ 1 1 0 1 0		
1 0 1 1 0 1		

(iii)   Octal

0	← Carries	A carry of 1 to the next higher position is generated when the sum in any column exceeds 7.
2 3		
3 2		
5 5		

(iv)   Hexadecimal

0	← Carries	A carry of 1 to the next higher position is generated when the sum in any column exceeds F (15).
1 3		
1 A		
2 D		

The answer in all cases is equivalent to 45_{10}.

6.3.1 Binary Addition

Though people are used to decimal addition and might find binary addition rather unusual, it should be apparent that binary addition is the easiest since one needs only to remember four possibilities:

$$0 + 0 = 0$$
$$0 + 1 = 1$$
$$1 + 0 = 1$$
$$1 + 1 = 0 \text{ plus a carry of 1 (10)}$$

If many digits are added, the answer is 1 or 0 depending on whether the number of 1s added is odd or even respectively.

Further, for every pair of 1s added, a 1 carry is generated.

Example (6.3)

Add the following binary numbers

$$011, 101, 011, 011$$

SOLUTION:

```
      1 1←
    1 1 1←  Carries
      0 1 1
      1 0 1
      0 1 1
  +   0 1 1
  ─────────
      1 1 1 0
```

Answer $1110_2 = 14_{10}$

6.3.2 Binary Multiplication

Multiplication in binary is easy. The main work, in fact, is the addition of the partial products. The partial product is equal to the multiplicand if the multiplier is 1, and equal to 0 if the multiplier is 0; it follows that

$$1 \times 1 = 1$$
$$0 \times 1 = 0$$
$$0 \times 0 = 0$$
$$1 \times 0 = 0$$

Example (6.4)

Using binary multiplication find the value of the product $25_{10} \times 5_{10}$

SOLUTION:

```
      1 1 0 0 1        Multiplicand
  ×       1 0 1        Multiplier
  ─────────────
      1 1 0 0 1  ⎫
      0 0 0 0 0  ⎬    Partial Products
  1 1 0 0 1      ⎭
  ───────────────
  1 1 1 1 1 0 1        Product
```

Answer $1111101_2 = 125_{10}$

6.3.3 Binary Division

The divisor is placed under the dividend so that the most significant digits coincide. The divisor is then compared to the portion of the dividend directly above the divisor.

If the portion of the dividend is equal to or greater than the divisor a 1 is written in the quotient and the divisor is subtracted from the dividend. Otherwise a 0 is written in the quotient and there is no subtraction. After the first step the divisor is shifted one position to the right and the procedure is repeated. When the unity digits line up, a binary point is placed to indicate the start of the fractional part.

Example (6.5)

Using binary methods perform the following division: $612_{10} \div 24_{10}$

Dividend   1 0 0 1 1 0 0 1 0 0
Divisor   1 1 0 0 0

SOLUTION:

		Quotient	
1 1 0 0 0 | 1 0 0 1 1 0 0 1 0 0			
1 1 0 0 0	No subtraction	0	Most significant digit
1 0 0 1 1 0 0 1 0 0			
1 1 0 0 0	Subtract	1	
0 0 1 1 1 0 0 1 0 0			
1 1 0 0 0	Subtract	1	
0 0 0 0 1 0 0 1 0 0			
1 1 0 0 0	No subtraction	0	
1 0 0 1 0 0			
1 1 0 0 0	No subtraction	0	
1 0 0 1 0 0			
1 1 0 0 0	Subtract	1	
0 0 1 1 0 0		.	
1 1 0 0 0 Subtract	1	
0 0 0 0 0 0 0			

Answer $1 1 0 0 1 \cdot 1_2 = 25 \cdot 5_{10}$

6.3.4 Binary Subtraction

As in binary addition, there are four possibilities to be remembered; these are:

$0 - 0 = 0$
$1 - 1 = 0$
$1 - 0 = 1$
$0 - 1 = 1$ and a borrow of 1

In the last case a 1 is borrowed from the next higher position whose value is twice the 1 in the present position.

Example (6.6)
Subtract the binary number 10110 from the binary number 11010.

SOLUTION:

$$
\begin{array}{rr}
1\,1\,0\,1\,0 & 2\,6 \\
-\,1\,0\,1\,1\,0 & -\,2\,2 \\
\hline
0\,0\,1\,0\,0 & 0\,4
\end{array}
$$

In the third column from the right a $0-1$ operation is encountered necessitating a borrow from the next higher position in column four, thus leaving a 0 in that position.

6.4   COMPLEMENTARY ARITHMETICS

6.4.1 Introduction

Towards the end of this chapter, it will be shown how to design binary adders and subtractors. It is known, however, that subtraction can be achieved by adding the minuend to the negative value of the subtrahend. We say that the minuend is added to the complement of the subtrahend. If this is possible, the same circuit can be used to achieve addition and subtraction.

Complements
For binary operations two types of complement are of interest to us; these are the 2's complement and the 1's complement.

For a positive integer number N having n binary digits (bits), the 2's complement is defined as

$$(N)_2 = 2^n - N$$

Similarly the 1's complement of N is defined as

$$(N)_1 = 2^n - N - 1$$

Example (6.7)
Find the 2's complements of the binary number 1010.

SOLUTION:

$$
\begin{aligned}
N &= 1010 \\
n &= 4 \\
(N)_2 &= 2^4 - 1010 \\
&= 10000 - 1010 \\
&= 0110
\end{aligned}
$$

Answer $(1010)_2 = 0110$

Example (6.8)
Find the 1's complement of the binary number 1010.

SOLUTION:
$$N = 1010 \quad , \quad n = 4$$
$$(N)_1 = 2^4 - 1010 - 1 = 0101$$

Answer $(1010)_1 = 0101$

It can be seen that the 1's complement is obtained by complementing all the bits in the number, which is an easy task for a digital circuit. In other words the 1's complement is obtained by changing all the 1s to 0s and all the 0s to 1s. Further, the 2's complement is obtained by adding a 1 to the 1's complement.

Example (6.9)
Find the 1's complement of the binary number 110101.
Answer $(110101)_1 = 001010$

Example (6.10)
Find the 2's complement of the binary number 1010.

SOLUTION:
$$N = 1010$$
$$(N)_1 = 0101$$
$$(N)_2 = (N)_1 + 1$$
$$= 0101 + 1$$
$$= 0110$$

Answer $(1010)_2 = 0110$

The rule for obtaining the 2's complement of a binary number can be written in a form more suitable for computation:

(*a*)   Inspect the bits one at a time starting from the right (least significant digit) keep the first 1 you meet and all 0s to its right.

(*b*)   Complement all other bits.

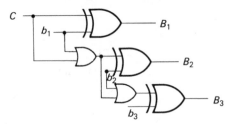

Fig. 6.4   Three-digit complement circuit

The circuit of Fig. 6.4 can be used for deriving either the 1's complement when the control signal $C = 1$, or the 2's complement when $C = 0$.

For a three-digit binary number, the three bits b_3 b_2 and b_1 are applied to the input of the circuit. The three-digit output is the complement of the input. In this circuit $B_1(b_1)$ is the least significant digit.
From Fig. 6.4

$$B_1 = b_1 \oplus C$$
$$B_2 = b_2 \oplus (C + b_1)$$
$$B_3 = b_3 \oplus (C + b_1 + b_2)$$

1's COMPLEMENTS: When $C = 1$, the outputs of the OR gates in Fig. 6.4 are always 1. All three EX-OR gates receive one high input and will act as inverters. The digits b_1, b_2 and b_3 will be inverted. Hence the circuit acts as a 1's complement circuit.

2's COMPLEMENTS:
$$C = 0$$
Then $B_1 = b_1$
$$\begin{aligned}
B_2 &= b_2 \oplus b_1 \\
&= b_2 \text{ if } b_1 \text{ is } 0 \\
&= \bar{b}_2 \text{ if } b_1 \text{ is } 1 \\
B_3 &= b_3 \oplus (b_1 + b_2) \\
&= b_3 \text{ if } b_1 \text{ AND } b_2 \text{ are both } 0 \\
&= \bar{b}_3 \text{ if } b_1 \text{ AND(OR) } b_2 \text{ are(is) } 1
\end{aligned}$$

All digits after the first binary 1 are inverted. Hence the circuit acts as a 2's complement circuit.

Example (6.11)
Find the 2's complement of the binary number 1010

SOLUTION:
Use the rule given after Example (6.10):

$$N = 1010$$
$$(N)_2 = 0110$$

6.4.2 Binary Arithmetic Using the 1's Complements

Consider two positive numbers N_1 and N_2 each having n-binary digits

(i) Assume $N_1 < N_2$

$$(N_1)_1 = 2^n - N_1 - 1$$
$$(N_2)_1 = 2^n - N_2 - 1$$

$$\begin{aligned}
N_1 - N_2 &= N_1 - N_2 + 2^n - 1 - 2^n + 1 \\
&= N_1 + (2^n - N_2 - 1) - 2^n + 1 \\
&= N_1 + (N_2)_1 - 2^n + 1 \\
&= -(2^n - (N_1 + (N_2)_1) - 1) \\
&= -(N_1 + (N_2)_1)_1
\end{aligned}$$

Therefore to subtract N_2 from N_1, add N_1 to the 1's complement of N_2, take the 1's complement of this addition to give the answer, which should be negative.

We note at this stage that the digital circuit is required only to add and complement, thus eliminating the need for subtraction.

Example (6.12)

Calculate $1001 - 11010$ using the 1's complements.

SOLUTION: Add zero to the left so as to make the number of digits equal. Then

$$N_1 = 01001$$
$$N_2 = 11010$$
$$(N_2)_1 = 00101$$
$$N_1 + (N_2)_1 = 01001 + 00101$$
$$= 01110$$
$$N_1 - N_2 = -(N_1 + (N_2)_1)_1$$
$$= -(10001) \; Answer$$

In decimal notation the problem is $9 - 26$, which equals -17. To be able to see how the sign is recognized we rewrite the solution in the following manner:

$$
\begin{array}{llll}
N_1 & = & 0\ 1\ 0\ 0\ 1 \\
(N_2) & = & 1\ 1\ 0\ 1\ 0 \\
(N_2)_1 & = & \lceil 0\ 0\ 1\ 0\ 1 \\
N_1 & = & \lfloor 0\ 1\ 0\ 0\ 1 \\
\hline
& & 0\lvert 0\ 1\ 1\ 1\ 0 \\
& & \uparrow
\end{array}
$$

No overflow. This indicates a negative answer, which is in the 1's complement. The numerical value is obtained by taking the 1's complement of the answer. Thus

$$(01110)_1 = 10001$$

(ii) Assume $N_1 \geqslant N_2$

If N_1 is greater than or equal to N_2, the answer to calculating $N_1 - N_2$ is positive or zero. The rule for the calculation may be summarized as follows:

Add N_1 to the 1's complements of N_2, then add 1 and ignore the last carry.

Example (6.13)

Calculate $110101 - 100110$ using the 1's complements.

SOLUTION:
$$N_1 \quad = 110101$$
$$N_2 \quad = 100110$$
$$(N_2)_1 \quad = 011001$$

By adding N_1 and $(N_2)_1$ and then adding 1, we have:

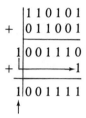

Discarded

Discarding the last carry, we obtain the answer:

$N_1 - N_2 = 001111$

In the decimal system the problem is $53 - 38 = 15$. The overflow indicates a positive answer. No complementing is required. The 1 added may be taken as the last carry brought forward (end-around carry).

Example (6.14)
Compute the following binary subtraction: $10110 - 10110$

SOLUTION:
$$N_1 \quad = 10110$$
$$N_2 \quad = 10110$$
$$(N_2)_1 = 01001$$

Using the above rule, we have:

```
 |10110
+|01001
─────────
 |11111      The overflow is zero at this stage. The answer may be
+|    1      interpreted as negative in complements form resulting in
─────────    minus zero, or a 1 is added and the overflow discarded.
1|00000
↑
```

Discarded

Answer $N_1 - N_2 = 00000$

6.4.3 Complementary Arithmetic Using 2's Complements

The main drawback in the 1's complement method of subtraction is that a 1 must be added. This can be overcome using the 2's complement, which is one greater than the 1's complement by definition.

Again consider two positive number N_1 and N_2 each having n bits. If the number of digits is different they must be made equal. To illustrate the procedure the same problem as in Example (6.12) will be considered.

(i) Assume $N_1 < N_2$

We have $N_1 - N_2 = -(\dot{N}_1 + (N_2)_2)_2$

Example (6.15)

Find $N_1 - N_2$ using 2's complements where

$N_1 = 1001$
$N_2 = 11010$

SOLUTION:

$(N_2)_2 = 00110$

$$
\begin{array}{r}
|0\ 1\ 0\ 0\ 1 \\
+|0\ 0\ 1\ 1\ 0 \\
\hline
|0\ 1\ 1\ 1\ 1 \\
\uparrow
\end{array}
$$

No overflow. The answer is negative and is shown in 2's complements.

Thus $N_1 - N_2 = -(01111)_2$
$\qquad\qquad = -10001 \qquad Answer$

The procedure can be summarized as follows:

Add N_1 to the 2's complement of N_2, take the 2's complement of the answer and remember it should be negative.

(ii) $N_1 \geqslant N_2$

The difference between N_1 and N_2 is the sum of N_1 and the 2's complement of N_2 with the carry to the 2^n position discarded (no need to add 1 as in the 1's complement method).

Example (6.16)

$N_1 = 110101$
$N_2 = 100110$

SOLUTION:

$(N_2)_2 = 011010$

$$
\begin{array}{r}
|1\ 1\ 0\ 1\ 0\ 1 \\
|0\ 1\ 1\ 0\ 1\ 0 \\
\hline
1|0\ 0\ 1\ 1\ 1\ 1 \\
\uparrow
\end{array}
$$

Discarded

Answer $N_1 - N_2 = 001111$

In the decimal system the problem is $57 - 38 = 15$
The overflow indicates a positive answer. No complementing is required.

Example (6.17)

$N_1 = 10110$

$N_2 = 10110$

Find $N_1 - N_2$ using the 2's complements.

$(N_2)_2 = 01010$

$$
\begin{array}{r}
1\,0\,1\,1\,0 \\
0\,1\,0\,1\,0 \\
\hline
1\,|0\,0\,0\,0\,0 \\
\end{array}
$$
+

↑

Discarded

Compared to the 1's complement, the answer is obtained without adding a 1, provided that the last carry is discarded.

6.4.4 2's Complement Arithmetic Including Sign Digit

So far N_1 and N_2 were assumed positive numbers. Further when $N_1 < N_2$ a minus sign was included in the answer. In digital computers this is overcome by using what is called 2's complement representation which takes care of the sign.

In this system, a positive number is represented by its binary equivalent with a 0 for the sign digit. A negative number is represented by its 2's complement with a 1 for the sign digit. For example:

$+10_{10} = 0\ 1010$
$-10_{10} = 1\ 0110$

$+15_{10} = 0\ 1111$
$-15_{10} = 1\ 0001$

$+23_{10} = 0\ 10111$
$-23_{10} = 1\ 01001$

It can be seen that a negative number is the 2's complement of the positive number including the sign (the sign digit is automatically taken care of).

Now the subtraction process becomes addition. Hence

$$N_1 - N_2 = N_1 + (-N_2)$$

where $-N_2$ is represented by its 2's complement.

From now on addition and subtraction are the same.

Example (6.18)

Calculate the value of $1110 - 1001$.

SOLUTION: This is equivalent to $14 - 9$ in the decimal system.

$$N_1 = + 14 = 0\ 1110$$
$$N_2 = + 9 = 0\ 1001$$
$$- 9 = 1\ 0111$$

$$N_1 - N_2 = N_1 + (-N_2)$$

Hence

$$
\begin{array}{cccccc}
C_2 & C_1 & & & & \\
\downarrow & \downarrow & & & & \\
1 & 1\ 1\ 1 & & & \leftarrow & \text{Carries} \\
& 0|1\ 1\ 1\ 0 & & & & N_1 \\
+ & 1|0\ 1\ 1\ 1 & & & & -N_2 \\
\hline
C_2' \rightarrow 1 & 0|0\ 1\ 0\ 1 & & & & \\
& \uparrow & & & & \\
& \text{Sign} & & & &
\end{array}
$$

The sign digit is 0, indicating a positive answer whose value is $0101_2 = 5_{10}$.
Then $N_1 - N_2 = 0\ 0101$
$$= + 5 \qquad \textit{Answer}$$

Note N_1, N_2 and the answer can all be represented by four digits (five digits if the sign digit is included). The last digit C_2' is discarded.
     If N_1 and N_2 are both positive or both negative the answer might require five digits to represent overflow. In such a case the fifth digit will be part of the answer and C_2' will represent the sign.

Example (6.19)
$N_1 = 14_{10}$
$N_2 = 9_{10}$
Find $N_1 + N_2$

SOLUTION:
$$N_1 = + 14 = 0\ 1110$$
$$N_2 = + 9 = 0\ 1001$$

$$
\begin{array}{ccccccc}
C_2 & C_1 & & & & & \text{Carries} \\
\downarrow & \downarrow & & & & & \\
0 & 1 & & & & & \\
& & 0|1 & 1 & 1 & 0 & \\
+ & & 0|1 & 0 & 0 & 1 & \\
\hline
0 & & 1\ 0 & 1 & 1 & 1 & \\
& \uparrow & & & & & \\
& \text{Sign} & & & & &
\end{array}
$$

The answer (numerically) is $10111 = 23$. $C_2' = 0$ represents the sign digit.

Therefore the answer is $0\ 10111 = +23_{10}$

Example (6.20)

$N_1 = -14_{10}$

$N_2 = + 9_{10}$

Find $N_1 - N_2$.

SOLUTION:

$N_1 - N_2 = N_1 + (-N_2)$

$N_1 = -14_{10} = 1\ 0010$

$-N_2 = -9_{10} = 1\ 0111$

$$
\begin{array}{ccccc}
C_2 & C_1 & & & \\
\downarrow & \downarrow & & & \\
\end{array}
$$

```
      C₂  C₁
      ↓   ↓
      1  0 1 1              Carries
        1│0 0 1 0
    +   1│0 1 1 1
    ─────────────
      1  0│1 0 0 1
Sign     ↑
```

Again C_2' represents the sign.

The answer is 1 01001

         ↑Sign

Remember that negative numbers are represented by their 2's complement, that

is

$+ 23 = 0\ 10111$

$- 23 = 1\ 01001$

It can be seen that the answer is -23.

6.4.5 General Rule

Looking back at the last three examples, we can state the following:

If $C_1 = C_2$ (Both 0 or Both 1), C_2' is discarded.

If $C_1 \neq C_2$, C_2' is the sign digit.

Example (6.21)

$N_1 = +29_{10}$, $N_2 = +6_{10}$

Find $N_1 - N_2$

SOLUTION:

$+29 = 0\quad 11101$

$+ 6 = 0\quad 00110$

$- 6 = 1\quad 11010$

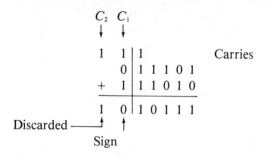

$$C_1 = C_2 = 1$$

Therefore C_2' is discarded.

The sign digit is 0 indicating a positive answer whose value is 10111.

Answer   0   10111 $= +23$

Note 1. In all cases addition is carried out, thus eliminating the need for subtraction.

Note 2. Overflow occurs only if the two operands have the same sign ($-A-B$ or $+A+B$); otherwise the sign of the answer is at the same position as the sign of the two operands.

6.5   BINARY ADDITION CIRCUITS

Having discussed the various binary arithmetic operations we must look at some basic arithmetic circuits. An important part of the central processor of any computer is the arithmetic unit in which binary addition, subtraction, division and multiplication are carried out.

Subtraction, however, can be performed by adding complemented numbers. Multiplication can also be performed by repeated addition. Division can be achieved by repeated subtraction. This makes the adder the centre piece of the arithmetic unit.

6.5.1 Half Adder (HA)

This is a circuit capable of adding two bits. It has two inputs and two outputs (Fig. 6.5(*a*).)

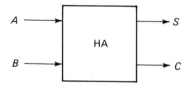

Fig. 6.5(a)   Block diagram for a half adder

A and B are two one-bit binary numbers.
S is the sum.
C is the carry.
The operation of a half adder can be expressed in the form of a truth table, as shown in Table 6.5.

Input		Output	
A	B	S	C
0	0	0	0
0	1	1	0
1	0	1	0
1	1	0	1

Table 6.5   Truth table for a binary half adder

From the truth table expressions for the sum and carry can be derived:

$$S = A\bar{B} + \bar{A}B$$
$$= A \oplus B$$
$$C = AB$$

The sum output is 1 when either A or B is 1 but not both. The carry output is 1 when both A and B are 1.

The half adder can be implemented in many different ways depending on available devices. For example, it can be implemented using AND/OR gates, assuming that the inputs are available in their true and complemented form, as in Fig. 6.5(b). Note also that the sum can be realized using one EXclusive-OR gate.

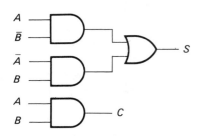

Fig. 6.5(b)   AND/OR circuit for a half adder

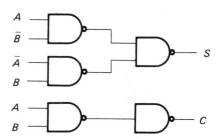

Fig. 6.5(c)   NAND circuit for a half adder

Nowadays NAND/NOR gates are becoming standard components for logic circuits and are available in integrated circuit form. Fig. 6.5(c) shows a NAND realization and Fig. 6.5(d) shows a NOR realization of the same half adder.
In Fig. 6.5(c) the output S is given by

$$S = \overline{\overline{A\bar{B}} \cdot \overline{\bar{A}B}} = A\bar{B} + \bar{A}B$$

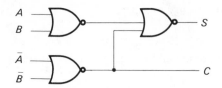

Fig. 6.5(d)   NOR circuit for a half adder

In Fig. 6.5(*d*) the outputs *S* and *C* can be expressed as

$$S = \overline{\overline{A + B} + \overline{\overline{A} + \overline{B}}}$$
$$= (A + B) \cdot (\overline{A} + \overline{B})$$
$$= A\overline{B} + \overline{A}B$$
$$C = \overline{\overline{A} + \overline{B}} = AB$$

In the above circuits it was assumed that $\overline{A}$ and $\overline{B}$ are available or are generated using additional NAND, NOR, or NOT gates.

The circuit in Fig. 6.5(*e*) requires only true inputs:

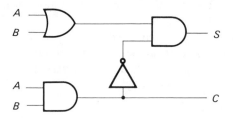

Fig. 6.5(e)   A half-adder circuit using true inputs

In Fig. 6.5(*e*), the expression for *S* may be best seen by writing the expression for $\overline{S}$ from the truth table.

$$\overline{S} = \overline{A}\,\overline{B} + AB$$
$$S = \overline{\overline{A}\,\overline{B} + AB}$$
$$= (A + B)\,(\overline{AB})$$

6.5.2 Full Adder

A half adder is not very useful on its own, and a third input is often required for carries. An adder with three inputs and two outputs as shown in Fig. 6.6 is called a *full adder*.

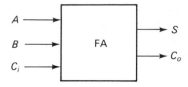

Fig. 6.6   A block diagram for a full adder

C_i is the carry-in from the previous addition.
C_o is the carry-out to the next addition.

The truth table for a full adder (FA) is given in Table 6.6

Inputs			Outputs	
A	B	C_i	S	C_o
0	0	0	0	0
0	0	1	1	0
0	1	0	1	0
0	1	1	0	1
1	0	0	1	0
1	0	1	0	1
1	1	0	0	1
1	1	1	1	1

Table 6.6   The truth table for a full adder

From the truth table logical expressions for the sum and carry out can be derived:

$$S = \bar{A}\bar{B}C_i + \bar{A}B\bar{C}_i + A\bar{B}\bar{C}_i + ABC_i$$
$$C_o = \bar{A}BC_i + A\bar{B}C_i + AB\bar{C}_i + ABC_i$$

The expressions can be entered on a K-map (Figs 6.7 and 6.8) for simplification.

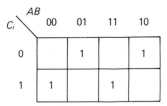

Fig. 6.7   K-map for the sum S

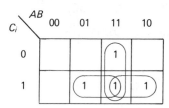

Fig. 6.8   K-map for the carry-out C_o

It can be seen that S cannot be simplified, but C_o can be simplified as follows:

$$C_o = AB + AC_i + BC_i$$

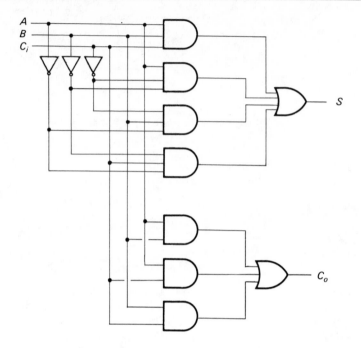

Fig. 6.9   Logic circuit for a binary full adder

The next step is to implement the expressions for S and C_o. This is done in Fig. 6.9.

It should be clear by now that other types of logic gate can be used and since the full adder is a basic circuit this possibility will be briefly discussed.

The expression for S may be written as follows:

$$S = \overline{\overline{\overline{A}\overline{B}C_i} + \overline{\overline{A}B\overline{C_i}} + \overline{ABC_i} + \overline{A\overline{B}\overline{C_i}}}$$

$$= \overline{\overline{\overline{A}\overline{B}C_i} \cdot \overline{\overline{A}B\overline{C_i}} \cdot \overline{ABC_i} \cdot \overline{A\overline{B}\overline{C_i}}}$$

Similarly C_o may be written as:

$$C_o = \overline{\overline{AB} + \overline{AC_i} + \overline{BC_i}}$$

$$= \overline{\overline{AB} \cdot \overline{AC_i} \cdot \overline{BC_i}}$$

The two equations can be implemented using NAND gates.

From the Os of the K-maps for S and C_o we may write

$$\overline{S} = \overline{A}\overline{B}\overline{C_i} + AB\overline{C_i} + \overline{A}BC_i + A\overline{B}C_i$$

Inverting the expression for $\bar{S}$, we get

$$S = (A + B + C_i)(\bar{A} + \bar{B} + C_i)(A + \bar{B} + \bar{C_i})(\bar{A} + B + \bar{C_i})$$
$$\bar{C_o} = \bar{A}\bar{B} + \bar{A}\bar{C_i} + \bar{B}\bar{C_i}$$

Inverting again, we get

$$C_o = (A + B)(A + C_i)(B + C_i)$$

These expressions for S and C_o are suitable for OR/AND realization as compared to the AND/OR realization shown in Fig. 6.9.

Again by inverting twice these can be converted to a form suitable for NOR gate realization

$$S = \overline{\overline{(A + B + C_i)(\bar{A} + \bar{B} + C_i)(A + \bar{B} + \bar{C_i})(\bar{A} + B + \bar{C_i})}}$$

$$= \overline{\overline{(A + B + C_i)} + \overline{(\bar{A} + \bar{B} + C_i)} + \overline{(A + \bar{B} + \bar{C_i})} + \overline{(\bar{A} + B + \bar{C_i})}}$$

Similarly

$$C_o = \overline{\overline{(A + B)} + \overline{(A + C_i)} + \overline{(B + C_i)}}$$

An important realization of a full adder employs two half adders as shown in Fig. 6.10.

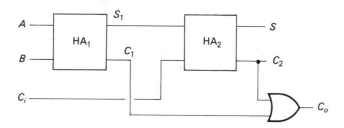

Fig. 6.10   Full adder using two half adders and one OR gate

$$S_1 = A\bar{B} + \bar{A}B \qquad \text{The EX-OR of the two inputs}$$
$$C_1 = AB \qquad\qquad\; \text{The AND of the two inputs}$$

Similarly

$$S = S_1\bar{C_i} + \bar{S_1}C_i$$

For a half adder, we have

$$\bar{S_1} = AB + \bar{A}\bar{B}$$

giving

$$S = (A\bar{B} + \bar{A}B)\bar{C}_i + (AB + \bar{A}\bar{B})C_i$$
$$= A\bar{B}\bar{C}_i + \bar{A}B\bar{C}_i + ABC_i + \bar{A}\bar{B}C_i$$

It can be seen that this is the expression for the sum of a full adder.

$$C_2 = C_i(A\bar{B} + \bar{A}B)$$
$$C_o = AB + C_2$$
$$= AB + A\bar{B}C_i + \bar{A}BC_i$$

If this is entered on a K-map, the pattern is the same as the carry-out of a full adder:

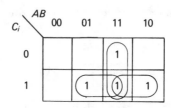

From the K-map

$$C_o = AB + AC_i + BC_i$$

the same expression for C_o as that for a full adder.

Therefore two half adders and one OR gate connected as in Fig. 6.10 can be used as a full adder.

Example (6.22)

Draw a truth table and design a binary full adder using two 2-input EXclusive-OR gates and three 2-input NAND gates

SOLUTION:

A	B	C_i	S	C_o
0	0	0	0	0
0	0	1	1	0
0	1	0	1	0
0	1	1	0	1
1	0	0	1	0
1	0	1	0	1
1	1	0	0	1
1	1	1	1	1

$$S = \bar{A}\bar{B}C_i + \bar{A}B\bar{C}_i + A\bar{B}\bar{C}_i + ABC_i$$

$$= (\bar{A}\bar{B} + AB)C_i + (\bar{A}B + A\bar{B})\bar{C}_i$$

$$= \overline{(A \oplus B)}C_i + (A \oplus B)\bar{C}_i$$

$$= (A \oplus B) \oplus C_i$$

$$C_o = \bar{A}BC_i + A\bar{B}C_i + AB\bar{C}_i + ABC_i$$

$$= (\bar{A}B + A\bar{B})C_i + AB$$

$$= (A \oplus B)C_i + AB$$

$$= \overline{((A \oplus B)C_i) \cdot (\overline{AB})}$$

These two equations can be implemented as shown in Fig. 6.11.

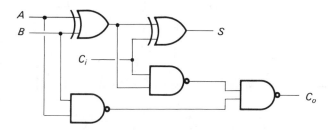

Fig. 6.11   Logic circuit for the binary full adder of Example (6.22)

6.5.3 Multi-Bit Adders

The full adder described is capable of adding two numbers A and B and a carry-in C_i assuming each number is made up of 1 bit. If A and B contain more than one bit, full adders can be used in serial or parallel to add the multi-bit numbers.

Serial Adders

In a serial adder one full adder circuit is required and addition is caried out one bit at a time. The carry-out C_o resulting from any addition must be stored and applied as a carry-in C_i one unit time later. The storage device may be a clocked D flip flop controlled by the clock of the system and cleared at the start of the addition.

A simplified block diagram for an eight-bit serial adder is shown in Fig. 6.12. Two eight-bit storage registers are used to hold the two binary numbers A and B. The data are shifted, using a shift pulse, into the adder starting with the least significant digit from both registers. In the example given in Fig. 6.12 the digits from registers A and B are both 1 while the carry-in is 0. These are added, resulting in a sum of 0 and a carry-out of 1.

After the first clock pulse the sum of 0 is shifted into the most significant position in the sum register. The carry-out of 1 appears now as the carry-in to the full adder. The data in registers A and B move one position to the right.

The inputs to the full adder are now 1(from A), 1 (from B) and 1 (C_i). These are added resulting in a sum of 1 and carry-out of 1.

After the second pulse the data in all registers are shifted one place to the right. The new sum (1) is now placed in the most significant position in the sum register.

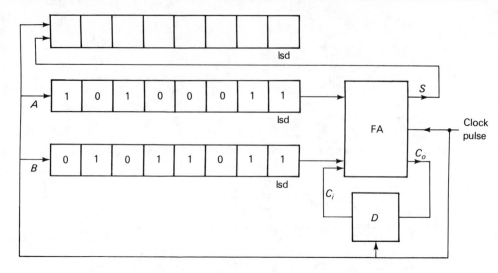

Fig. 6.12   A block diagram of a serial adder
lsd = least significant digit

The carry-out (1) now appears as a carry-in and the third digits from A and B now appear as inputs to the full adder, and so on.

After two shift pulses the sum register appears as

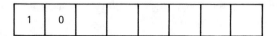

After the eight pulses the sum register contains eight digits corresponding to the sum of the two numbers in A and B.

We should add that, in some situations, the addition of two eight-bit numbers could result in a nine-bit answer. The overflow of 1 can be detected by the presence of a 1 on the flip-flop at the end of the addition cycle.

The system is economical since only one full adder is used. The disadvantage, however, is that it is slow since it would take n units of time to add n digits.

Parallel Adders

If speed is an important factor (it usually is), then parallel adders can be used. In this case, as the name implies, the data are presented in a parallel form as shown in Fig. 6.13.

Though this system is faster, it is more costly since n full adders are required to add two n-bit numbers. That means one full adder for each bit, though a half adder can be used for the first stage which has no carry-in.

Ideally, the sum should be available after one unit of time. In reality, each stage has to wait for the carry-in from the previous stage. In the worst case the carry will propagate through all stages as in, say, a four-bit adder adding 0001 and 0111. For

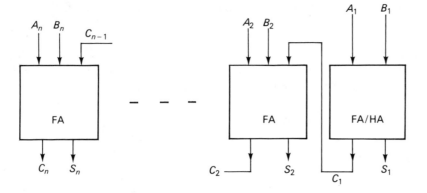

Fig. 6.13 A block diagram for an n-bit parallel adder

this reason a fixed time is allowed for the addition to be completed or a logic is added to detect the completion before the sum is gated into the register. The second alternative is usually faster though uses more logic.

6.5.4 Carry-Look-Ahead Adders

The main disadvantage of parallel adders is that the carry signal requires time to propagate through the various stages. In this section a new technique is discussed which overcomes this problem. Consider the truth table for a full adder given in Table 6.7.

	A	B	C_i	S	C_o
1	0	0	0	0	0
2	0	0	1	1	0
3	0	1	0	1	0
4	0	1	1	0	1
5	1	0	0	1	0
6	1	0	1	0	1
7	1	1	0	0	1
8	1	1	1	1	1

Table 6.7   Truth table for a full adder

If the carry-in and carry-out columns are examined it can be seen that in rows 4 and 6, the carry is propagated. That means that the carry of stage $(n-1)$, say, is passed to the nth stage.

In rows 7 and 8 a carry is generated. The part ('carry-propagate') of the carry-out concerned with propagating the carry can be expressed as

$$P = \bar{A}B + A\bar{B}$$

Similarly, the carry-generating part ('carry-generate') can be written as

$$G = AB$$

In the first two rows, no carry is generated.

The idea is therefore to test the inputs to each stage and then carries are computed for all stages simultaneously. This technique is known as *carry-look-ahead*.

In general, for any stage j

$$G_j = A_j B_j$$
$$P_j = A_j \bar{B}_j + \bar{A}_j B_j$$
$$\quad = A_j \oplus B_j$$

The unity output carry of any stage n is C_n. This can be expressed in terms of G_n, P_n and C_{n-1}, which is the unity output carry of the $(n-1)$ stage; hence

$$C_n = G_n + P_n C_{n-1}$$

Substituting for C_{n-1}, we get

$$C_n = G_n + P_n (G_{n-1} + P_{n-1} C_{n-2})$$
$$\quad = G_n + P_n G_{n-1} + P_n P_{n-1} C_{n-2}$$

Now we can substitute for C_{n-2} just as we substituted for C_{n-1}, and so on.

For four stages, we have

$$C_1 = G_1 + P_1 C_i$$
$$C_2 = G_2 + P_2 C_1$$
$$\quad = G_2 + P_2 G_1 + P_2 P_1 C_i$$
$$C_3 = G_3 + P_3 C_2$$
$$\quad = G_3 + P_3 G_2 + P_3 P_2 G_1 + P_3 P_2 P_1 C_i$$
$$C_4 = G_4 + P_4 C_3$$
$$\quad = G_4 + P_4 (G_3 + P_3 G_2 + P_3 P_2 G_1 + P_3 P_2 P_1 C_i)$$
$$\quad = G_4 + P_4 G_3 + P_4 P_3 G_2 + P_4 P_3 P_2 G_1 + P_4 P_3 P_2 P_1 C_i$$

C_i is the carry-in to the first stage (normally 0).

It has already been shown that a full adder can be implemented using two half adders and one OR gate, as in Fig. 6.14(*a*).

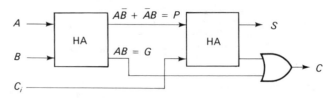

Fig. 6.14(a)   A full adder stage

It can be seen that the adder produces its own carry-propagate P and carry-generate G. For three stages then, the circuit will be along the lines of Fig. 6.14(*b*).

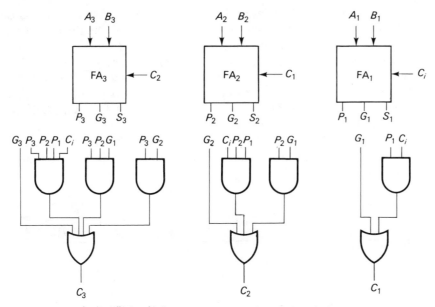

Fig. 6.14(b)   A three-stage carry-look-ahead circuit

Example (6.23)

Explain how a carry-look-ahead adder generates the carries for the two three-bit numbers 001 and 011.

SOLUTION:

C_3	C_2	C_1	C_i				
	A_3	A_2	A_1	$=$	0	1	1
	B_3	B_2	B_1	$=$	0	0	1
	S_3	S_2	S_1		0	1	1
					1	0	0

$A_1 = 1, B_1 = 1$

$$G_1 = A_1B_1 = 1$$
$$P_1 = A_1 \oplus B_1 = 0$$
$$C_1 = G_1 + P_1C_i$$
$$= 1 + 0$$
$$= 1$$

$A_2 = 0, B_2 = 1$

$$G_2 = A_2B_2 = 0$$
$$P_2 = A_2 \oplus B_2 = 1$$
$$C_2 = G_2 + G_1P_2 + P_1P_2C_i$$
$$= 0 + 1 + 0$$
$$= 1$$

$A_3 = 0, B_3 = 0$

$$G_3 = A_3B_3 = 0$$
$$P_3 = A_3 \oplus B_3 = 0$$
$$C_3 = G_3 + P_3G_2 + P_3P_2G_1 + P_3P_2P_1C_i$$
$$= 0 + 0 + 0 + 0$$
$$= 0$$

It can be seen that the carries are functions of the carry propagate and carry generate, which can be produced simultaneously if the inputs are available simultaneously. The problem, however, is that the logic required to produce the carries becomes impractical for large numbers, requiring more gates, some of which must have a large number of inputs. Hence the trade-off between speed and cost.

The usual practice is to use carry-look-ahead in groups of four or five stages and to use ripple carry between the levels. In this system each level must wait for the carry from the last level. The computation time is therefore proportional to the number of levels and the propagation time for one level. The block diagram in Fig. 6.15 illustrates this technique, where three levels of carry look-ahead groups are used to add two 12-bit numbers.

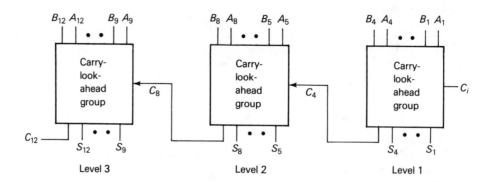

Fig. 6.15   Carry-look-ahead/ripple-through adder

6.5.5 Binary Coded Decimal Addition

In binary coded decimal (BCD) technique, each decimal number (of one digit) is represented by its four-bit binary equivalent.

To illustrate this technique consider the addition of 3 and 6:

$$
\begin{array}{r}
3 \\
+\ 6 \\
\hline
9
\end{array}
\qquad
\begin{array}{cccc}
0 & 0 & 1 & 1 \\
+\ 0 & 1 & 1 & 0 \\
\hline
1 & 0 & 0 & 1
\end{array}
$$

In the above example the two decimal numbers are replaced by their BCD codes which are then added. The answer is also in BCD code.

The same method is used if the decimal numbers contain more than one digit. Thus:

$$
\begin{array}{r}
2\ \ 6 \\
+\ 5\ \ 2 \\
\hline
7\ \ 8
\end{array}
\qquad
\begin{array}{cccc}
0\ 0\ 1\ 0 & \quad & 0\ 1\ 1\ 0 \\
+\ 0\ 1\ 0\ 1 & & 0\ 0\ 1\ 0 \\
\hline
0\ 1\ 1\ 1 & & 1\ 0\ 0\ 0
\end{array}
$$

Problems arise if the sum of any two digits exceeds decimal nine. Consider the addition of 8 + 6

$$
\begin{array}{r}
8 \\
+ \quad 6 \\
\hline
1 \ 4
\end{array}
\qquad
\begin{array}{r}
1 \ 0 \ 0 \ 0 \\
+ \ 0 \ 1 \ 1 \ 0 \\
\hline
1 \ 1 \ 1 \ 0
\end{array}
\qquad \text{NOT VALID}
$$

The answer for the last addition is not valid because in the BCD technique the code 1110 is not used. The answer is corrected by subtracting 10(1010) or adding 6(0110) to allow for the six forbidden codes.

Note that the 2's complements of 1010 is 0110. This can be shown as follows:

$$
\begin{array}{r}
8 \\
+ \quad 6 \\
\hline
1 \ 4
\end{array}
\qquad
\begin{array}{r}
\left| 1 \ 0 \ 0 \ 0 \right. \\
+ \left| 0 \ 1 \ 1 \ 0 \right. \\
\hline
\left| 1 \ 1 \ 1 \ 0 \right. \\
+ \left| 0 \ 1 \ 1 \ 0 \right. \\
\hline
0 \ 0 \ 01 \left| 0 \ 1 \ 0 \ 0 \right.
\end{array}
$$

NOT VALID
Add 6

BCD for 14

$$\uparrow \qquad\qquad \uparrow$$
BCD for 1   BCD for 4

The answer is 14_{10}, which is $00001 \quad 0100_{BCD}$

Now consider the addition of two-digit numbers:

$$
\begin{array}{r}
2 \ \ 6 \\
+ \ 5 \ \ 5 \\
\hline
8 \ \ 1
\end{array}
\qquad
\begin{array}{r}
0 \ 0 \ 1 \ 0 \ | \ 0 \ 1 \ 1 \ 0 \\
+ \ 0 \ 1 \ 0 \ 1 \ | \ 0 \ 1 \ 0 \ 1 \\
\hline
0 \ 1 \ 1 \ 1 \ | \ 1 \ 0 \ 1 \ 1 \\
+ \qquad\qquad | \ 0 \ 1 \ 1 \ 0 \\
\hline
\qquad\qquad 1 \ | \\
\hline
1 \ 0 \ 0 \ 0 \ | \ 0 \ 0 \ 0 \ 1
\end{array}
$$

← NOT VALID
← Add 6
← Carry

← BCD for 81

The procedure can be summarized as follows:

(i) Add the BCD code for each digit.
(ii) If the sum for any position $\leqslant 9$ the answer is correct.
(iii) If the sum > 9, add 6 to that position. A carry will be produced and must be added to the next position.

To generalize the procedure consider the two four-bit binary codes *A* and *B*, each consisting of four bits.

$$
\begin{array}{ccccc}
A_4 & A_3 & A_2 & A_1 \\
B_4 & B_3 & B_2 & B_1 \\
\hline
S_5 & S_4 & S_3 & S_2 & S_1
\end{array}
$$

Let the carry into the next stage be C *if the sum* > 9.
It can be seen that $C = 1$ if $S_5 = 1$ (sum > 15).
         Also $C = 1$ if $S_4 = 1$ and S_3, S_2 or both are 1.

Hence we may have

$$C = S_5 + S_4 (S_3 + S_2)$$
$$= S_5 + S_4 S_3 + S_4 S_2$$

Obviously if $C = 1$ the correction code 0110 must be added. This can be implemented as in Fig. 6.16

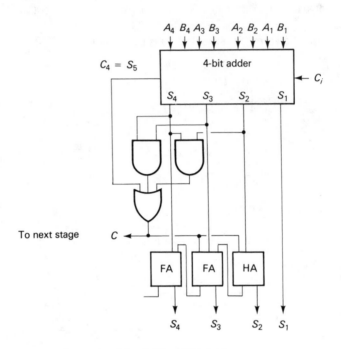

Fig. 6.16   BCD Adder

Example (6.24)

Two four-bit binary adders and a variety of logic gates are available.
Design a single-stage 8421 BCD adder. Using a block diagram show how this can be extended to a four-stage 8421 BCD adder.

SOLUTION: The adder and its extension are shown in Fig. 6.17(*a*) and (*b*).

6.5.6 Carry-Save Adders

To add a sequence of numbers, several full adders are connected. This can be achieved in more than one way, remembering that a full adder (FA) receives three inputs (3-bit) of equal significance and produces two outputs.
The outputs are:

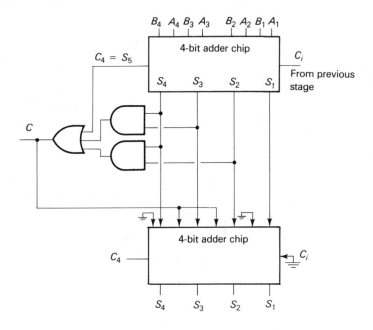

Fig. 6.17(a)   Four-bit BCD adding stage

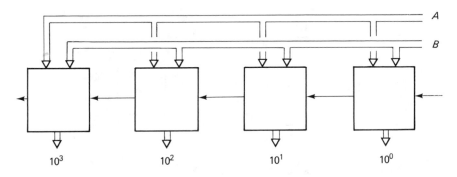

Fig. 6.17(b)   Four-stage BCD adder

(a)   A sum of the same significance as the inputs

(b)   A carry of double the significance of the inputs

For the three 3-bit numbers X, Y, Z the circuit may be connected as in Fig. 6.18.

A faster connection is possible if the carries are saved and added separately as shown in Fig. 6.19.

Example (6.25)

Add the decimal numbers $10 + 7 + 9 + 3 + 5 + 11$ using carry-save technique.

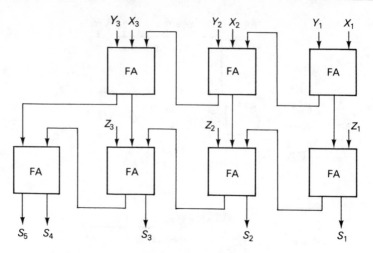

Fig. 6.18   Adding a sequence of numbers

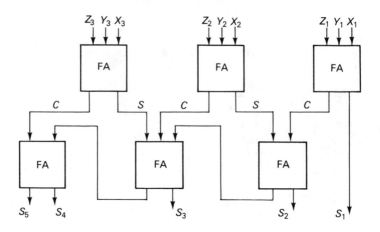

Fig. 6.19   Carry-save adder

SOLUTION:

Step (1). Add 1, 2, and 3 and 456 (simultaneously)

	Weights 8 4 2 1		Weights 8 4 2 1	
1 0	1 0 1 0	3	0 0 1 1	
7	0 1 1 1	5	0 1 0 1	
9	1 0 0 1	1 1	1 0 1 1	
Pseudo-sum (1)	0 1 0 0		1 1 0 1	Pseudo-sum (2)
Pseudo-carry (1)	1 0 1 1		0 0 1 1	Pseudo-carry (2)

Pseudo-carries are always shifted one bit.

Step (2). Add pseudo-carry (1) and pseudo-sums (1) and (2)

	0 1 0 0	Pseudo-sum (1)
	1 1 0 1	Pseudo-sum (2)
	1 0 1 1	Pseudo-carry (1)
Pseudo-sum (3)	1 1 1 1 1	
Pseudo-carry (3)	0 1 0 0 ↑	

$$\text{lsb} = 2^0$$

The least significant bit (lsb) of the results is generated here and is dropped from pseudo-sum (3)

Step (3). Add pseudo-carries (2) and (3) and pseudo-sum (3)

	0 0 1 1	Pseudo-carry (2)
	1 1 1 1	Pseudo-sum (3)
	0 1 0 0	Pseudo-carry (3)
Pseudo-sum (4)	1 0 0 0	
Pseudo-carry (4)	0 1 1 1 ↑	

$$\text{lsb} = 2^1$$

The 2^1 bit generated here is dropped from the pseudo-sum.

Step (4). Add pseudo-sum (4) and pseudo-carry (4) in a conventional ripple-through adder

0 1 0 0	Pseudo-sum (4)
0 1 1 1	Pseudo-carry (4)
0 1 0 1 1	Result bits

Therefore the final answer is $0\ 1\ 0\ 1\ 1\ 0\ 1\ = 45_{10}$

Dropped in step (3) ↑ ↑ Dropped in step (2)

This technique is useful in adding the partial products in fast multipliers.

6.6   BINARY SUBTRACTION CIRCUITS

6.6.1 Half Subtractor

The subtraction of two binary numbers can also be represented by a truth table as in the binary adder.

From the truth table

$$D = A\bar{B} + \bar{A}B$$
$$= A \oplus B$$
$$B_o = \bar{A}B$$

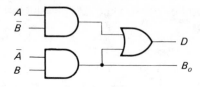

Inputs		Difference	Borrow
A	B	D	B_o
0	0	0	0
0	1	1	1
1	0	1	0
1	1	0	0

Fig. 6.20   Binary half subtractor   **Table 6.8   Truth table for half subtraction**

Looking back at the expression for the half adder, we can see that the expression for D is identical to the sum S.

One possible realization for D and B_o is shown in Fig. 6.20.

As for binary adders a half subtractor is not very useful on its own. A full subtractor is required to take care of the borrow-in (borrow from previous stage).

6.6.2 Full Subtractor

Inputs			Outputs	
Minuend	Subtrahend	Borrow-in	Difference	Borrow-out
A	B	B_i	D	B_o
0	0	0	0	0
0	0	1	1	1
0	1	0	1	1
0	1	1	0	1
1	0	0	1	0
1	0	1	0	0
1	1	0	0	0
1	1	1	1	1

Table 6.9   Truth table for a full subtractor

When deriving the table one should remember that the borrow-in is added to the subtrahend. The borrow-out is, therefore, 1 when $A < B$ plus B_i. The difference is 1 when the number of inputs in the 1 state is odd.

$$D = \bar{A}\bar{B}B_i + \bar{A}B\bar{B_i} + A\bar{B}\bar{B_i} + ABB_i$$
$$B_o = \bar{A}\bar{B}B_i + \bar{A}B\bar{B_i} + \bar{A}BB_i + ABB_i$$

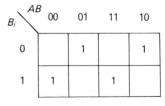

K-map for D

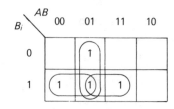

K-map for B_o

The K-maps show that D is the EX-OR of the inputs and cannot be simplified (see Fig. 6.7 and Example (6.22)).

The borrow B_o can be simplified to the following:

$$B_o = \bar{A}B + BB_i + \bar{A}B_i$$

It seems, in this case, that it is more economical to implement the expanded form of B_o than the minimized form. One possible circuit using AND/OR logic is shown in Fig. 6.21.

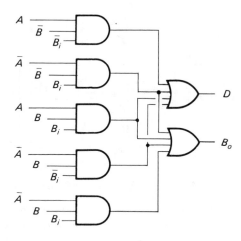

Fig. 6.21 Logic circuit for a binary full subtractor

6.6.3 Adder/Subtractor

The purpose of Section 6.6.2 is to show that it is possible to design and build a binary subtractor. It is better, however, if we can use the same adder to do both addition and subtraction.

Earlier in this chapter, it was shown that one can subtract by adding complements. This idea will be used in a four-bit adder/subtractor, which will now be explained.

Consider the two EX-OR gates shown below:

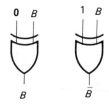

The EX-OR of (B and 0) is B
The EX-OR of (B and 1) is $\bar{B}$

This means that it is possible to obtain B unchanged or its 1's complement at the output of an EX-OR gate by having a 0 or a 1 on the second (control) input of the gate.

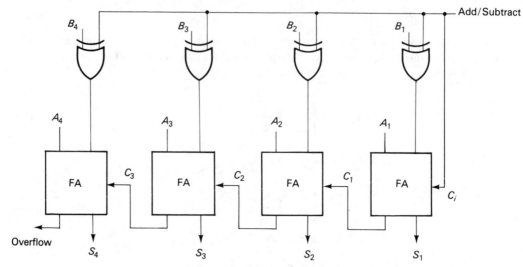

Fig. 6.22   A four-bit binary adder/subtractor

From the logic diagram a logic 0 on the add/subtract line (Fig. 6.22) results in addition mode. But a logic 1 on the add/subtract line inverts B and adds 1 to the first stage, thus resulting in a subtraction mode.

Carry-ripple-through is used in Fig. 6.22. Carry-look-ahead can be used between the adders to speed up the operation, and avoid having to wait for the carry to ripple through.

Example (6.26)

The 2's complement of a binary number can be obtained by serially testing all the bits, and complementing every bit after the first one has passed starting with the least significant digit. Draw any circuit for achieving this.

Explain, with the aid of a block diagram, how this circuit can be used together with one binary full adder, and any available logic circuits, to construct an eight-bit adder/subtractor which functions as follows:

> If the ADD control line is high (SUBTRACT is low), the contents of the input register are added to the contents of the accumulator. If the SUBTRACT control line is high (ADD is low), the contents of the input register are subtracted from the contents of the accumulator. The answer is stored in the accumulator, whose contents are in 2's complements form.

If the eighth bit of the operands is used for the sign digit, design the necessary logic for detecting the overflow.

SOLUTION: The three requirements of the question are met by Fig. 6.23(a), (b) and (c) and the associated explanations.

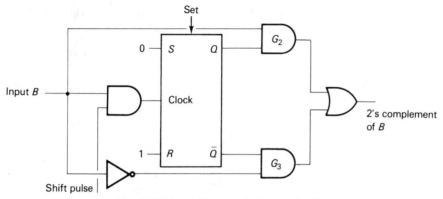

Fig. 6.23(a)   A 2's complement circuit

In Fig. 6.23(a) the flip-flop is initially set, gate G_3 is disabled, gate G_2 is enabled, and the input B appears at the output.

When the first binary 1 appears, the shift pulse will trigger the flip-flop. Because $S = 0$, $R = 1$ this pulse will reset the flip-flop.

Then G_2 is disabled, G_3 is enabled, and the complement ($\bar{B}$) of B will appear at the output. The flip-flop will stay in the reset condition.

Note. See Fig. 6.4 for an alternative circuit.

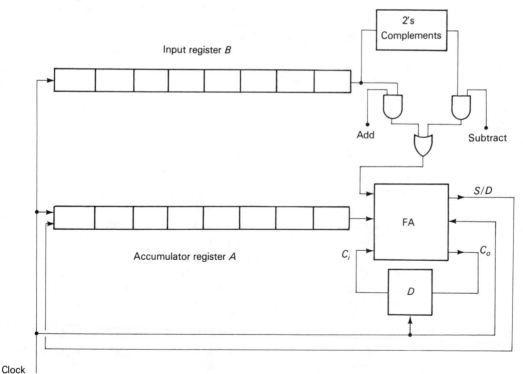

Fig. 6.23(b)   A 2's complements adder/subtractor

The circuit operation is the same as in Section 6.5.3. The overflow can be detected in more than one way:

(a) By testing C_7 and C_8 (if available); there is overflow if $C_7 \neq C_8$, i.e. the overflow $= C_7\bar{C}_8 + \bar{C}_7C_8$

(b) By testing the sign bits of the two operands A and B and the sign of the sum, i.e. the overflow $= A_8B_8\bar{S}_8 + \bar{A}_8\bar{B}_8S_8$.

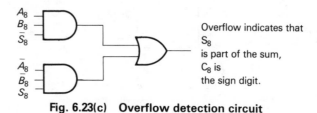

Overflow indicates that
S_8
is part of the sum,
C_8 is
the sign digit.

Fig. 6.23(c)   Overflow detection circuit

6.7   BINARY MULTIPLICATION CIRCUITS

6.7.1 Pencil and Paper Multiplication

It was stated earlier that multiplication can be achieved by repeated additions, a process very convenient for digital computers.

Consider the following multiplication:

$$
\begin{array}{l}
1\,0\,1\,1 \quad \text{Multiplicand} \\
1\,0\,0\,1 \quad \text{Multiplier} \\
\hline
1\,0\,1\,1 \\
0\,0\,0\,0 \\
0\,0\,0\,0 \\
1\,0\,1\,1 \\
\hline
1\,1\,0\,0\,0\,1\,1
\end{array}
$$

The pencil and paper multiplication method can be summarised as follows. Inspect the multiplier one bit at a time. If the bit under consideration is 1, add the multiplicand and shift; otherwise add zero and shift (i.e. shift only).

In pencil and paper multiplication the partial products are shifted left and addition is carried out at the end. The same effect can be obtained by carrying out addition (if required) after every bit inspection and then shifting the sum to the right.

Registers are required to hold the multiplicand, the multiplier, the sign and the product. It should be noted that if the operands have n bits each, the product requires a $2n$-bit register.

The end of the computation is indicated by a counter which holds the number of digits of the multiplier. Assume the multiplicand of the last example (1011) is held in register A. The sum (initially 0000) is held in register B. The multiplier (1001) is held in register C.

At the end of the multiplication process, the product (double length) will occupy registers B and C as illustrated below. The multiplier is lost in this case. If the multiplier is to be retained, register B is replaced by a double-length register. The bits of the multiplier can be returned to C using a circulating shift register.

Counter   A   B   C

$\boxed{4}$ $\boxed{1\,0\,1\,1}$ $\boxed{0\,0\,0\,0}$ $\boxed{1\,0\,0\,1}$

$\overline{\underline{|1\,0\,1\,1|}}$

Initial conditions
Counter not zero
lsd in C is 1

Action
Add contents of A to that of B. Shift content of B and C by one position. Decrease the counter by one.

$\boxed{3}$ $\boxed{1\,0\,1\,1}$ $\boxed{0\,1\,0\,1}$ $\boxed{1\,1\,0\,0}$

Counter not zero
lsd in C is 0

Action
Shift content of B and C by one position. Decrease the counter by one.

$\boxed{2}$ $\cdot\boxed{1\,0\,1\,1}$ $\boxed{0\,0\,1\,0}$ $\boxed{1\,1\,1\,0}$

Counter not zero
lsd in C is 0

Action
Shift content of B and C by one position. Decrease the counter by one.

$\boxed{1}$ $\boxed{1\,0\,1\,1}$ $\boxed{0\,0\,0\,1}$ $\boxed{0\,1\,1\,1}$

Counter not zero $\overline{\underline{|1\,1\,0\,0|}}$
lsd in C is 1

Action
Add content of A to that of B. Shift the content of B and C by one position. Decrease the counter by one.

$\boxed{0}$ $\boxed{1\,0\,1\,1}$ $\boxed{0\,1\,1\,0}$ $\boxed{0\,0\,1\,1}$

Counter zero
End of computation.

Answer   0 1 1 0 0 0 1 1

6.7.2 Speed-up Techniques for Multiplication

An important factor affecting the design of arithmetic circuits is the speed. The speed depends on the speed of additions and the frequency with which other operations, like multiplication, are carried out.

Possible speeding-up techniques are:

(a) Shifting Across

To speed up the process, no addition cycles are initiated for zeros but shift cycles are initiated immediately. Further improvement is possible if we allow for variable shift length. This can be done if a single shift could shift across a whole string of zeros.

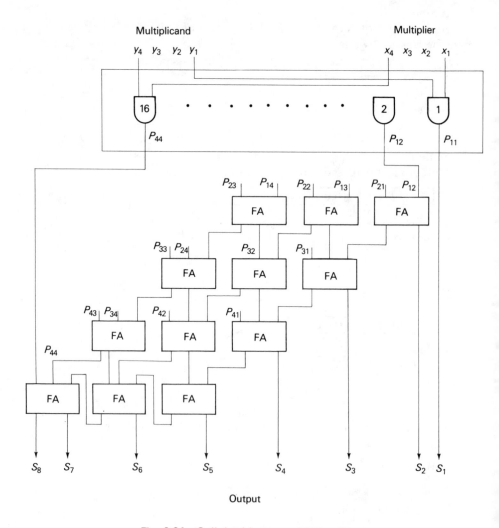

Fig. 6.24   Cellular binary multiplication

(b) Multiplication by Addition and Subtraction

If 0111100 is part of the number, let n be the position number of the highest order 1 in the group.

Let x be the number of successive 1s.

The value of the number N is given by

$$N = 2^n - 2^{n-x}$$

In this example $n = 6$
$$x = 4$$

$$N = 2^6 - 2^{6-4}$$
$$= 64 - 4 = 60$$

Check. $111100 = 1 \times 2^5 + 1 \times 2^4 + 1 \times 2^3 + 1 \times 2^2 + 0 \times 2^1 + 0 \times 2^0$
$$= 32 + 16 + 8 + 4 = 60$$

Therefore, for a string of ones in a multiplier, we need one addition and one subtraction, with some extra control, instead of one addition for each bit.

This can be achieved as follows:

(i) Inspect the bits of the multiplier starting with the least significant digit.
(ii) When you meet the first 1 in a string of 1s, subtract the multiplicand from the accumulator.
(iii) When you meet the first 0 after a series of 1s add the multiplicand to the accumulator.
(iv) For a single 1, add the multiplicand to the accumulator.
(v) Remember that the multiplicand must be shifted an appropriate number of places as usual.

6.7.3 Cellular Binary Multiplication

Consider the following multiplication of two 4-bit binary numbers x and y.

				y_4	y_3	y_2	y_1	Multiplicand
				x_4	x_3	x_2	x_1	Multiplier
				P_{14}	P_{13}	P_{12}	P_{11}	
			P_{24}	P_{23}	P_{22}	P_{21}		Partial
		P_{34}	P_{33}	P_{32}	P_{31}			products
	P_{44}	P_{43}	P_{42}	P_{41}				
S_8	S_7	S_6	S_5	S_4	S_3	S_2	S_1	Answer

The multiplication process can therefore be carried out in two steps:

(a) Developing the partial products
(b) Summing these partial products

For two n-bit numbers, the partial products can be developed in a single gate time using an array of n^2 AND gates. The partial products are added using an array of full adders as shown in Fig. 6.24. The speed of this cellular array multiplication depends on the type of logic and speed of the adder.

Note. Binary adders add two numbers at a time except for specially designed circuits.

The worst case delay in the array will be along the furthest right diagonal and the lowest row.

For $n \times n$ multiplication, the delay is along $(n - 1) + (n - 1) = 2(n - 1)$ full adders.

If the full adder is made of two levels of gate, then the delay through it is two units (two gate delays). The partial products take one unit time to be formed.

Total time $T = 1 + 2(n - 1) \times 2$
$$= 1 + 4(n - 1) \text{ gate delays}$$

In the arrangement above, the worst case for carry propagation is through six adders, and the total time is 13 gate delays.

The speed can be further improved if carry-look-ahead is employed in the last stage (row) instead of ripple carry.

The hardware requirement is rather high. For two n-bit numbers, n^2 AND gates are required for the generation of the partial products and $n(n - 1)$ full adders for adding the partial products.

In the example of two four-bit numbers shown in Fig. 6.24, 16 AND gates and 12 full adders are used.

6.7.4 Combinational Logic Multiplier

One easy way of designing a fast parallel multiplier is to treat it as a combinational logic problem. Table 6.10 shows all possible products for the multiplication of two 2-bit numbers:

Multiplier		Multiplicand		Product			
a_2	a_1	b_2	b_1	P_4	P_3	P_2	P_1
0	0	0	0	0	0	0	0
0	0	0	1	0	0	0	0
0	0	1	0	0	0	0	0
0	0	1	1	0	0	0	0
0	1	0	0	0	0	0	0
0	1	0	1	0	0	0	1
0	1	1	0	0	0	1	0
0	1	1	1	0	0	1	1
1	0	0	0	0	0	0	0
1	0	0	1	0	0	1	0
1	0	1	0	0	1	0	0
1	0	1	1	0	1	1	0
1	1	0	0	0	0	0	0
1	1	0	1	0	0	1	1
1	1	1	0	0	1	1	0
1	1	1	1	1	0	0	1

Table 6.10   **Truth table for a two-bit multiplier**

From the table expressions can be developed for P_1, P_2, P_3 and P_4, which are simplified and implemented using suitable logic:

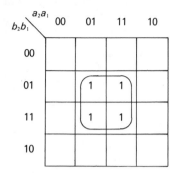

K-map for P_1

$$P_1 = a_1 b_1$$

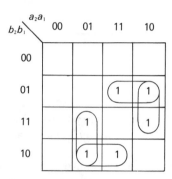

K-map for P_2

$$P_2 = a_2 \bar{b}_2 b_1 + a_2 \bar{a}_1 b_1 + \\ \bar{a}_2 a_1 b_2 + a_1 b_2 \bar{b}_1$$

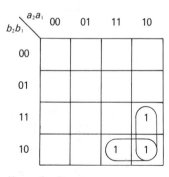

K-map for P_3

$$P_3 = a_2 \bar{a}_1 b_2 + a_2 b_2 \bar{b}_1$$

From the table
$$P_4 = a_2 a_1 b_2 b_1$$

This sort of multiplier accepts the inputs in parallel and hence can be fast. The obvious disadvantage is the logic complexity for large operands.

6.7.5 Serial-Parallel Multipliers

As the name suggests, one of the operands is applied in series (the multiplicand in this case) and the other (multiplier) is applied in parallel.

In Fig. 6.25 two four-bit numbers are multiplied.

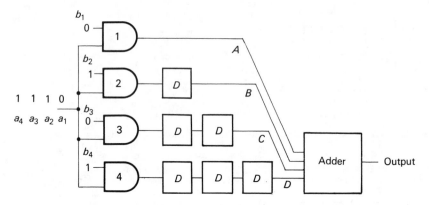

Fig. 6.25   A simple serial-parallel multiplier

The multiplier $(b_4 b_3 b_2 b_1)$ is applied in parallel and controls the AND gates. If the multiplier digit is 1, the AND gate allows the multiplicand digits to pass to the adder (as in gates 2 and 4). If the multiplier digit is 0, the output of the AND gates is also 0 (as in gates 1 and 3).

The multiplicand is applied in series with the least significant digit first. The delays provide the necessary shifting operation. The adder adds the partial products. Consider the multiplication of 10×14. The multiplier is 1010. The multiplicand is 1110. The multiplication process can be explained as follows:

	P_8	P_7	P_6	P_5	P_4	P_3	P_2	P_1
A	0	0	0	0	0	0	0	0
B	0	0	0	1	1	1	0	0
C	0	0	0	0	0	0	0	0
D	0	1	1	1	0	0	0	0
Output	1	0	0	0	1	1	0	0

$10001100_2 = 140_{10}$

It can be seen that the output at A is always 0 since the least significant digit of the multiplier is 0. The output at B is the multiplicand delayed by one unit delay (one clock pulse). The output at $C = 0$ (for the same reason as A).

Finally the output at D is the multiplicand delayed by three unit delays. The adder adds all these partial products.

Compare this to pencil and paper multiplication:

$$
\begin{array}{ll}
1\,1\,1\,0 & \text{Multiplicand} \\
1\,0\,1\,0 & \text{Multiplier} \\
\hline
0\,0\,0\,0 & A \\
1\,1\,1\,0\,0 & B \\
0\,0\,0\,0\,0\,0 & C \\
1\,1\,1\,0\,0\,0\,0 & D \\
\hline
1\,0\,0\,0\,1\,1\,0\,0 & \text{Product}
\end{array}
$$

The arrangement in Fig. 6.25 requires a multi-input adder and a large number of delays. This can be overcome by using the arrangement of Fig. 6.26.

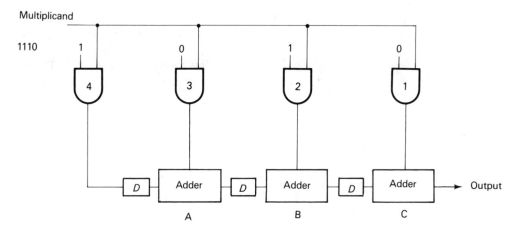

Fig. 6.26   Alternative serial-parallel multiplier

This arrangement requires $(n - 1)$ adders and delays for an n-bit multiplier. D represents clocked delays.

Once again the AND gates allow the multiplicand digits through if the multiplier digit is 1. The multiplicand is applied in series with the least significant digits first. The multiplicand may have any number of digits.

The output of gate 4 after four clock pulses is 1110. The output of gate 3 is always 0.

Adder A adds the output of gate 3 to the output of gate 4 delayed by one unit delay.

Addition at A:
$$
\begin{array}{l}
0\,0\,0\,0 \\
1\,1\,1\,0\,0 \\
\hline
1\,1\,1\,0\,0
\end{array}
$$

The output of gate 2 after four clock pulses is 1110.

Adder B then adds the output of gates 2 to the output of adder A delayed by one unit delay.

Addition at B
```
      1 1 1 0
  1 1 1 0 0 0
  _____
  1 0 0 0 1 1 0
```

Similarly the adder C adds the output of gate 1 to the output of adder B delayed by one unit delay.

Addition at C
```
        0 0 0 0
  1 0 0 0 1 1 0 0
  _____
  1 0 0 0 1 1 0 0
```

The output of adder C is the product which appears with the least significant digit first.

Note that the adders in Fig. 6.26 did not allow for the carry-out signal. If single full adders are used, the carry-out of each stage must be fed to the next (higher order) stage as shown in Fig. 6.27.

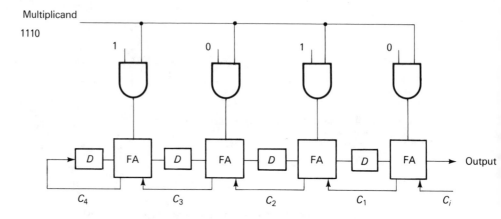

Fig. 6.27 A practical serial-parallel multiplier

In practice, a four-bit parallel adder may be used instead of the four full adders. The carries are internally connected except for the carry-in C_i, which is zero (in this case), and the carry-out C_4, which must be delayed by one unit delay.

For the four delay units, one four-bit parallel-in/parallel-out shift register may be used. Also a quad-two input AND gate chip may be used. This can be arranged as in Fig. 6.28.

The output (product) can be stored in a double-length (eight-bit in this case) storage register, but this is not shown.

Signed Numbers

In practice, the two binary numbers may contain sign digits. Usually logic 0 indicates positive and logic 1 indicates negative. With this convention, the sign digit

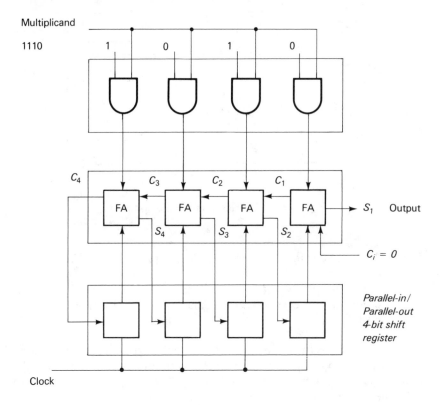

Fig. 6.28   IC version of a serial-parallel multiplier

of the product is 0 if the signs of the multiplier and multiplicand are the same; other-wise the sign of the product is 1. This can be implemented using a single EXclusive-OR gate. The two inputs to the gate are the sign digits of the multiplier and multipli-cand, and its output is the sign digit of the product.

6.7.6 Look-up-Table Multiplication

The increasing availability of low-cost, large and fast-memory devices like ROMs justifies the look-up-table approach to binary arithmetic. In this technique, the product value of two digits x and y is stored in the ROM and can be accessed by addressing the correct location. In other words, the multiplication table stored in the ROM contains results of all possible combinations of the two-input operands as shown in Fig. 6.29.

Let the multiplicand contains m bits.
Let the multiplier contain n bits.
Then the total number of inputs to the ROM is $m + n$.
The product will have up to $(m + n)$ bits.

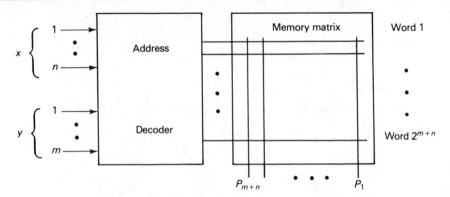

Fig. 6.29   ROM used as a look-up table

The ROM must have $(m + n)$ inputs addressing 2^{m+n} words. Each word must have $(m + n)$ bits.

Total storage capacity $M = (m + n)2^{m+n}$ bits.

As an example, consider two eight-bit binary numbers, i.e. $m = n = 8$.

Then the ROM capacity $M = 16 \times 2^{16} = 2^{20}$ bits

This means that over a million bits are required to store the whole of the multiplication table for two eight-bit numbers. This is becoming feasible, but still large for present-day devices. On the other hand only a 256×8 (2k) ROM is required to implement the 4×4 table.

Though only ROMs have been mentioned, other similar devices like PROMs and EPROMs can be used for storing look-up tables.

Storage reduction techniques

(*a*) ROUNDING: Many small computers employ rounding to reduce the storage requirement at the expence of accuracy. Instead of the product containing $(m + n)$ bits the product is rounded to w bits where $w < m + n$. Memory requirement is now $Mr = w \cdot 2^{m+n}$.

Table 6.11 shows the memory requirement with and without rounding.

(*b*) PARTITIONING: The multiplier and multiplicand are partitioned into *disjoint**
sections, thus enabling large tables to be implemented using small memories.

To illustrate the procedure consider two binary numbers A (eight bits) and B (four bits).

The eight-bit number can be partitioned into two parts, one part containing bits 5 to 8 and the other containing bits 1 to 4. Thus

---

* See Section 8.1 for the meaning of 'disjoint'.

Input	Word length	Memory bits
$m = n$	w	
4	8	2 048
4	4	1 024
8	16	1 048 576
8	12	786 432
8	8	524 288
10	20	20 971 520
10	16	16 777 216
10	10	10 485 760

Table 6.11 Look-up-table memory requirement

where $A_{5-8} = a_8\, a_7\, a_6\, a_5\, 0\,0\,0\,0$
$A_{1-4} = 0\,0\,0\,0\, a_4\, a_3\, a_2\, a_1$

The product $A_8 B_4 = (A_{5-8} + A_{1-4}) \times B_{1-4}$
$= A_{5-8}\, B_{1-4} + A_{1-4}\, B_{1-4}$

Similarly if B contains eight bits, then

$B_8 = B_{5-8} + B_{1-4}$

and the product is given by

$A_8 B_8 = (A_{5-8} + A_{1-4}) \times (B_{5-8} + B_{1-4})$
$= A_{5-8}\, B_{5-8} + A_{5-8}\, B_{1-4} + A_{1-4}\, B_{5-8} + A_{1-4}\, B_{1-4}$

The leading and trailing zeros are not connected but are used to decide the significance of the partial products.

The product can now be implemented using readily available 8×2^8 (2k) bit memories and four-bit adders as shown in Fig. 6.30.

(c) LOGARITHMIC MULTIPLICATION: Another way of reducing the storage capacity of a ROM multiplication table is to convert multiplication into addition and division into subtraction, by using logarithms:

$A \times B = $ Antilog $(\log A + \log B)$

The logarithm and antilogarithm require less storage, but speed is reduced. Addition can be done by hard-wired logic or by table look-up if desired.

Example (6.27)
Design a parallel binary multiplier for the multiplication of two eight-bit numbers, assuming that three four-bit binary full adders and a sufficient number of two-input AND gates are available.

SOLUTION: AND gates can be used to generate the partial product. These are added using the four-bit adders. To simplify the connection, we shall assume that the eight inputs are one one side of the adder chip and the four outputs together with the

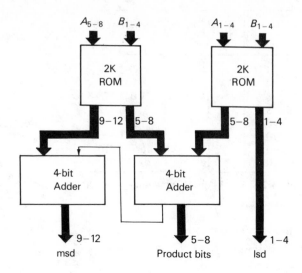

Fig. 6.30(a)   Multiplication of a four-bit by an eight-bit number

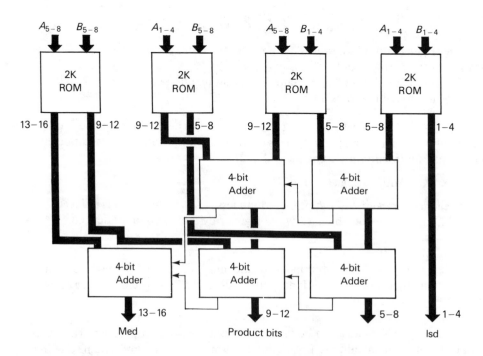

Fig. 6.30(b)   Multiplication of two eight-bit numbers

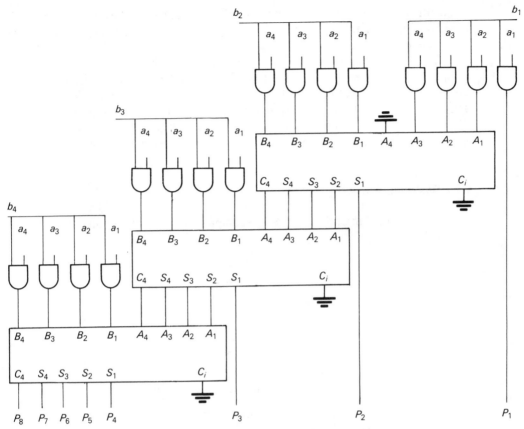

Fig. 6.31   A 4 × 4 parallel binary multiplier for Example (6.27)

carry-in and carry-out are on the other side. The 7483 chip given in Appendix 6.1 may be used for this purpose. See Fig. 6.31.

Note. On the chip in Appendix 6.1 C_o is used to indicate carry-in.
In this book C_i is used to indicate carry-in to avoid confusion with C_o for carry-out.

6.8   BINARY DIVISION CIRCUITS

6.8.1 Introduction

It was mentioned earlier that multiplication can be reduced to a series of additions. Similarly division can be reduced to a series of subtractions. In fact even the comparison process can be achieved by subtracting and noting the sign of the answer. Division can also be accomplished by multiplying the dividend by the reciprocal of the divisor. The reciprocals can be read from a reference table. Division occurs less frequently than multiplication, and therefore a slow speed can be tolerated. Roughly speaking, division takes four times as long as multiplication, depending on the technique used.

In the binary division described in Section 6.3.3, the divisor is compared with the part of the dividend (or current partial remainder) above it. If the divisor is greater, a 0 is placed in the quotient, no subtraction occurs, and the dividend is shifted one place left (or the divisor is shifted right). If the divisor is smaller, a 1 is placed in the quotient, the divisor is subtracted, and the dividend is shifted one place to the left. This can be shown by means of a flow diagram as in Fig. 6.32.

6.8.2 Restoring Division

In the algorithm of Section 6.8.1, a comparison of the divisor and dividend was assumed.

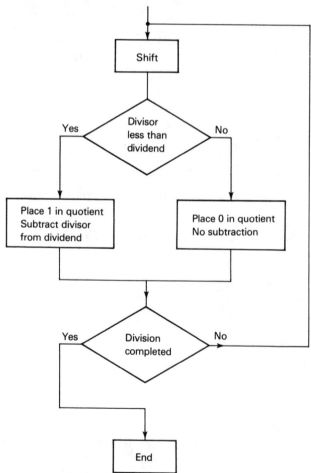

Fig. 6.32   Flow diagram for the division procedure

To avoid using expensive comparators, computers usually carry out the test by subtraction. In this case the divisor is subtracted from the part of the dividend under consideration. A positive answer indicates that the divisor is smaller and a 1 is placed in the quotient. A negative answer, however, indicates that the divisor is larger and subtraction was not necessary; then the divisor is added back to the dividend. The addition process restores the original value of the dividend, which gives rise to the name *restoring division*.

This can be represented by the flow chart of Fig. 6.33.

To implement the restoring procedure the divisor is placed in the divisor register. The dividend or partial dividend is placed in the accumulator register, as shown in Fig. 6.34. At each clock pulse subtraction is performed using the 2's complement method. The answer determines whether restoring is necessary and whether a 1 or 0 is to be placed in the quotient register.

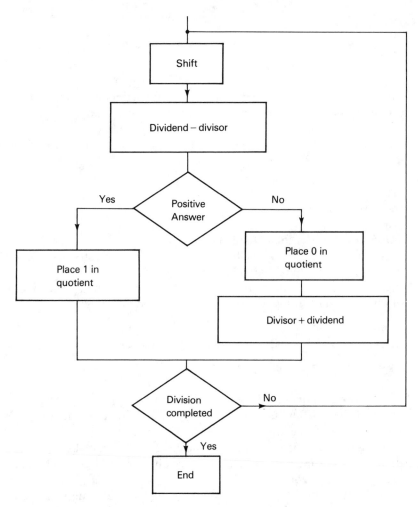

Fig. 6.33   **Flow diagram representation of restoring division**

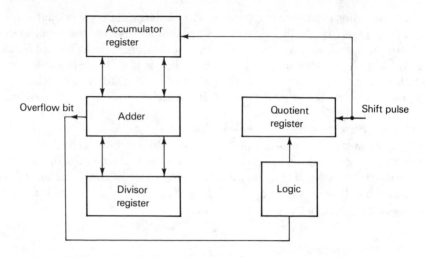

Fig. 6.34 Block diagram for the restoring division procedure

Division, like multiplication, is carried out on fractional operands. These are then scaled by the program. The decimal number 981, for exampe, is taken as 0.981 with a scale factor of 1000.

When compilers are written for fixed-point arithmetic machines, scaling is provided and the user is not aware of it.

6.8.3 Non-Restoring Division

(*a*) Divisor is Greater than Dividend

In restoring division, each time the divisor is added to restore the dividend it is subtracted during the next cycle after the shift operation.

A right-hand shift of the divisor reduces its value by half. The right-hand branch of the flow chart in Fig. 6.33 can be interpreted as follows:

Add present divisor.
Subtract half present divisor (subtract shifted divisor).

These two steps can be replaced by 'add shifted divisor'.

(*b*) Divisor is Less than Dividend

In this case the next instruction is 'subtract shifted divisor'. The two parts can be combined together in the modified flow chart of Fig. 6.35.

This method avoids the restoration of the dividend which makes it faster compared with the previous method. As usual, the penalty is more logic, and more complex control circuits are required.

It should be observed that not all computers incorporate multiply and divide logic. In some cases, the processes of add, complement and shift are used to achieve

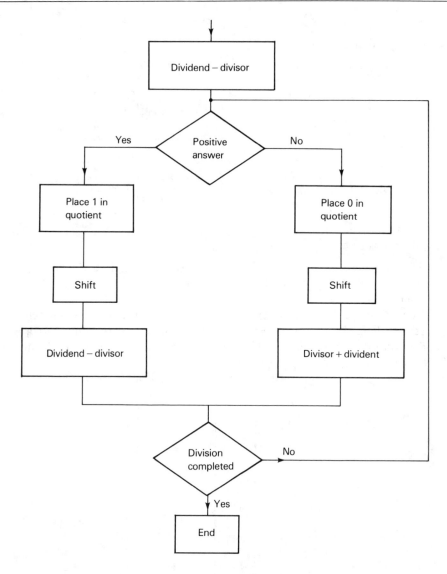

Fig. 6.35 Flow diagram representation of non-restoring division

multiplication and division using software techniques. With the advent of LSI, however, more computers are incorporating specialized multiplication and division circuits, especially when speed is important.

To illustrate the non-restoring-division procedure we shall assume a six-bit dividend and a four-bit divisor given in 2's complement representation. For simplicity, the two operands are assumed positive (msd = 0). Subtraction is achieved by adding the 2's complements. Let us first start with a numerical example and divide 21 by 7, using the procedure outlined in Fig. 6.35:

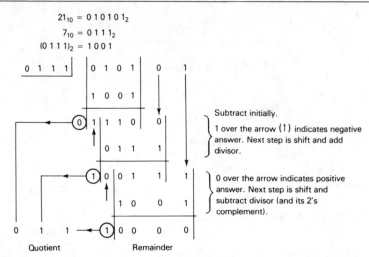

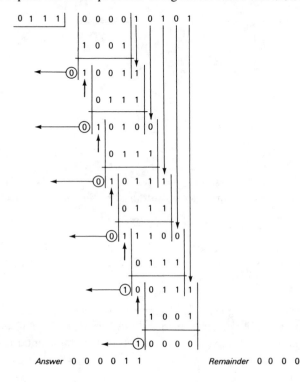

The quotient is obtained from the encircled overflow digits giving a quotient of 011 and a remainder of 0000. In the decimal system the answer is 3 without remainder.

The answer in this case is obtained in three levels of addition/subtraction. If however, we divide 21 by 1, six levels of addition/subtraction are required to produce a six-digit quotient. Generally speaking, the number of levels is equal to the number of digits in the dividend.

Let us repeat the same problem using six levels of addition:

The circuit to achieve this is given in Fig. 6.36. In the first level at the top, the control line feeding the EX-OR gates is 1. This inverts the divisor digits and adds 1 to the least significant digit (connected to the carry-in C_0, of the four-bit adder). The effect is a 2's complement subtraction.

After the first level, the fourth digit of the adder (S_4) decides whether the next level will add the shifted divisor ($S_4 = 1$) or subtract it ($S_4 = 0$). The carry-out (C_4) digits from the six levels are taken as the answer (quotient). To understand the function of the last four-bit adder, we consider the following two examples:

(i) $21 \div 6 = 3$, Remainder 3

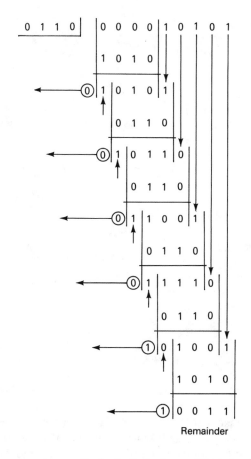

Answer   0 0 0 0 1 1   Remainder 0   0 1 1

The msd of the remainder (sign digits) is zero. This disables the four AND gates. The last four-bit adder will have no effect in this case.

(ii) $21 \div 5 = 4$, Remainder 1

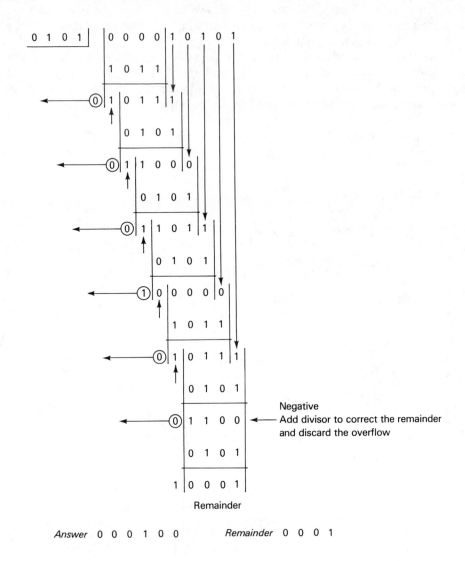

Answer 0 0 0 1 0 0   Remainder 0 0 0 1

Note that S_4 of the last stage, which is the sign digit of the remainder, is 1. This indicates a negative remainder, which must be corrected. The logic 1 is used to enable the four AND gates and allows the divisor to be added to the remainder and produce the correct answer.

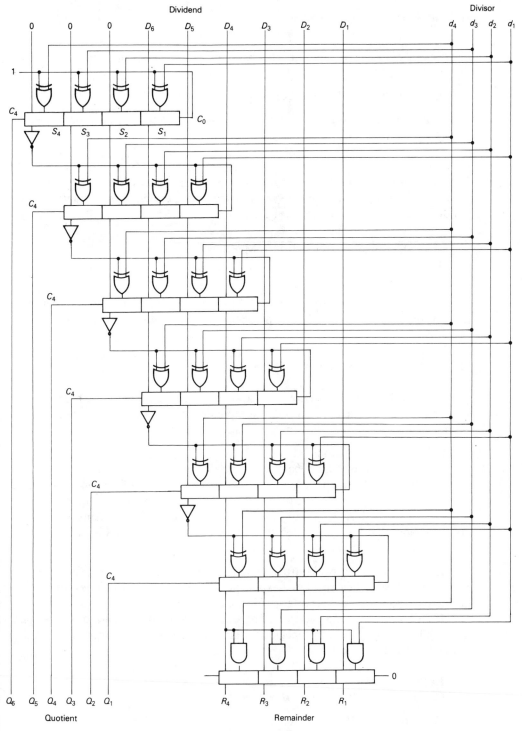

Fig. 6.36   Non-restoring division

6.9   FLOATING POINT ARITHMETIC

Like decimal numbers, binary numbers can be expressed in a floating point format as follows:

$$1010.01 = .101001 \times 2^4 = .101001 \times 2^{100}$$

If M is the mantissa and E is the exponent, any binary number m can be written as follows:

$$m = M \times 2^E$$

There are many ways of arranging the format in a digital computer. In a 48-bit machine, the exponent occupies 7–14 bits. The rest are used for the mantissa. The first digit is usually reserved for the sign.

Exponent	Mantissa
0 10	11 47

The usual practice is to make the first bit of the mantissa 1. M is therefore a fraction, such that $1 > M \geqslant \frac{1}{2}$, as shown above. This is called the *normalized form*.

6.9.1 Addition and Subtraction

For addition and subtraction, the exponents must be equal before the mantissa can be added or subtracted. To make the exponents equal and yet keep the mantissa in a normalized form the following steps are followed:

(i)   Compare the exponents.
(ii)   Subtract the smaller exponent from the larger.
(iii) Increase the smaller exponent, and simultaneously shift the binary point of the corresponding mantissa to the left (shift the mantissa to the right), by a number of bits equal to the difference between the exponents.
(iv)   Now that the binary points are aligned, the mantissas are added or subtracted as required.

If the results overflow, shift the mantissa one place to the right and increase the exponent by 1. Then check the exponent. If the exponent overflows because of this increase, the number is too large for the machine.

If there is no mantissa overflow, check that the result is normalized. If not, shift the result left until a non-zero appears in the most significant position, decreasing the exponent by one for every shift.

Example (6.28)
Calculate $A + B$ and $A - B$ where

$$A = .10010 \times 2^7 \quad\; = .10010 \times 2^{111}$$
$$B = .10100 \times 2^6 \quad\; = .10100 \times 2^{110}$$

SOLUTION:

$E_a = 111$ $E_b = 110$
E_b is adjusted to 111 (increased by 1)

The mantissa M_b must be adjusted accordingly, by shifting it one position to the right.

This results in the following:

$$A = .10010 \times 2^{111}$$
$$(+)\ B = .01010 \times 2^{111}$$
$$\text{The sum } S = .11100 \times 2^{111}$$

Similarly for subtraction:

$$A = .10010 \times 2^{111}$$
$$B = .01010 \times 2^{111}$$
$$\text{The difference } D = .01000 \times 2^{111}$$

The difference must be normalized by shifting the binary point one position to the right (shifting the mantissa one position to the left), and decreasing the exponent by 1.
Hence

$$D = .10000 \times 2^{110}$$

Now consider the addition of the following two numbers

$$A = .10010 \times 2^{7}$$
$$B = .10101 \times 2^{1}$$

Obviously B must be adjusted to make its exponent equal to that of A. This results in the following

$$B = .00000 \times 2^{7}$$

The shifting of the binary point of the mantissa for B has resulted in a zero mantissa. In such a case the larger operand is taken as the answer.

6.9.2 Multiplication and Division

(i) When floating point numbers are multiplied their mantissas are multiplied and their exponents are added.

(ii) When the two numbers are divided, their mantissas are divided and their exponents are subtracted.

For a change, multiplication and division seem easier than addition and subtraction as no 'pre-shifting' is required. The result must be represented in the standard form to bring the mantissa within the accepted range $1 > M \geqslant \frac{1}{2}$.

Example (6.29)

Calculate $\dfrac{A}{B}$ and $A \times B$

where

$$A = .1010 \times 2^{101}$$
$$B = .1000 \times 2^{-001}$$

SOLUTION:

The quotient $Z = \dfrac{A}{B} = \dfrac{.1010}{.1000} \times 2^{101-(-001)}$

$$= 1.0100 \times 2^{110}$$
$$= .1010 \times 2^{111}$$

The product $Y = A \times B$
$$= .1010 \times 2^{101} \times .1000 \times 2^{-001}$$
$$= .1010 \times .1000 \times 2^{101+(-001)}$$
$$= .0101 \times 2^{100}$$
$$= .1010 \times 2^{011}$$

It is obvious that the floating point arithmetic requires more complex hardware since the mantissa and exponents are treated differently.

The main advantage is that the floating point technique extends the range of the computer and enables it to handle much wider ranges of very large and very small numbers.

Finally, we should indicate that only the large expensive computers do have the hardware circuitry required for floating point arithmetic. Others perform the operation using slower, but less expensive, software. Again, VLSI technology is going to change the situation in the near future.

Exercises

Q(6.1) Convert the following to binary:

1984_{10} ; 68.32_{10} ; 53.61_8 ; $B37.6C_{16}$

Q(6.2) Convert the decimal numbers 67 and 36 into eight-bit binary numbers. Using these binary numbers, perform the following subtractions:

(i) $67 - 36$ using the 1's complement arithmetic.
(ii) $36 - 67$ using the 1's complement arithmetic.
(iii) $67 - 36$ using the 2's complement arithmetic.
(iv) $36 - 67$ using the 2's complement arithmetic.

Q(6.3) Design and minimize a logic circuit to convert the 8421 binary code into the reflected code given in Table 6.2. Implement the circuit using suitable logic.

Q(6.4) Draw the truth table for a binary full adder and derive logical expressions for the sum and carry signals. Implement the circuit using a minimum number of NAND gates. Using a four-bit binary adder and NAND gates, design a four-bit adder/subtractor.

Q(6.5) Using a 1024×10 bit ROM, implement the sines function of angles between 0 and 90 degrees, using the look-up-table principle. Estimate the error compared with a four-digit handbook value.

Q(6.6) If 256×8 bit ROMs and four-bit adders are available, show how the division of an eight-bit number by a four-bit number can be achieved. The quotient has 12 bits, with accuracy of $\pm 1/16$. Compare the storage required with a direct ROM implementation using 12 outputs and 12 inputs.

Appendix 6.1 A sample of available ICs

4 × 4 parallel binary multiplier
74284

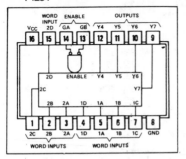

2 × 4 parallel binary multiplier
74261

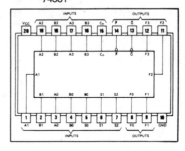

4-bit binary full adder
7483

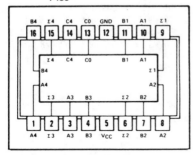

Arithmetic Logic Unit
74381

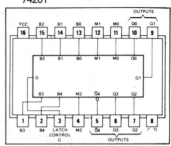

4-bit magnitude comparator
7485

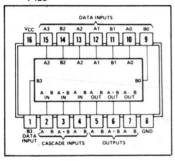

FUNCTION TABLE

SELECTION			ARITHMETIC/LOGIC
S2	S1	S0	OPERATION
L	L	L	CLEAR
L	L	H	B MINUS A
L	H	L	A MINUS B
L	H	H	A PLUS B
H	L	L	A $\oplus$ B
H	L	H	A + B
H	H	L	AB
H	H	H	PRESET

Appendix 6.2   Table for powers of 2

2^n	n	2^{-n}
1	0	1.0
2	1	0.5
4	2	0.25
8	3	0.125
16	4	0.062 5
32	5	0.031 25
64	6	0.015 625
128	7	0.007 812 5
256	8	0.003 906 25
512	9	0.001 953 125
1 024	10	0.000 976 562 5
2 048	11	0.000 488 281 25
4 096	12	0.000 244 140 625
8 192	13	0.000 122 070 312 5
16 384	14	0.000 061 035 156 25
32 768	15	0.000 030 517 578 125
65 536	16	0.000 015 258 789 062 5
131 072	17	0.000 007 629 394 531 25
262 144	18	0.000 003 814 697 265 625
524 288	19	0.000 001 907 348 632 812 5
1 048 576	20	0.000 000 953 674 316 406 25
2 097 152	21	0.000 000 476 837 158 203 125
4 194 304	22	0.000 000 238 418 579 101 562 5
8 388 608	23	0.000 000 119 209 289 550 781 25
16 777 216	24	0.000 000 059 604 644 775 390 625
33 554 432	25	0.000 000 029 802 322 387 695 312 5
67 108 864	26	0.000 000 014 901 161 193 847 656 25
134 217 728	27	0.000 000 007 450 580 596 923 828 125
268 435 456	28	0.000 000 003 725 290 298 461 914 062 5
536 870 912	29	0.000 000 001 862 646 149 230 957 031 25
1 073 741 824	30	0.000 000 000 931 322 574 615 478 515 625
2 147 483 648	31	0.000 000 000 465 661 287 307 739 257 812 5
4 294 967 296	32	0.000 000 000 232 830 643 653 869 628 906 25

PART TWO

Optimization of Logic Systems

7

Combinational Logic Techniques

Up to now only small logic functions have been considered. For larger logic functions, one may need to use more than one module, and more than one level. In such a situation, it becomes necessary to look at some optimization techniques that are relevant to design at the sub-system level. Four techniques will be described in this chapter.

7.1   COLUMN PARTITIONING

7.1.1   Partitioning Techniques ⟨66⟩

In extending the number of words by cascading memory modules, the number of modules doubles for every additional input variable.

In many situations this is not necessary, and some modules may be saved by a technique we shall call *column partitioning*.

Consider the switching function given in Example (7.1).

Example (7.1)
The logic function F given below is used to illustrate the partitioning technique:

$$F = \Sigma \ (41, \ 49, \ 56, \ 63, \ 65, \ 94, \ 97, \ 105, \ 113, \ 120, \ 121, \ 124, \ 125, \ 126, \ 127)$$

The truth table for this function is given in Table 7.1.

To illustrate the point of multi-module design, we assume that only small modules are available, say 16×4 and 32×4 ROMs and three-variable ULMs. From Equation (3.13) the number Mo of ROM modules required for the function in this example is given by

$$Mo = 2^{v-l}$$
$$= 2^{7-5}$$
$$= 4 \qquad \text{(four } 32 \times 4 \text{ ROM modules)}$$
or $\quad Mo = 2^{7-4}$
$$= 8 \text{ (eight } 16 \times 4 \text{ ROM modules)}$$

If, on the other hand, ULMs are used, then, from Equation (3.6), the number of levels l is:

$$l = \frac{7-1}{2} = 3$$

From equation (3.8) the maximum number of modules is:

$$Mo = \frac{1-4^3}{1-4} = 21 \text{ (21 three-variable ULMs)}$$

Inspecting Table 7.1, we observe that it is possible to partition and recode some of the variables by making use of the common variables, or minterms, to achieve a reduction in the number of modules required to implement the same expression.

Tables 7.2 and 7.3 show how b, c, d, e and f are partitioned and recoded and finally implemented using a cascade of two (32×4) ROM modules, as shown in Fig. 7.1. Note that in the second level, a 32×1 module could have been used for F if available.

a	b	c	d	e	f	g
0	1	0	1	0	0	1
0	1	1	0	0	0	1
0	1	1	1	0	0	0
0	1	1	1	1	1	1
1	0	0	0	0	0	1
1	0	1	1	1	1	0
1	1	0	0	0	0	1
1	1	0	1	0	0	1
1	1	1	0	0	0	1
1	1	1	1	0	0	0
1	1	1	1	0	0	1
1	1	1	1	1	0	0
1	1	1	1	1	0	1
1	1	1	1	1	1	0
1	1	1	1	1	1	1

Table 7.1 Elementary products for $F_i = 1$ Example (7.1)

b	c	d	e	f	Z_1	Z_2	Z_3
0	0	0	0	0	0	0	0
1	0	1	0	0	0	0	1
1	1	0	0	0	0	1	0
1	1	1	0	0	0	1	1
1	0	0	0	0	1	0	0
0	1	1	1	1	1	0	1
1	1	1	1	0	1	1	0
1	1	1	1	1	1	1	1

Table 7.2 Recoding of partitioned variables in Example (7.1)

a	Z_1	Z_2	Z_3	g
1	0	0	0	1
0	0	0	1	1
0	0	1	0	1
0	0	1	1	0
1	1	0	0	1
1	0	0	1	1
1	0	1	0	1
1	0	1	1	0
1	1	0	1	0
1	0	1	1	1
1	1	1	0	0
0	1	1	1	1
1	1	1	0	1
1	1	1	1	0
1	1	1	1	1

Table 7.3 Example (7.1), after partitioning and recoding

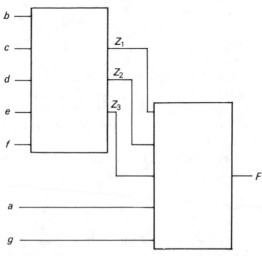

Fig. 7.1 Final implementation of Example (7.1) using two ROMs

Let v_1 be the number of paritioned variables applied to the first stage (coder).

v_2 be the reduction in the number of inputs to the second stage due to partitioning.

The total number *Mop* of ROM modules required by the partitioned function is given by:

$$Mop = 2^{(v_1 - I)} + 2^{(v - v_2 - I)} \tag{7.1}$$

For a given module, I the number of inputs is fixed and v is also fixed, since it represents the number of variables of the original function.

Substituting numerical values in Equations (3.13) and (7.1) shows that a reduction in the number of modules is possible even for values of v_2 as low as 1. If a choice is possible, a high value of v_2 and a low value of v_1 give a better saving.

In Example (7.1), the number of modules *Mop* required for the partitioned function is given by:

$$Mop = 2^{5-5} + 2^{7-2-5}$$
$$= 2 \quad \text{(two 32} \times \text{4 ROM modules)}$$

or $Mop = 2^{5-4} + 2^{7-2-4}$
$$= 4 \quad \text{(four 16} \times \text{4 ROM modules)}$$

The partitioning has therefore resulted in a 50 per cent reduction in the number of modules, even though storage locations used are 120 bits both before and after partitioning.

For ULMs, this technique is not very useful if the function requires more than one output, since ULM modules are one-output devices. In fact in this case we have four expressions F, Z_1, Z_2, Z_3. Each is a function of five variables, and together they need a maximum of 20 modules, as compared with the original requirement of 21 modules.

7.1.2 A Digital Computer Program

It is obvious that the partitioning technique is very attractive for memory devices. Unfortunately there is no systematic algorithm to find the best way to partition the function, if any. This makes it necessary to search through all the possibilities, which is a very laborious job, especially for large functions. A digital computer program was therefore developed to search through all the combinations, as specified by the designer, and to print out both the function and the most suitable columns to partition and recode.

The number of distinguishable partitions on a set of N distinct objects, splitting them into two classes of $N_x \leqslant N$, and $N - N_x$ objects, respectively, is given by:

$$\binom{N}{N_x} = \frac{N!}{(N - N_x)! \ N_x!} \tag{7.2}$$

The computation is repeated for all combinations specified by the user, normally from $N_x = 2$ to $N_x = N - 1$.

In Example (7.1), $N = 7$. When $N_x = 3$ say, we have to test the following:

abc,   abd,   abe,   abf,   abg,   acd,   ace,   acf,   acg,   ade,
adf,   adg,   aef,   aeg,   afg,   bcd,   bce,   bcf,   bcg,   bde,
bdf,   edg,   bef,   beg,   bfg,   cde,   cdf,   cdg,   cef,   ceg,
cfg,   def,   deg,   dfg,   and   efg. Similarly for $N_x = 2, 4$, etc.

The program is very general and economical in terms of time and storage. It can test any number of functions, and each can have any number of variables.

Using the ICL 1900 computer all combinations of four different functions were tested in only 35 s, and using less than 20K of core size. The program is written in 1900 Fortran and is given in Appendix 7.1.

How to Use the Program:

To use the program, one needs to specify the following:

(i)   Number of functions to be tested (IW)
(ii)   For each function, the dimensions of the function, i.e. the number of rows and number of columns of the table of elementary products (N_1, N_2), are given followed by the upper and lower limits for N_x as given in Equation (7.2) (IG, IH). Then the matrix of elementary products, i.e. the binary numbers associated with the configuration i for which $F_i = 1$ must be given. For example, suppose that two functions are tested and that say, the first one is Example (7.1) given in Table 7.1. Then the layout is as follows:

Document data
2   IW (number of functions)
15 7 2 6 N_1 N_2 IG IH (for first function)
0 1 0 1 0 0 1
0 1 1 0 0 0 1   Minterms of first function
.
.
.

12 6 2 4 N_1 N_2 IG IH (for second function)
0 0 0 1 0 1
0 0 1 1 1 1   Minterms of second function
.
.
.

The computer will print each function, followed by the number of columns that may be partitioned. The number of unique occurrences is given, thus providing an indication of the number of variables required for recoding.

In Example (7.1) the output of the computer indicated that b, c, d, e, and f occur in only eight different combinations out of a total of 32 possible combinations. These are therefore coded with $\log_2 (8) = 3$ variables Z_1, Z_2 and Z_3.

7.2   FUNCTIONAL DECOMPOSITION

The procedure outlined in Section 7.1, though very useful, cannot be successfully applied to every logic function. The function which is given in Example (7.2) below and whose truth table appears in Table 7.4 has no useful partitions, and cannot be simplified using this method.

The decomposition technique developed by Ashenhurst and Curtis proved very useful in saving discrete components. Here its applicability to the present-day devices is considered.

Consider a logic function $f(x_1, x_2, \ldots, x_n)$ of n variables, where $n \geq 3$. If this function can be expressed as a composition of a number of functions, each depending on fewer than n-variables, the process is refered to as *functional decomposition*.

If there are two functions F and ϕ such that

$$f(x_1, x_2, \ldots, x_n) = F\left[\phi\left(y_1, y_2, \ldots, y_t\right), Z_1, Z_2, \ldots, Z_r\right] \tag{7.3}$$

where $1 \leq t \leq n - 1$, and both $(y_1, y_2, \ldots, y_t)$ and $(Z_1, Z_2, \ldots, Z_r)$ are subsets of $(x_1, x_2, \ldots, x_n)$, such that their union is equal to the original set of $(x_1, x_2, \ldots, x_n)$. If the subsets $(y_1, \ldots, y_t)$ and $(Z_1, \ldots, Z_r)$ are *disjoint*, that is $r = n - t$, then the decomposition is refered to as disjoint. The expressions given in Equation (7.3), F of $r + 1$, and ϕ of t variables represent a simple disjunctive decomposition of the logic function f.

If the two disjoint subsets $y_1, \ldots, y_t$ and $Z_1, \ldots, Z_r$ are named the bound and free variables respectively, the K-map with the free variables running down one side, and the bound variables across the top, is called the *partition matrix*. Further, the number of distinct row vectors is called its *row multiplicity*.

It has been shown that a switching function $f(x_1, \ldots, x_n)$ possesses a simple disjunctive decomposition given by:

$F(\phi, Z_1, Z_2, \ldots, Z_{n-t})$ and
$\phi(y_1, y_2, \ldots, y_t)$ if and only if its $Z_1, Z_2, \ldots, Z_{n-t} \mid y_1, y_2, \ldots, y_t$

Partition matrix has column multiplicity less than or equal to two.

To any n-variable function, there are 2^n partition matrices. Since the transpose of a partition matrix is also a partition matrix, any partition matrix when looked at sideways corresponds to a partition matrix. This means that only 2^{n-1} matrices need to be tested. This is done by using the decomposition chart which consists of 2^{n-1} sub-charts, each of which has 2^n entries, corresponding to the integers 0 to $2^n - 1$. The integers in each entry represent the decimal value of corresponding minterms.

When a logic function is to be tested, all the entries corresponding to minterms for which the function is unity are circled on all the sub-charts. The sub-charts are viewed normally and sideways to check for column multiplicity that is equal to or less than 2. The case when the multiplicity equals 1 means that the function is independent of ϕ.

In a book by Curtis ⟨20⟩, other forms of decomposition are given, and may be used to simplify complicated functions. Examples of a more complex decomposition are:

$$f(x_1, x_2, \ldots, x_n) = F\left[\phi(y_1, \ldots, y_t)\, \psi\,(u_1, \ldots, u_s), Z_1, \ldots, Z_{n-s-t}\right]$$

or may be

$$f(x_1, x_2, ..., x_n) = F[\phi\{y_1, ..., y_t, \psi(u_1, ..., u_s)\}, Z_1, ..., Z_{n-s-t}]$$

Example (7.2)

The logic function f is used to illustrate the decomposition technique. A logic function $f(u, v, w, x, y, z,)$ is expressed by the numerical sum of products form as below. [This example is taken from Curtis ⟨20⟩ where it is implemented using discrete components.]

$$f = \Sigma(0, 1, 6, 7, 11, 13, 16, 17, 22, 23, 27, 29, 36, 37, 38, 39, 41, 43, 50, 51, 57, 61, 63)$$

These minterms are given in their binary form in Table 7.4, and show no useful partition of the form described in the last section. The function is entered on the decomposition charts shown in Appendix 7.2.

u	v	w	x	y	z
0	0	0	0	0	0
0	0	0	0	0	1
0	0	0	1	1	0
0	0	0	1	1	1
0	0	1	0	1	1
0	0	1	1	0	1
0	1	0	0	0	0
0	1	0	0	0	1
0	1	0	1	1	0
0	1	0	1	1	1
0	1	1	0	1	1
0	1	1	1	0	1
1	0	0	1	0	0
1	0	0	1	0	1
1	0	0	1	1	0
1	0	0	1	1	1
1	0	1	0	0	1
1	0	1	0	1	1
1	1	0	0	1	0
1	1	0	0	1	1
1	1	1	0	0	1
1	1	1	1	0	1
1	1	1	1	1	1

Table 7.4   Elementary products for $F_i = 1$ [Example (7.2)]

The first matrix in Appendix 7.2(*a*) corresponds to the trivial partition where $t = 0$, and is included only to record the function. The rest of the matrices in Appendix 7.2(*a*) are only viewed normally, since the sideway view corresponds to a trivial decomposition for which $t = 1$.

The $wz \mid uvxy$ sub-chart in Appendix 7.2(*d*) shows a column multiplicity of two when viewed sideways and indicates a simple disjunction decomposition resulting in the following:

$$\phi(u, v, x, y) = \bar{u}\,\bar{v}\bar{x}\bar{y} + \bar{u}vxy + \bar{u}v\bar{x}\bar{y} + \bar{u}vxy + u\bar{v}x\bar{y} + u\bar{v}xy + uv\bar{x}y \qquad (7.4)$$

$$F(\phi, wz) = \bar{\phi}wz + \phi\bar{w}\bar{z} + \phi\bar{w}z \qquad (7.5)$$

These two expressions are shown in Tables 7.5 and 7.6 respectively.

u	v	x	y
0	0	0	0
0	0	1	1
0	1	0	0
0	1	1	1
1	0	1	0
1	0	1	1
1	1	0	1

ϕ	w	z
0	1	1
1	0	0
1	0	1

Table 7.5 Elementary products for $\phi_i = 1$ [Example (7.2)]

Table 7.6 Elementary products for $F_i = 1$ [Example (7.2)]

(i) ROM IMPLEMENTATION: In terms of actual storage required, the original function would use $1.2^6 = 64$ bits.

After decomposition, the storage is
$1.2^4 + 1.2^3 = 24$ bits. (A saving of over 60 per cent).

If we choose to think in terms of modules, which is justified by the fact that ROMs are made in standard sizes, and that bits of storage do not always give a realistic picture, and if we assume that 16×1 ROM modules are used, the original functions would need four (16×1) ROM modules as compared with two (16×1) ROM modules required by the decomposed function, $-$ a 50 per cent reduction in modules (Fig. 7.2).

It is obvious that a good saving is possible, both in bits of storage and in modules, due to decomposition.

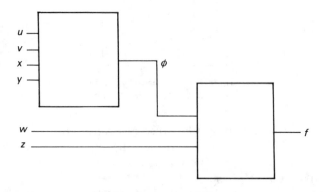

Fig. 7.2 Implementation of Example (7.2) after decomposition, using two ROM modules

(ii) ULM IMPLEMENTATION Assuming three-variable modules are used, the original function would need l levels, where l is given by Equation (3.6) as

$$l = \left\lceil \frac{6-1}{2} \right\rceil = 3 \text{ levels}$$

The maximum number of modules Mo that might be required is given by Equation (3.9):

$$Mo = 1 + 2^1 \left(\frac{1-4^2}{1-4} \right)$$

$$= 11 \text{ modules}$$

Whether 11 modules are actually used depends on the function, and on the control variables. Table 7.7 gives some of the many possible permutations of the control variables.

Control variables			Front input variable	Number of modules
1st level	*2nd level*	*3rd level*		
v	wx	uz	y	11
u	yz	vw	x	11
v	wy	uz	x	11
v	yz	uw	x	11
u	vw	yz	x	9
$\vdots$	$\vdots$	$\vdots$	$\vdots$	$\vdots$
u	vw	xy	z	8
u	vx	wy	z	6

Table 7.7   Showing the dependence of the number of modules on the choice of control variables for Example (7.2)

A technique based on pattern recognition, and a general computer program will be introduced in Section 7.4. For the time being we depend on trying a few permutations and we are prepared to use up to the maximum of 11 modules, as shown in Fig. 7.3, which is based on the following expansion:

$$f = \bar{u} \begin{bmatrix} \bar{y}\bar{z}\,[\,\bar{v}\bar{w}(\bar{x}) + \bar{v}w(0) + v\bar{w}(\bar{x}) + vw(0)] + \\ \bar{y}z\,[\,\bar{v}\bar{w}(\bar{x}) + \bar{v}w(x) + v\bar{w}(\bar{x}) + vw(x)] + \\ y\bar{z}\,[\,\bar{v}\bar{w}(x) + \bar{v}w(0) + v\bar{w}(x) + vw(0)] + \\ yz\,[\,\bar{v}\bar{w}(x) + \bar{v}w(\bar{x}) + v\bar{w}(x) + vw(\bar{x})] \end{bmatrix}$$

$$+ u \begin{bmatrix} \bar{y}\bar{z}\,[\,\bar{v}\bar{w}(x) + \bar{v}w(0) + v\bar{w}(0) + vw(0)] + \\ \bar{y}z\,[\,\bar{v}\bar{w}(x) + \bar{v}w(\bar{x}) + v\bar{w}(0) + vw(1)] + \\ y\bar{z}\,[\,\bar{v}\bar{w}(x) + \bar{v}w(0) + v\bar{w}(\bar{x}) + vw(0)] + \\ yz\,[\,\bar{v}\bar{w}(x) + \bar{v}w(\bar{x}) + v\bar{w}(\bar{x}) + vw(x)] \end{bmatrix}$$

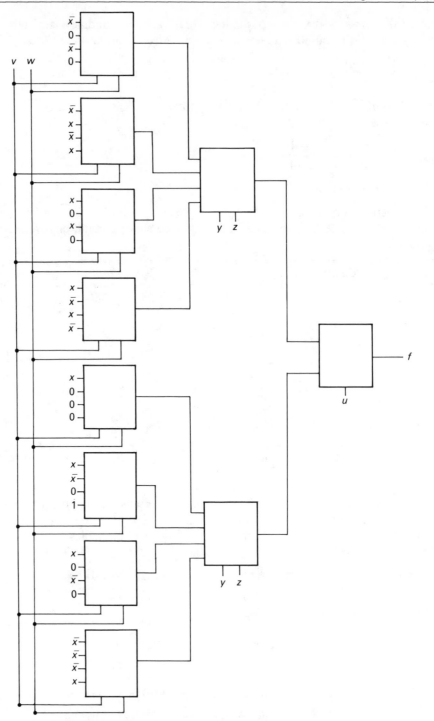

Fig. 7.3 Implementation of Example (7.2) before decomposition, using a maximum of 11 three-variable ULM modules

After decomposition we have two functions to implement as given in Equations (7.4) and (7.5), i.e.

$\phi(u, v, x, y)$ and $F(\phi, w, z)$

The first requires two levels using a maximum of three modules.
The second requires one three-variable module.
The maximum total is four three-variable modules.

Thus, after decomposition, the maximum number of modules has been reduced from 11 to 4, a saving of over 60 per cent. Figs 7.3 and 7.4 show the two circuits using the maximum number of modules before and after decomposition respectively.

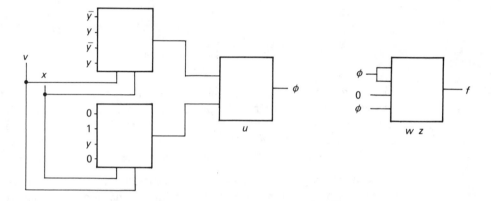

Fig. 7.4   Implementation of Example (7.2) after decomposition, using a maximum of 4 three-variable ULM modules

7.3   GROUP CODING ⟨3, 19⟩

7.3.1 Application to Memory Devices — Bit Packing

The bit packing technique can be useful in some cases to reduce the word length of memory devices like ROMs. As the name suggests, the technique helps in condensing the ROM word into a smaller one. The main points to note are summarized as follows:

(i)   Output lines that are the same for every address are equivalent and can be combined in one line.

(ii) Two output lines producing different outputs for every address are complementary, and can be combined by eliminating one line and complementing the other.

(iii) Output lines that are never '1' at the same time, are called exclusive; they can be replaced by fewer lines by coding the outputs. Exclusive groups are detected with the aid of the conflict matrix which lists all the outputs that conflict in pairs. From the matrix, a conflict diagram is constructed. Groups of non-conflicting

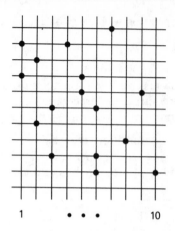

1   • • •   10

Fig. 7.5   Original ROM

outputs are grouped and coded, which implies that decoders have to be used. The procedure is best illustrated by packing the loosely packed information in the matrix given in Fig. 7.5 into the smaller matrix in Fig. 7.6(d). The technique is useful in cases where the ROM is very inefficiently used as in the example shown in Fig. 7.5 in which the word length is reduced from 10 to 5 bits.

Fig. 7.5 shows only the memory matrix. The ten functions are compared with each other. Function 1 (column 1) is compared with all other nine functions. It conflicts with 4 and 5 because they have common terms indicated by • on the same row. This is shown also in Fig. 7.6(a) by placing 1 in the appropriate boxes. Note that the matrix is symmetrical about the diagonal. Hence only half the conflict matrix needs to be drawn.

The same information is shown in Fig. 7.6(b) by joining functions that are conflicting (not exclusive). For example 1 is connected to 4 and 5. This means that 1 cannot be in the same group as 4 or 5. There is no conflict however between 4 and 5.

By inspecting the conflict diagram we recognise two groups of functions that are non-conflicting. These are 4, 5, 6 and 1, 2, 3, 7, 8, 9, 10. The first group has three members and can be coded using two binary digits. The larger group has seven members and requires three binary digits, as shown in Fig. 7.6(c).

The binary code can be stored in a smaller memory matrix and additional logic is used for decoding. Two coders are shown in Fig. 7.6(d) in order to obtain the original ten functions.

7.3.2 Application to ULM Devices

In this section we explain how the technique outlined in Section 7.3.1 can be extended to cover situations where universal logic modules are used.

Consider Example (7.3), where there are nine functions $f_1, \ldots, f_9$, each being a function of four variables $x_1, \ldots, x_4$.

	1	2	3	4	5	6	7	8	9	10
1	X			1	1					
2		X								
3			X			1				
4	1			X						
5	1				X			1		
6			1			X				1
7							X			
8								X		
9					1				X	
10						1				X

Fig. 7.6(a)   Conflict matrix

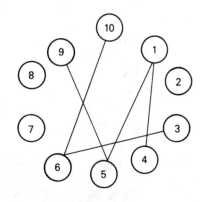

Fig. 7.6(b)   Conflict diagram gives 1,2,3,7,8,9,10 and 4,5,6 and two non-conflicting groups

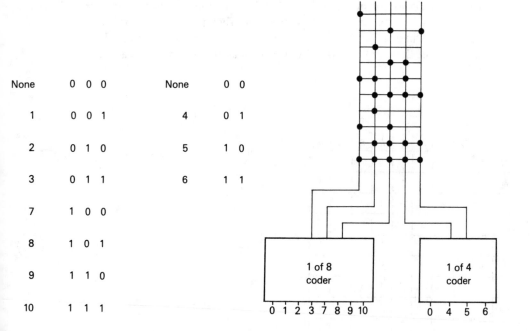

None	0 0 0
1	0 0 1
2	0 1 0
3	0 1 1
7	1 0 0
8	1 0 1
9	1 1 0
10	1 1 1

None	0 0
4	0 1
5	1 0
6	1 1

1 of 8 coder

0 1 2 3 7 8 9 10

1 of 4 coder

0 4 5 6

Fig. 7.6(c)   Groups coding   Fig. 7.6(d)   The packed version of the ROM in Fig. 7.5

Example (7.3)

$f_1 = \Sigma\ (2, 5, 9, 11)$
$f_2 = \Sigma\ (3, 5, 7, 8)$
$f_3 = \Sigma\ (6, 8)$
$f_4 = \Sigma\ (4, 13)$
$f_5 = \Sigma\ (1, 7, 15)$
$f_6 = \Sigma\ (1, 4, 9, 14)$
$f_7 = \Sigma\ (12, 14)$
$f_8 = \Sigma\ (0, 2, 11, 12)$
$f_9 = \Sigma\ (0, 3, 10)$

This example is used to illustrate the use of group coding to ULM circuits.

Each of the above functions can be implemented independently. The function f_1, for example, can be expanded as shown below, and implemented using three 3-variable ULMs.

$$f_1 = \bar{x}_1\ [\bar{x}_2\bar{x}_3(0) + \bar{x}_2x_3(\bar{x}_4) + x_2\bar{x}_3(x_4) + x_2x_3(0)] +$$
$$x_1\ [\bar{x}_2\bar{x}_3(x_4) + \bar{x}_2x_3(x_4) + x_2\bar{x}_3(0) + x_2x_3(0)]$$

In the following, we call any two functions having a common minterm *conflicting* or *non-exclusive* functions. By comparing every one of the logic functions in our example with every other function, we can draw a conflict matrix as shown in Fig. 7.7(a).

As an example f_1 conflicts with f_2 (having the common minterm 5), with f_6 (having the common minterm 9), and with f_8 (having the common minterm 11). A 1 is placed in those positions. Similarly in the conflict diagram f_1 is connected to f_2, f_6 and f_8 as shown in Fig. 7.7(b). We note that a function compared to itself is indicated by *x*s along the diagonal matrix. The conflict matrix is symmetrical about this diagonal.

	f_1	f_2	f_3	f_4	f_5	f_6	f_7	f_8	f_9
f_1	X	1				1		1	
f_2	1	X	1		1				1
f_3		1	X						
f_4				X		1			
f_5		1			X	1			
f_6	1			1	1	X	1		
f_7						1	X	1	
f_8	1						1	X	1
f_9		1						1	X

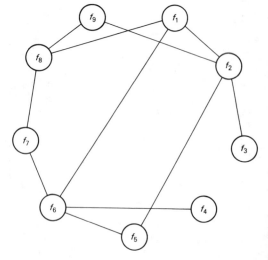

Fig. 7.7(a)   Conflict matrix for Example (7.3)   **Fig. 7.7(b)   Conflict diagram for Example (7.3)**

As in Section 7.3.1, we can now formulate suitable non-conflicting groups, and re-code them as in Fig. 7.7(c). The two groups in this case are f_2, f_6, f_8 and f_1, f_3, f_4, f_5, f_7, f_9.

The logic shown in Fig. 7.7(d) can be implemented using suitable devices. In the following, ULM multiplexers are suggested. Demultiplexers are used for the coders to obtain the original nine functions.

From the original expressions and the code used in Fig. 7.7(c) we derive the logical expression for the five variables $\psi_1, \ldots, \psi_5$, and expand them to a form suitable for direct ULM implementation.

	ψ_1	ψ_2	ψ_3
None	0	0	0
f_1	0	0	1
f_3	0	1	0
f_4	0	1	1
f_5	1	0	0
f_7	1	0	1
f_9	1	1	0

	ψ_4	ψ_5
None	0	0
f_2	0	1
f_6	1	0
f_8	1	1

Fig. 7.7(c)   Coding of the two exclusive groups in Example (7.3)

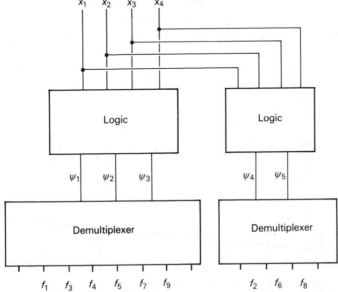

Fig. 7.7(d)   A block diagram implementation of Example (7.3)

$$\psi_1 = \bar{x}_1 \left[\bar{x}_2\bar{x}_3(x_4) + \bar{x}_2 x_3(x_4) + x_2\bar{x}_3(0) + x_2 x_3(x_4) \right] + x_1 \left[\bar{x}_2\bar{x}_3(0) + \bar{x}_2 x_3(\bar{x}_4) + x_2\bar{x}_3(\bar{x}_4) + x_2 x_3(1) \right]$$

$$\psi_2 = \bar{x}_1 \left[\bar{x}_2\bar{x}_3(\bar{x}_4) + \bar{x}_2 x_3(x_4) + x_2\bar{x}_3(\bar{x}_4) + x_2 x_3(\bar{x}_4) \right] + x_1 \left[\bar{x}_2\bar{x}_3(\bar{x}_4) + \bar{x}_2 x_3(\bar{x}_4) + x_2\bar{x}_3(x_4) + x_2 x_3(0) \right]$$

$$\psi_3 = \bar{x}_1 \left[\bar{x}_2\bar{x}_3(0) + \bar{x}_2 x_3(\bar{x}_4) + x_2\bar{x}_3(1) + x_2 x_3(0) \right] + x_1 \left[\bar{x}_2\bar{x}_3(x_4) + \bar{x}_2 x_3(x_4) + x_2\bar{x}_3(1) + x_2 x_3(\bar{x}_4) \right]$$

$$\psi_4 = \bar{x}_1 \left[\bar{x}_2\bar{x}_3(1) + \bar{x}_2 x_3(\bar{x}_4) + x_2\bar{x}_3(\bar{x}_4) + x_2 x_3(0) \right] + x_1 \left[\bar{x}_2\bar{x}_3(x_4) + \bar{x}_2 x_3(x_4) + x_2\bar{x}_3(\bar{x}_4) + x_2 x_3(\bar{x}_4) \right]$$

$$\psi_5 = \bar{x}_1 \left[\bar{x}_2\bar{x}_3(\bar{x}_4) + \bar{x}_2 x_3(1) + x_2\bar{x}_3(x_4) + x_2 x_3(x_4) \right] + x_1 \left[\bar{x}_2\bar{x}_3(\bar{x}_4) + \bar{x}_2 x_3(x_4) + x_2\bar{x}_3(\bar{x}_4) + x_2 x_3(0) \right]$$

Like the original nine functions, all the above five expressions are functions of four variables. Each can be implemented using a maximum of 3 three-variable ULMs (or 1 four-variable ULM).

The total maximum number of modules required is 15 three-variable ULMs (or 5 four-variable ULMs), plus two demultiplexers. This should be compared with a maximum of 27 three-variable ULMs, or 9 four-variable ULMs, required by the original functions.

7.3.3 Procedure for Computing the Maximum Exclusive Groups (MEGs)

A quick look at the conflict diagram in Fig. 7.8 can easily show how difficult it would be to use this method for large problems. Further, most problems have many possible groups and there does not appear to be any systematic procedure for computing these groups. It seems inevitable, therefore, to look for a systematic design technique that can be solved by a digital computer. This procedure will be illustrated with the aid of the following design example which consists of 13 logic functions.

Example (7.4)

$F_1 = \Sigma\ (0, 3, 7, 10, 17, 31)$ $F_8 = \Sigma\ (11, 19, 22, 24, 26, 30)$

$F_2 = \Sigma\ (0, 2, 10, 15)$ $F_9 = \Sigma\ (9, 12, 19, 28)$

$F_3 = \Sigma\ (1, 4, 6, 11)$ $F_{10} = \Sigma\ (3, 6, 7, 21, 23)$

$F_4 = \Sigma\ (1, 14, 17, 20, 25)$ $F_{11} = \Sigma\ (4, 13, 29)$

$F_5 = \Sigma\ (5, 16, 18, 30)$ $F_{12} = \Sigma\ (5, 9, 18, 31)$

$F_6 = \Sigma\ (2, 8, 13, 20)$ $F_{13} = \Sigma\ (14, 21, 27, 29)$

$F_7 = \Sigma\ (8, 12, 27, 28)$

The conflict diagram for this problem is given in Fig. 7.9.

DEFINITION 1   An exclusive or non-conflicting group is a group of logic functions that have no common minterms. Every function is, by itself, exclusive.

DEFINITION 2   A maximum exclusive group (MEG) is an exclusive group that is not a subset of any larger exclusive group.

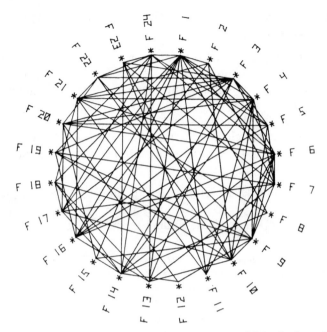

Fig. 7.8   A computer plot of the conflict diagram for 24 logic function having 96 minterms

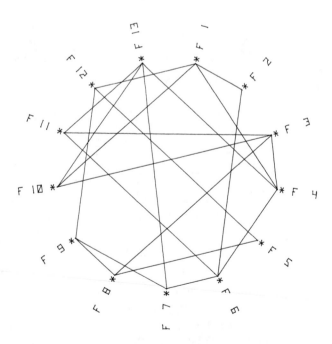

Fig. 7.9   A computer plot of the conflict diagram for Example (7.4)

DEFINITION 3 A set of functions constitute an exclusive group, if every pair of functions in the set is exclusive.

If we can formulate suitable MEGs of functions, we can code them, implement the coding functions using suitable logic like ROMs or ULMs, and then use decoders to obtain the original logic functions. This may be achieved as follows:

A table is drawn as in Fig. 7.10 and a mark X is placed in every position where the functions conflict. Blank entries signify exclusive functions.

Let S_i be the set containing all the functions whose entries in column i, of the conflict table, have blank entries. The procedure for computing the MEGs can be summarized as follows:

(*a*) An exclusive list (E-list) is initiated with one group containing the first function. This is possible according to Definition 1 above.

(*b*) Starting from left to right, column by column, the set S_i is formed and tested.

(*c*) If S_i is an empty set, a new group consisting of one function i is added to the E-list before moving to the next column. (Since function i conflicts with functions F_1 to F_{i-1}, it is placed in a one member group.)

	2	3	4	5	6	7	8	9	10	11	12	13
12												
11												X
10												X
9											X	
8								X				
7								X				X
6						X				X		
5							X				X	
4					X							X
3			X				X		X	X		
2					X							
1	X		X						X		X	

Fig. 7.10 Table of conflicting functions for Example (7.4)

(*d*) If S_i is not an empty set, it is intersected with every member of the current E-list, and, depending on the intersection, the next step is one of the following:

(i) If there is no intersection, the groups are unchanged.
(ii) If the intersection is equal to the group, function i is added to the group, since every member of the group can make an exclusive group of two with function i.
(iii) If the intersection has more than one member, the set consisting of the union of i and the intersection makes a new group. These new groups are arranged accord-

ing to their size and are added to the E-list only if they are not included in any larger group.

(e) Pairs consisting of i and members of S_i that did not appear in any previous intersection are added to the list. (Function i can still make exclusive groups of two with each of the remaining members of S_i.)

The above procedure generates, exhaustively and systematically, all the MEGs. Redundancy and repetition are eliminated in the process.

MEGs are equivalent to the maximal compatible states in sequential switching functions ⟨111⟩

Applying this procedure to Example (7.4), we obtain the following:

S_1 = $E = 1$
S_2 = $E = 1, 2$
S_3 = 1 2 $E = 1\ 3, 2\ 3$
S_4 = 2 $E = 1\ 3, 2\ 3, 2\ 4$
S_5 = 1 2 3 4 $E = 1\ 3\ 5, 2\ 3\ 5, 2\ 4\ 5$
S_6 = 1 3 5 $E = 1\ 3\ 5\ 6, 2\ 3\ 5, 2\ 4\ 5$
S_7 = 1 2 3 4 5 $E = 1\ 3\ 5\ 6, 2\ 3\ 5\ 7, 2\ 4\ 5\ 7, 1\ 3\ 5\ 7$
S_8 = 1 2 4 6 7 $E = 1\ 3\ 5\ 6, 2\ 3\ 5\ 7, 2\ 4\ 5\ 7,$
                                              $1\ 3\ 5\ 7, 2\ 4\ 7\ 8, 1\ 6\ 8, 1\ 7\ 8$
S_9 = 1 2 3 4 5 6 $E = 1\ 3\ 5\ 6\ 9, 2\ 3\ 5\ 7, 2\ 4\ 5\ 7,$
                                              $1\ 3\ 5\ 7, 2\ 4\ 7\ 8, 1\ 6\ 8, 1\ 7\ 8,$
                                              $2\ 3\ 5\ 9, 2\ 4\ 5\ 9$
S_{10} = 2 4 5 6 7 8 9 $E = 1\ 3\ 5\ 6\ 9, 2\ 3\ 5\ 7, 2\ 4\ 5\ 7\ 10,$
                                              $1\ 3\ 5\ 7, 2\ 4\ 7\ 8\ 10, 1\ 6\ 8, 1\ 7\ 8,$
                                              $2\ 3\ 5\ 9, 2\ 4\ 5\ 9\ 10, 5\ 6\ 9\ 10,$
                                              $6\ 8\ 10$
S_{11} = 1 2 4 5 7 8 9 10 $E = 1\ 3\ 5\ 6\ 9, 2\ 3\ 5\ 7, 2\ 4\ 5\ 7\ 10\ 11,$
                                              $1\ 3\ 5\ 7, 2\ 4\ 7\ 8\ 10\ 11, 1\ 6\ 8,$
                                              $1\ 7\ 8\ 11, 2\ 3\ 5\ 9, 2\ 4\ 5\ 9\ 10\ 11,$
                                              $5\ 6\ 9\ 10, 6\ 8\ 10, 1\ 5\ 9\ 11,$
                                              $1\ 5\ 7\ 11$
S_{12} = 2 3 4 6 7 8 10 11 $E = 1\ 3\ 5\ 6\ 9, 2\ 3\ 5\ 7, 2\ 4\ 5\ 7\ 10\ 11,$
                                              $1\ 3\ 5\ 7, 2\ 4\ 7\ 8\ 10\ 11\ 12, 1\ 6\ 8,$
                                              $1\ 7\ 8\ 11, 2\ 3\ 5\ 9, 2\ 4\ 5\ 9\ 10\ 11,$
                                              $5\ 6\ 9\ 10, 6\ 8\ 10\ 12, 1\ 5\ 9\ 11,$
                                              $1\ 5\ 7\ 11, 2\ 3\ 7\ 12, 3\ 6\ 12$
S_{13} = 1 2 3 5 6 8 9 12 $E = 1\ 3\ 5\ 6\ 9\ 13, 2\ 3\ 5\ 7,$
                                              $2\ 4\ 5\ 7\ 10\ 11, 1\ 3\ 5\ 7,$
                                              $2\ 4\ 7\ 8\ 10\ 11\ 12, 1\ 6\ 8\ 13,$
                                              $1\ 7\ 8\ 11, 2\ 3\ 5\ 9\ 13,$
                                              $2\ 4\ 5\ 9\ 10\ 11, 5\ 6\ 9\ 10,$
                                              $6\ 8\ 10\ 12, 1\ 5\ 9\ 11, 1\ 5\ 7\ 11,$
                                              $2\ 3\ 7\ 12, 3\ 6\ 12\ 13, 2\ 8\ 12\ 13,$
                                              $6\ 8\ 12\ 13, 2\ 3\ 12\ 13$

The last E-list contains all the MEGs.

Returning to Example (7.3), we can redraw the conflict matrix as shown in Fig. 7.11. In fact this table contains the same information as that of Fig. 7.7(*a*) in a compact form, with X used instead of 1.

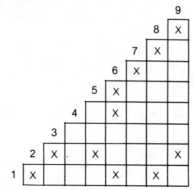

Fig. 7.11   Conflict table for Example (7.3)

By using the procedure outline earlier, the following list of MEGs is obtained:

247, 248, 268, 369, 3458, 134579, 368

The two groups found earlier from the conflict diagram, namely 134579 and 268, do cover all the functions with no repetition, but it leaves us in doubt whether we made the best choice. This problem will be tackled next.

Using Boolean Algebra for Calculating the MEGs ⟨74⟩

This is another method that may be used for calculating the MEGs (or in fact maximal compatibles in sequential circuits).

Starting from the conflict table in Fig. 7.11 above, the following steps are required:

(*a*) Inspect each row and write a Boolean product of sum for all boxes marked by X. This can be done in one of two ways:

(i) $(8 + 9)(7 + 8)(6 + 7)(4 + 6)(2 + 9)(2 + 5)(2 + 3)(1 + 8)(1 + 6)(1 + 2)$
(ii) $(8 + 9)(7 + 8)(6 + 7)(4 + 6)(2 + 359)(1 + 268)$

(*b*) Multiply out the expressions and eliminate all redundancies. The sum of products is obtained. This is rather lengthy and error prone, but is suitable for computer programming.

Applying this to the expression in (*i*), the following is obtained:

124578 + 124579 + 12679 + 134579 + 135678 + 135679 + 268

(*c*) The maximal compatibles are obtained by writing the missing terms. For example in 124578 the missing terms are 3, 6 and 9.

Applying this to the sum of products in (*b*), we obtain the following MEGs:

369, 368, 3458, 268, 247, 248, 134579

This agrees with the result obtained earlier.

7.3.4 Group Selection and Coding

As indicated earlier, the choice of a suitable set of MEGs to cover all the original functions is not always straightforward. More so, if an optimum design is required. Obviously if the design is to be carried out by a digital computer, some guide is required to help the designer or the computer to make a suitable choice.

Having obtained a list of the MEGs as given in the last E-list, we proceed to select the minimum number of these groups to cover all the logic functions. A procedure similar to the prime implicant table described in Chapter 1 may be used for the selection of the most suitable groups. For Example (7.3) the following list of MEGs were obtained: 247, 248, 268, 369, 3458, 134579 and 368. These are entered on the table shown in Fig. 7.12.

The two groups chosen intuitively, namely 268 and 134579, do cover all the functions. The choice of the optimum groups requires some practice.

	✓ 1	✓ 2	✓ 3	✓ 4	✓ 5	✓ 6	✓ 7	✓ 8	✓ 9
247	X		X			X			
248	X		X				X		
268	X				X		X		
369		X			X			X	
3458		X	X	X			X		
134579	X	X	X	X		X		X	
368		X			X		X		

Fig. 7.12   Covering table of exclusive groups for Example (7.3)

It is important to note that one must not include any function in more than one group, and that large groups can be subdivided into smaller groups since any subgroup of an exclusive group is also exclusive.

It is also advisable to keep the number of coding functions to a minimum. This might be achieved by choosing larger groups that make the fullest use of the code as given by the constraints imposed by Equation (7.7) below. For example, two groups of six functions each will require six coding functions, while four groups of three functions each will require eight coding functions.

In general if m_i is the number of elements (functions) in group i, and if we have j groups, the total number of coding functions is given by:

$$n = \sum_{i=1}^{j} \lceil \log_2(m_i + a_i) \rceil \tag{7.6}$$

where a_i is either 1 or 0, and $\lceil g \rceil$ is the smallest integer greater than or equal to g.

Note that $a = 1$ if it is necessary to reserve a code for the no-output condition as in Example (7.3) and $a = 0$ if all the minterms are within the group and no code is necessary for the no-output condition. For Example (7.4) given earlier, the following two groups are selected:

1 3 5 6 9 13,   2 4 7 8 10 11 12

$$n = \lceil \log_2(6 + 1) \rceil + \lceil \log_2(7 + 1) \rceil = 6$$

Therefore six coding functions, $R_1, \ldots, R_6$, are required. A binary code is then given to these groups as shown in Tables 7.8 and 7.9. Usually one code must be reserved for no output. Hence, any n-bit code can code a group of up to $2^n - 1$ functions. A two-bit code can code groups of two or three functions. The lower limit is trivial since there is no advantage in coding two-variables using a two-bit code. A three-bit code can code groups having 4–7 functions. In general, for any m functions in a group and n-bit code we have

$$\begin{aligned} 2^{n-1} \leqslant m \leqslant 2^n - a \quad & \text{for } n > 2 \\ m = 2^2 - a \quad & \text{for } n = 2 \end{aligned} \tag{7.7}$$

The case when $n = 1$ is trivial.

As before $a = 1$ if a code is reserved for no output; otherwise $a = 0$.

	R_1	R_2	R_3
None	0	0	0
F_2	0	0	1
F_4	0	1	0
F_7	0	1	1
F_8	1	0	0
F_{10}	1	0	1
F_{11}	1	1	0
F_{12}	1	1	1

Table 7.8   Coding of the first MEG

	R_4	R_5	R_6
None	0	0	0
F_1	0	0	1
F_3	0	1	0
F_5	0	1	1
F_6	1	0	0
F_9	1	0	1
F_{13}	1	1	0

Table 7.9   Coding of the second MEG (Example (7.4))

From Tables 7.8 and 7.9 expressions for the coding functions, $R_1, \ldots, R_6$, can be obtained in terms of the input variables $x_1, \ldots, x_5$. For example:

$$R_1 = \Sigma (3, 4, 5, 6, 7, 9, 11, 13, 18, 19, 21, 22, 23, 24, 26, 29, 30, 31)$$

This may be expanded and implemented as in Fig. 7.13 using ULMs, say:

$$\begin{aligned} R_1 = \; & \bar{x}_1\bar{x}_2[\bar{x}_3\bar{x}_4(0) + \bar{x}_3x_4(x_5) + x_3\bar{x}_4(1) + x_3x_4(1)] + \\ & \bar{x}_1x_2[\bar{x}_3\bar{x}_4(x_5) + \bar{x}_3x_4(x_5) + x_3\bar{x}_4(x_5) + x_3x_4(0)] + \\ & x_1\bar{x}_2[\bar{x}_3\bar{x}_4(0) + \bar{x}_3x_4(1) + x_3\bar{x}_4(x_5) + x_3x_4(1)] + \\ & x_1x_2[\bar{x}_3\bar{x}_4(x_5) + \bar{x}_3x_4(x_5) + x_3\bar{x}_4(x_5) + x_3x_4(1)] \end{aligned}$$

Similarly expressions for $R_2, \ldots, R_6$ can be obtained as shown in the computer outputs in Appendix 7.3 and implemented using suitable logic such as gates, ULMs or ROMs together with decoders to obtain the original functions as shown in Fig. 7.14. Programmed ROMs are shown in Fig. 7.15 for the original 13 logic functions

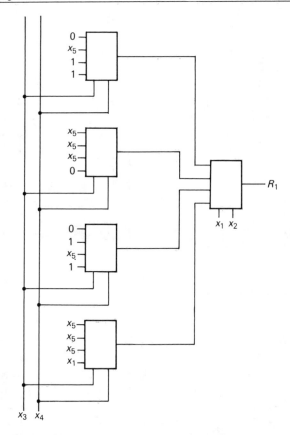

Fig. 7.13   ULM implementation of the coding function R_1

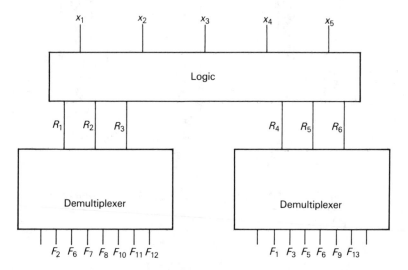

Fig. 7.14   A block diagram implementation of Example (7.4)

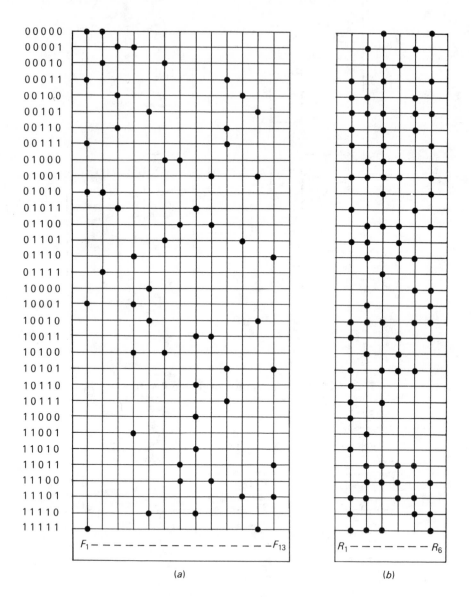

Fig. 7.15   ROM storage requirement for Example (7.4)
(a) Original programs
(b) After group coding

and the new six coding functions, demonstrating the amount of saving in memory locations.

If three-variable ULM modules are used, the maximum number of modules required for such an example would be 65 modules. After applying the group coding technique it would require at most 30 modules and two 1-of-8 demultiplexers, a saving of over 50 per cent.

7.3.5 Digital Computer Program

A digital computer program was developed to execute the design procedure outlined. This computes all the MEGs and selects covering groups, taking all the special cases mentioned earlier into consideration. These include the minimization of the total number of coding functions, making the best use of the code, testing whether a code is required for no output or not, which is particularly useful in cases where the number of elements in a group is an exact power of 2. As an illustration, two groups, of 7 members each, are preferred to groups of 6 and 8 members, but if only 6 and 8 are available the last test is carried out to see if it is necessary to break the large group into 7 and 1.

Once the groups are chosen, a binary code is given to the groups and then expressions are obtained for the coding functions. Finally the computer will display the output in a form suitable for either ROM or ULM realization as shown.

Because the minicomputer used was rather small with limited storage capacity, effort was made to minimize the storage requirements. In the program the dimensions of the matrices are given in terms of the number of minterms M. Only matrix D is given as $D(N, ?)$ where the unknown dimension refers to the number of MEGs expected. This may be given an arbitrary value in the trial run, say 20 for an average problem. The total memory for the data is usually less than 1 K bits.

The input data required by the program are as follows:

(i)   Number of functions N
(ii) Number of words M
(iii) For each function the number of minterms and the minterms, given in decimal notation
(iv) An integer I which is given the value 0, 1 or 2, depending on the output required.

   $I = 0$ gives the output up to the expressions for the coding functions.
   $I = 1$ the solutions up to and including the truth table for ROM.
   $I = 2$ all the solutions including the expressions to suit ULM implementation.

The program is given in Appendix 7.3.

For most problems, typical run time is of the order of 1−5 min including printing time using a HP9830A minicomputer. A FORTRAN copy of the program, run on a UNIVAC 9400/9480 computer for trial, solved the problem given here plus another four of similar size in 62 s. The solution of the larger problem, whose conflicts diagram is given in Fig. 7.8, took 112 s.

Computer Output for Example (7.4)

FUNCTIONS = 13   WORDS = 32

The data

Functions   Minterms

Functions						
F_1	0	3	7	10	17	31
F_2	0	2	10	15		
F_3	1	4	6	11		
F_4	1	14	17	20	25	
F_5	5	16	18	30		
F_6	2	8	13	20		
F_7	8	12	27	28		
F_8	11	19	22	24	26	30
F_9	9	12	19	28		
F_{10}	3	6	7	21	23	
F_{11}	4	13	29			
F_{12}	5	9	18	31		
F_{13}	14	21	27	29		

S Table

$S_2 =$
$S_3 = 1 \quad 2$
$S_4 = 2$
$S_5 = 1 \quad 2 \quad 3 \quad 4$
$S_6 = 1 \quad 3 \quad 5$
$S_7 = 1 \quad 2 \quad 3 \quad 4 \quad 5$
$S_8 = 1 \quad 2 \quad 4 \quad 6 \quad 7$
$S_9 = 1 \quad 2 \quad 3 \quad 4 \quad 5 \quad 6$
$S_{10} = 2 \quad 4 \quad 5 \quad 6 \quad 7 \quad 8 \quad 9$
$S_{11} = 1 \quad 2 \quad 4 \quad 5 \quad 7 \quad 8 \quad 9 \quad 10$
$S_{12} = 1 \quad 3 \quad 4 \quad 6 \quad 7 \quad 8 \quad 10 \quad 11$
$S_{13} = 1 \quad 2 \quad 3 \quad 5 \quad 6 \quad 8 \quad 9 \quad 12$

The possible groups

1	1	3	5	6	9	13		10	5	6	9	10
2	2	3	5	7				11	6	8	10	12
3	2	4	5	7	10	11		12	1	5	9	11
4	1	3	5	7				13	1	5	7	11
5	2	4	7	8	10	11	12	14	2	3	7	12
6	1	6	8	13				15	3	6	12	13
7	1	7	8	11				16	2	8	12	13
8	2	3	5	9	13			17	6	8	12	13
9	2	4	5	9	10	11		18	2	3	12	13

The chosen groups

1	2	4	7	8	10	11	12
2	1	3	5	6	9	13	

THE CODE

Group 1   Elements = 8   Degree = 3

	R_1	R_2	R_3
NON	0	0	0
F_2	0	0	1
F_4	0	1	0
F_7	0	1	1
F_8	1	0	0
F_{10}	1	0	1
F_{11}	1	1	0
F_{12}	1	1	1

Group2   Elements = 7   Degree = 3

	R_4	R_5	R_6
NON	0	0	0
F_1	0	0	1
F_3	·0	1	0
F_5	0	1	1
F_6	1	0	0
F_9	1	0	1
F_{13}	1	1	0

The final output

Functions   Minterms

R_1	3	4	5	6	7	9	11	13
	18	19	21	22	23	24	26	29
	30	31						
R_2	1	4	5	8	9	12	13	14
	17	18	20	25	27	28	29	31
R_3	0	2	3	5	6	7	8	9
	10	12	15	18	21	23	27	28
	31							
R_4	2	8	9	12	13	14	19	20
	21	27	28	29				
R_5	1	4	5	6	11	14	16	18
	21	27	29	30				
R_6	0	3	5	7	9	10	12	16
	17	18	19	28	30	31		

For ROMs

Word	R_1	R_2	R_3	R_4	R_5	R_6
0	0	0	1	0	0	1
1	0	1	0	0	1	0
2	0	0	1	1	0	0
3	1	0	1	0	0	1
4	1	1	0	0	1	0
5	1	1	1	0	1	1
6	1	0	1	0	1	0
7	1	0	1	0	0	1
8	0	1	1	1	0	0
9	1	1	1	1	0	1
10	0	0	1	0	0	1
11	1	0	0	0	1	0
12	0	1	1	1	0	1
13	1	1	0	1	0	0
14	0	1	0	1	1	0
15	0	0	1	0	0	0
16	0	0	0	0	1	1
17	0	1	0	0	0	1
18	1	1	1	0	1	1
19	1	0	0	1	0	1
20	0	1	0	1	0	0
21	1	0	1	1	1	0
22	1	0	0	0	0	0
23	1	0	1	0	0	0
24	1	0	0	0	0	0
25	0	1	0	0	0	0
26	1	0	0	0	0	0
27	0	1	1	1	1	0
28	0	1	1	1	0	1
29	1	1	0	1	1	0
30	1	0	0	0	1	1
31	1	1	1	0	0	1

```
10 REM DIM X[M + 1, N], S[N, N], P[N], Q[N], Z[M], H[M], G[M],
D[N, ?], R[N, N], E[N]
20 DIM X[33, 8], S[13, 9], P[13], Q[13], Z[32], H[18], G[18],
D[13, 7], R[6, 8], E[2]
7020 DATA 13, 32
7030 DATA 6, 0, 3, 7, 10, 17, 31, 4, 0, 2, 10, 15, 4, 1, 4, 6, 11, 5, 1, 14, 17, 20, 25
7040 DATA 4, 5, 16, 18, 30, 4, 2, 8, 13, 20, 4, 8, 12, 27, 28, 6, 11, 19, 22, 24,
26, 30
7050 DATA 4, 9, 12, 19, 28, 5, 3, 6, 7, 21, 23, 3, 4, 13, 29, 4, 5, 9, 18, 31
7060 DATA 4, 14, 21, 27, 29
7090 DATA 1
```

7.4 MINIMIZATION OF ULM TREES

7.4.1 Preliminaries

Following the discussion in Chapter 3, it will be shown how to use a universal logic module (ULM) in a multi-level array. Briefly, one decides what type of module to use, proceeds to write the function in full canonical form, and then selects the control variables for the various levels. The function is next repeatedly expanded with respect to the variables used at the control inputs of the first, second and subsequent levels, until the residues are functions of one variable only, which is used at the front inputs of the last level. A logic function of n variables is expanded using Shannon's expansion theorem as shown in Equation (7.8) below.

$$
\begin{aligned}
f(x_1, x_2, \ldots, x_n) = \bar{x}_1 \bar{x}_2 & \quad f(0, 0, x_3, \ldots, x_n) + \\
\bar{x}_1 x_2 & \quad f(0, 1, x_3, \ldots, x_n) + \\
x_1 \bar{x}_2 & \quad f(1, 0, x_3, \ldots, x_n) + \\
x_1 x_2 & \quad f(1, 1, x_3, \ldots, x_n)
\end{aligned}
\tag{7.8}
$$

The choice of control variables, as was hinted earlier, can substantially affect the total number of modules required for a particular implementation. In order to ensure an optimum solution, the designer has to try all possible permutations of control variables. This is a difficult individual job, even for small functions; for large functions of, say, ten variables, it is difficult even for a computer, as will be shown later in this chapter. It is therefore preferable to employ a more efficient technique which gives a 'good', though not necessarily an optimum, solution.

In this section, a technique based on level-by-level minimization is outlined; this gives an optimum or near optimum solution. The main advantage of the method is that it greatly reduces the amount of work required, compared with carrying out a complete enumeration of the various permutations of control variables.

Two versions of the method are presented. The first is essentially a pattern recognition technique, based on the relationships which exist between the variable ordering and the residue interchange pattern extracted from the expanded function. This can be used manually for functions of up to about seven or eight variables. For larger functions, the pattern becomes more difficult to detect, and a computer program is described which uses level-by-level minimization. It is shown that this is far more efficient in terms of storage and computation time than a completely enumerative algorithm.

In order to simplify subsequent development, it is necessary to impose one restriction, namely the assumption that all modules in the same level of implementation use the same control variables. This is in line with current practice in that it preserves the modularity of the system and the regularity of the interconnections. However, it should be clear that cases do arise where it is advantageous to use mixed variables within a level, or even use multi-variable functions as control inputs, as is so with logic functions decomposed according to Ashenhurst−Curtis techniques.

In order to implement an n-variable logic function $f(x_1, x_2, \ldots, x_n)$, let:

l be the number of levels of implementation.
c be the number of control inputs to a module.
$I = 2^c$ be the number of front inputs to a module.

c_i be the number of control inputs per module used in the ith level
$(1 \leqslant c_i \leqslant c, \quad \text{and } i = 1, 2 \ldots l)$.

Then the maximum number of modules Mo required to implement the expanded function is given by Equation (3.8), reproduced here:

$$Mo = \frac{1 - I^l}{1 - I} \tag{7.9(a)}$$

If c does not divide $(n - 1)$ exactly, the first level will have some of its control inputs left unused; or possibly smaller modules are used; therefore Equation (3.8) is modified to the following:

$$Mo = 1 + 2^{c1} \left(\frac{1 - I^{l-1}}{1 - I} \right) \tag{7.9(b)}$$

However, after fixing the number of control variables in each level, there is still a possibility of saving modules. This occurs when two or more modules having the same control variables happen to have the same front inputs. In this case the criterion for minimization is that as many modules as possible should have identical inputs, i.e. should have identical residue terms in the expanded function. Consequently, one module can do the job of others in that its output can be shared at the corresponding inputs of the preceding level. In addition, as will subsequently be seen, it is advantageous to have modules whose front inputs are independent of the function variables, i.e. residue terms which consist of the logical constants 0 and 1.

In view of the foregoing, the resulting number of modules required in an implementation can often be considerably less than Mo as calculated from Equation (7.9).

7.4.2 General Theory

Consider the n-variable logic function in Equation (7.8). If the expansion is continued into canonical form, the residue terms become functions of only one variable. This will be called the *residue variable* and can be denoted as R_{ij} in that the functions can be considered as elements of a matrix which we shall call a *residue matrix*. In the basic form of the residue matrix, i will be taken as the decimal equivalent of $x_1, x_2 \ldots x_{c1}$ expressed as a binary number, and j as the decimal equivalent of the remaining control variables $x_{c1+1}, x_{c1+2}, \ldots x_{n-1}$.

As an example, consider any four-variable function $f(x_1, x_2, x_3, x_4)$. This may be written as:

$$f(x_1, x_2, x_3, x_4) = \bar{x}_1 \left[\bar{x}_2\bar{x}_3(R_{00}) + \bar{x}_2x_3(R_{01}) + x_2\bar{x}_3(R_{02}) + x_2x_3(R_{03}) \right] +$$
$$x_1 \left[\bar{x}_2\bar{x}_3(R_{10}) + \bar{x}_2x_3(R_{11}) + x_2\bar{x}_3(R_{12}) + x_2x_3(R_{13}) \right] \tag{7.10}$$

In this case, the first-level expansion is about x_1, the second-level expansion is about x_2 and x_3, and the residue terms R_{ij} are functions only of x_4 and take on values of $\bar{x}_4$, x_4, 0, or 1.

$$R = \begin{bmatrix} R_{00} & R_{01} & R_{02} & R_{03} \\ R_{10} & R_{11} & R_{12} & R_{13} \end{bmatrix}$$

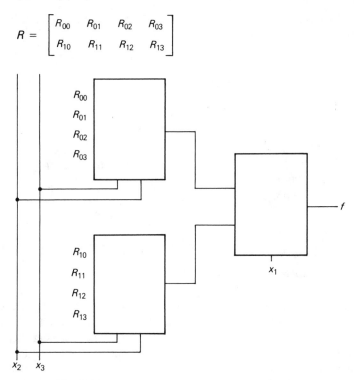

Fig. 7.16   Implementation of Equation (7.10)

The function expressed in the form of Equation (7.10) might then, for example, be implemented using three-variable multiplexer modules as shown in Fig. 7.16, where x_1 is the control variable for the first level, x_2 and x_3 are the control variables for the second level, and the resulting residue terms are the front inputs for the second-level multiplexers. These residue terms can be written in matrix form as shown in the residue matrix R in Fig. 7.16.

To simplify subsequent notations, let v refer to the variable of binary weight 2^v in the first level of the canonical expansion, where the first-level control variables are expressed as a binary number. Also, let v' refer to the variable of binary weight 2^v in the succeeding levels, where all the remaining control variables are expressed as a binary number. Then the ordering of the variables in Equation (7.10) is:

$$x_1 \ [x_2 \ x_3 \ (R_{ij}) + \ldots] \qquad \longrightarrow \quad 0 \ [1'0' \ (R_{ij}) + \ldots]$$

Alternatively the ordering may also be seen from the following representation, where the variable ordering is shown above the corresponding variables:

x_1 \ $\overset{0}{}$	$\overset{1'}{x_2}$	$\overset{0'}{x_3}$		
	R_{00}	R_{01}	R_{02}	R_{03}
	R_{10}	R_{11}	R_{12}	R_{13}

For three levels of three-variable modules, the ordering when minimizing at the first level is shown by the following correspondence:

$$x_1\, x_2\, [x_3\, x_4\, [x_5\, x_6\, (R_{ij}) + \ldots]] \qquad \longrightarrow \quad 10\, [3'2'1'0'\, (R_{ij}) + \ldots]$$

Then the ordering of the expanded terms can generally be represented by a string of numbers as:

$$(c_R - 1) \ldots 210\, [(c_C - 1)' \ldots (c_1 - 1)' \ldots 2'1'0'\, (R_{ij}) + \ldots] \tag{7.11}$$

where c_R is the number of row variables, i.e. the number of variables defining the rows of the residue matrix, and c_C is the number of column variables, i.e. the number of variables defining the columns of the residue matrix.

In the above expression x_1 and x_2 are row variables, and x_3, x_4, x_5 and x_6 are column variables.

7.4.3 Two-Level Implementation

For two or more modules to have the same front inputs requires the residue matrix to have two or more identical rows. Interchange of various elements in the residue matrix specifically so as to make two or more of the rows identical can lead to correlations between various interchange patterns and different permuations of control variables from different levels.

For example, if in Equation (7.10) x_1 and x_3 are interchanged, the effect on the matrix elements is a diagonal interchange as indicated by the diagonally connected boxes in Equation (7.12).

$$R = \begin{bmatrix} R_{00} & \boxed{R_{01}} & R_{02} & \boxed{R_{03}} \\ \boxed{R_{10}} & R_{11} & \boxed{R_{12}} & R_{13} \end{bmatrix} \tag{7.12}$$

Furthermore, if x_1 and x_2 are interchanged, the result is now a single diagonal shift, this time involving two elements as shown in Equation (7.13).

$$R = \begin{bmatrix} R_{00} & R_{01} & \boxed{R_{02}} & R_{03} \\ \boxed{R_{10}} & R_{11} & R_{12} & R_{13} \end{bmatrix} \tag{7.13}$$

The residue interchange patterns in Equations (7.12) and (7.13) are similar in the sense that, starting at the top right-hand corner, alternate sub-matrices which are diagonally adjacent and of equal size are interchanged. That this is true in general, for any residue matrix may be readily seen as due to the binary ordering of the control variables. The dimensions of these sub-matrices can be related to the ordering of the variables which are interchanged as follows:

Interchanging row variable number 0, and column variable number 0' (decimal numbers), corresponds to interchange of 1×1 (i.e. $2^0 \times 2^0$) sub-matrices in the residue matrix in the diagonal pattern as described.

Interchange of the variables numbered 0 and $1'$ corresponds to interchange of 1×2 (i.e. $2^0 \times 2^1$) sub-matrices. In general if two variables numbered e and t' are interchanged, the corresponding sub-matrices are of dimensions $u \times w$ where:

$$u = 2^e$$
$$\text{and} \quad w = 2^t \tag{7.14}$$

It is possible therefore to write the logic function in its expanded form, and see visually if by interchanging the elements of the residue matrix in a pattern similar to that described above, two or more rows can be made similar; if so, the corresponding control variables are interchanged using Equation (7.14). With a little practice, it should be possible to exchange the variables without reference to Equation (7.14). This, in a way, is a prediction of the form of a residue matrix for other permutations of the control variables, and saves the designer the trouble of rewriting and expanding the logic function for each permutation of the control variables. Fig. 7.17 gives further illustrations of the pattern and its relation to the control variables.

The described diagonal residue interchange patterns do not, of course, cover all possibilities of permutations of control variables and, to complete the picture, the following summary should be noted:

(i)   The interchange pattern holds for any size module, regardless of the number of control variables used, provided that only two variables (one from each level) are interchanged at any one time. This covers $c_1 \times c_2$ permutations.

(ii)   Where equal numbers of control variables are employed in both levels, it is possible to exchange rows and columns in the residue matrix (matrix transposition), by exchanging all the control variables in both levels in the same order. This point is illustrated in a later example.

(iii)   Exchanging more than two variables at a time, apart from the special case (ii), can result in different useful interchange patterns in the residue matrix. However, the patterns are more complex and, to make full use of these, one would need to keep a set of all possible interchange patterns; alternatively one may use (i) above more than once.

(iv)   Exchanging variables within the same level is not useful, since this merely permutes the rows or columns of the residue matrix leaving their elements unchanged.

The procedure is illustrated by Example (7.5).

Example (7.5)
Implement the function $f_1(x_1, x_2, x_3, x_4, x_5)$ expressed in expanded form as:

$$f_1(x_1, x_2, x_3, x_4, x_5) =$$

$$
\begin{aligned}
&= \bar{x}_1\bar{x}_2 \left[\bar{x}_3\bar{x}_4(1) && + \bar{x}_3x_4(x_5) && + x_3\bar{x}_4(1) && + x_3x_4(x_5) \right] + \\
&\quad \bar{x}_1x_2 \left[\bar{x}_3\bar{x}_4(1) && + \bar{x}_3x_4(\bar{x}_5) && + x_3\bar{x}_4(1) && + x_3x_4(1) \right] + \\
&\quad x_1\bar{x}_2 \left[\bar{x}_3\bar{x}_4(x_5) && + \bar{x}_3x_4(0) && + x_3\bar{x}_4(x_5) && + x_3x_4(1) \right] + \\
&\quad x_1x_2 \left[\bar{x}_3\bar{x}_4(\bar{x}_5) && + \bar{x}_3x_4(0) && + x_3\bar{x}_4(1) && + x_3x_4(1) \right]
\end{aligned}
$$

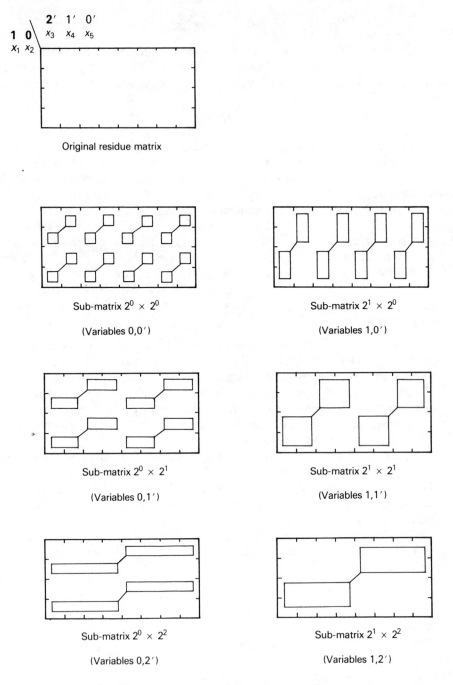

Fig. 7.17 **Typical patterns illustrating the relationship between the dimensions of the sub-matrices and the variable ordering**

SOLUTION: This function requires two levels of three-variable modules with a maximum of five modules for implementation, as shown in Fig. 7.18(*a*).

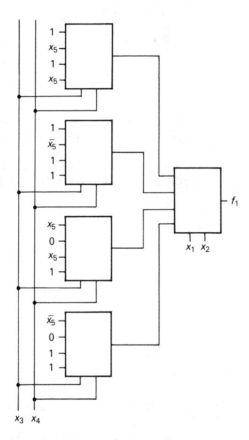

Fig. 7.18(a)   Original implementation of Example (7.5)

The basic residue matrix is of the form:

$$
\begin{bmatrix}
\begin{array}{cc|cc}
1 & x_5 & 1 & x_5 \\
1 & \bar{x}_5 & 1 & 1 \\
\hline
x_5 & 0 & x_5 & 1 \\
\bar{x}_5 & 0 & 1 & 1
\end{array}
\end{bmatrix}
$$

A useful pattern may be observed by interchange of the sub-matrices shown. This results in a matrix of the form:

$$
\begin{bmatrix}
1 & x_5 & x_5 & 0 \\
1 & \bar{x}_5 & \bar{x}_5 & 0 \\
1 & x_5 & x_5 & 1 \\
1 & 1 & 1 & 1
\end{bmatrix}
$$

From the expression (7.11) the variable ordering of the terms in the expansion is:

$$x_1 x_2 [x_3 x_4 (R_{ij}) + \ldots] \rightarrow 1\,0[1'0'(R_{ij}) + \ldots]$$

From Equation (7.14) the dimensions of the sub-matrices are related to this ordering by:

$$2 = 2^e \quad \text{giving} \quad e = 1$$
$$\text{and} \quad 2 = 2^t \quad \text{giving} \quad t' = 1'$$

i.e. interchange of 1 and 1' (or x_1 and x_3). Then by the column-row interchange rule, a matrix having two identical rows results from the use of variables $x_1 x_4$ at the first level and $x_3 x_2$ at the second level.

As an alternative, this solution might have been arrived at in a single step by interchange of sub-matrices of unit dimension as shown in the following.

$$
\begin{bmatrix}
1 & \boxed{x_5} & 1 & \boxed{x_5} \\
\boxed{1} & \bar{x}_5 & \boxed{1} & 1 \\
x_5 & \boxed{0} & x_5 & \boxed{1} \\
\boxed{\bar{x}_5} & 0 & \boxed{1} & 1
\end{bmatrix}
$$

The new matrix is then of the form:

$$
\begin{bmatrix}
1 & 1 & 1 & 1 \\
x_5 & \bar{x}_5 & x_5 & 1 \\
x_5 & \bar{x}_5 & x_5 & 1 \\
0 & 0 & 1 & 1
\end{bmatrix}
$$

giving the same solution as before. The implementation based on this requires only two modules and is shown in Fig. 7.18(*b*). This example also illustrates the usefulness of having residue terms consisting of logical constants, as pointed out earlier.

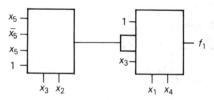

Fig. 7.18(b) Modified implementation of Example (7.5)

7.4.4 General Multi-Level Implementation

The same basic concepts can be extended to general multi-level networks, provided that the variables in the terms of the canonical expansion are numbered in the order previously described by Expression (7.11). For example, for a seven-variable function implemented by using three levels of three-variable modules, as previously described, the function is expanded with respect to x_1x_2, x_3x_4 and x_5x_6, which are used as the control variables of the first, second and third levels respectively. The residue terms are then functions of x_7 only.

When minimizing at the first level, the correspondence between the terms of the expansion is of the form:

$$x_1x_2[x_3x_4[x_5x_6(R_{ij}) + ...]] \to 10[3'2'1'0'(R_{ij}) + ...] \qquad (7.15)$$

for $i = 0, 1, 2, 3$
      $j = 0, 1, 2, ..., 15$

The first-level variables are x_1x_2 (the row level), whilst the remaining control variables $x_3x_4x_5x_6$ are considered equivalent to one level (the column level), and the two-level minimization technique is then applied. The same basic procedure as described before may then be employed at each level except the last, since the permutation of control variables at the last level will change only the front connection of the modules in that level. This means that the residue matrix requires to be expressed in $l - 1$ different forms to construct a general l-level network.

Starting with the basic residue matrix, if a useful interchange pattern is observed, resulting in two or more identical rows, all but one of these rows may be deleted and a modified residue matrix constructed with rows defined by both first- and second-level variables together. For example, with reference to expression (7.15), if it is found that interchange of x_1 and x_5 results in rows 0 and 3 of the basic residue matrix being identical, then x_5 and x_2 would be used at the first-level control inputs and, arbitrarily deleting row 0 from the matrix, the resulting correspondence for the second step of the procedure would be:

$$x_5x_2[x_3x_4[x_1x_6(R_{ij}) + ...]] \to 3\,2[1\,0[1'0'(R_{ij}) + ...]]$$

for $i = 4, 5, 6, ... 15$, and $j = 0, 1, 2, 3$ $\qquad (7.16)$

Choosing to delete row 3 from the basic residue matrix, the correspondence would be:

$$x_5x_2[x_3x_4[x_1x_6(R_{ij}) + ...]] \to 3\,2[1\,0[1'0'(R_{ij}) + ...]]$$

for $i = , 1, 2, ... 11$ and $j = 0, 1, 2, 3$ $\qquad (7.17)$

The final results in terms of module count however will be unaffected, whichever row is chosen to be deleted.

If, on the other hand, no useful interchange pattern can be observed, for the second step of the procedure, both first- and second-level variables together define the rows of the modified residue matrix, and the variable correspondence becomes:

$$x_1x_2[x_3x_4[x_5x_6(R_{ij}) + ...]] \to 32[1\,0[1'\,0'\,(R_{ij}) + ...]]$$

for $i = 0, 1, 2, ..., 15$ and $j = 0, 1, 2, 3$ $\qquad (7.18)$

In this context, the first and second levels might then be referred to as the 'row levels' of the matrix, and the third level as the 'column level'. It is suggested that minimization of multi-level ULM trees should start from the first level, since finding two similar rows at an early level results in saving a whole branch as opposed to one module in the last level, as will be shown.

In general, this basic procedure can be adopted at each level of implementation except the last, with the rules given in the preceding subsection generalized as:

(i)   Exchange of any two variables, one from the 'row levels' and one from the 'column levels' results in a residue interchange pattern as for two levels, with the dimensions of the sub-matrices given by Equation (7.14).

(ii)   Interchange of rows and columns in the matrix by an ordered interchange of variables in the row and column levels is applicable only where the row and column levels have equal numbers of variables.

(iii) Interchange of more than two variables from the 'row levels' and the 'column levels' of the residue matrix results in a residue interchange pattern which does not follow the general rule, and may or may not be useful. However step (i) may be used more than once to produce the same effect.

(iv)   As for two levels, interchange of variables within the same level permutes only the front connections of that level.

With these points in mind, the procedure described earlier may be used to determine whether by interchange of residues in a recognizable pattern two or more similar rows may be obtained. If so, this simply means the interchange of the corresponding control variables determined from the described ordering, in conjunction with Equation (7.14).

Using this basic procedure, the work involved in logic design using ULMs can be greatly reduced. For example, to implement a function of seven variables using two levels of four-variable modules, it is possible to write the function in expanded form just once and then determine what happens for another ten permutations of control variables without rewriting and expanding the function for each permutation.

With a little practice, rearranging the residue matrix to give identical rows will be found much easier than re-expanding the function and comparing for identical terms. However it should be noted that one variable is assigned to the front input from the start of the procedure, and this might have to be changed for an optimum solution.

Example (7.6)

Show how multi-level implementation of a seven-variable function using three levels of three-variable modules can be effected, when the function has a basic residue matrix of the form:

$3'2'1'0'$

$X_3 X_4 X_5 X_6$

$\begin{array}{cc} 1 & 0 \\ X_1 & X_2 \end{array}$
$\begin{bmatrix} 1 & 0 & x_7 & x_7 & x_7 & \bar{x}_7 & \bar{x}_7 & 1 & 0 & 1 & \bar{x}_7 & 1 & x_7 & \bar{x}_7 & x_7 & 0 \\ x_7 & x_7 & 1 & 0 & 1 & \bar{x}_7 & x_7 & x_7 & 0 & x_7 & 1 & 0 & 1 & 1 & 1 & 0 \\ 1 & 0 & 0 & x_7 & x_7 & \bar{x}_7 & \bar{x}_7 & \bar{x}_7 & 0 & 1 & x_7 & 0 & x_7 & \bar{x}_7 & 0 & 0 \\ x_7 & x_7 & 0 & 0 & 1 & \bar{x}_7 & x_7 & 0 & 0 & x_7 & 1 & 1 & 1 & 1 & 0 & x_7 \end{bmatrix}$

SOLUTION:

In accordance with Equation (7.9), a direct implementation of this function requires 21 modules. By interchange of residues as shown, the first and third rows are seen to be identical, resulting in the matrix below:

$$\begin{bmatrix} 1 & 0 & x_7 & x_7 & x_7 & \bar{x}_7 & 1 & \bar{x}_7 & 0 & 1 & 0 & x_7 & x_7 & \bar{x}_7 & 1 & 1 \\ x_7 & x_7 & 1 & 0 & \bar{x}_7 & 1 & x_7 & x_7 & \bar{x}_7 & 1 & 1 & 0 & x_7 & 0 & 1 & 0 \\ 1 & 0 & x_7 & x_7 & x_7 & \bar{x}_7 & 1 & \bar{x}_7 & 0 & 1 & 0 & x_7 & x_7 & \bar{x}_7 & 1 & 1 \\ 0 & x_7 & 0 & 0 & \bar{x}_7 & \bar{x}_7 & x_7 & 0 & x_7 & 0 & 1 & 1 & 0 & 0 & 0 & x_7 \end{bmatrix}$$

Equation (7.14) indicates that this matrix is produced by interchange of variables x_2 and x_5. Delete row 0 from this matrix; then the new variable ordering for the choice of the second-level variables becomes:

$$x_1 x_5 [x_3 x_4 [x_2 x_6 (R_{ij}) + \dots]] \rightarrow 32[1\,0[1'0'(R_{ij}) + \dots]]$$

for $i = 4, 5, 6, \dots 15$, and $j = 0, 1, 2, 3$.

The modified residue matrix based on this ordering is then:

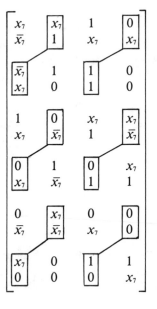

No identical rows are present in this matrix. However, interchange of residues as shown results in the following matrix which has two identical rows:

$$
\begin{bmatrix}
x_7 & \bar{x}_7 & 1 & 1 \\
\bar{x}_7 & x_7 & x_7 & 1 \\
x_7 & 1 & 0 & 0 \\
1 & 0 & x_7 & 0 \\
1 & 0 & x_7 & 0 \\
x_7 & x_7 & 1 & 1 \\
0 & 1 & x_7 & x_7 \\
\bar{x}_7 & \bar{x}_7 & \bar{x}_7 & 1 \\
0 & x_7 & 0 & 1 \\
\bar{x}_7 & 0 & x_7 & 0 \\
x_7 & 0 & 0 & 1 \\
\bar{x}_7 & 0 & 0 & x_7
\end{bmatrix}
$$

From this, Equation (7.14) shows that interchange of variables x_6 and x_3 is required. The final ordering of variables is thus:

$$x_1 x_5 [x_6 x_4 [x_2 x_3 (R_{ij}) + \ldots]]$$

The implementation resulting from this final choice of variables is shown in Fig. 7.19.

Notice that a whole branch of modules is saved by finding two identical rows in the matrix at the first level, whereas just a single module is saved by a similar finding at the second level. In general, as the level of implementation increases, module savings become proportionately less, although the rows of the matrix being shorter, the probability of a saving is considerably increased.

Another feature of the method which is brought out by this example is the relationship between the structure of the residue matrix and the actual circuit structure. For example, at the first stage of minimization, the matrix has four rows, and the first level of implementation has four front inputs corresponding to these. In the modified matrix with the first and third rows equal, deletion of the first row removes the branch feeding the first front input from the circuit as in Fig. 7.19. In a similar way, at the second stage of minimization, the modified matrix has its fourth and fifth rows equal. In this case, deletion of the fourth row results in the removal of the module feeding the fourth front input in the second level of implementation. This is also shown in Fig. 7.19.

7.4.5 Symmetry and Other Properties of Logic Functions (22, 81, 116)

For economical implementation of switching functions, it is necessary to know whether the function exhibits total or partial symmetry. Yau and Tang proposed a special module to implement such functions economically, but the detection of symmetry in logic functions is useful whether ULMs are used or not. The identifica-

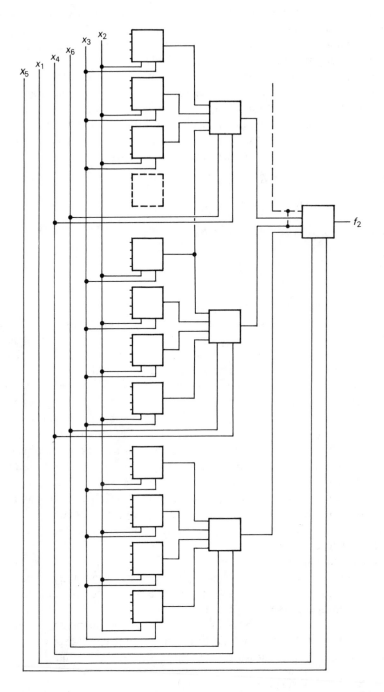

Fig. 7.19   Implementation of Example (7.6) using three-variable ULMs

tion of total or partial symmetry has been considered by many authors and only a brief recapitulation is given here.

A switching function $f(x_1, x_2, \ldots, x_n)$ is totally symmetric if the function remains unchanged for any permutations of its n variables, while it is described as partially symmetric in, say, m variables if there exists a subset of m variables, where $2 \leqslant m < n$, such that any permutations of these variables leaves the function unchanged. These m variables are called the variables of partial symmetry. A switching function that is symmetric with respect to two variables x_p and x_q is denoted by $f{:}x_p \sim x_q$.

Consider a switching function of n variables expanded as shown in Equation (7.8). If the four rows of residues are labelled $R_0, \ldots, R_3$ then:

$$f(x_1, x_2, \ldots, x_n) = \bar{x}_1\bar{x}_2 R_0 + \bar{x}_1 x_2 R_1 + x_1\bar{x}_2 R_2 + x_1 x_2 R_3$$

It has been previously shown that:

$$
\left.
\begin{array}{llll}
\text{(i)} & f{:}x_1 \sim x_2 & \text{if and only if} & R_1 = R_2 \\
\text{(ii)} & f{:}\bar{x}_1 \sim \bar{x}_2 & \text{if and only if} & R_1 = R_2 \\
\text{(iii)} & f{:}\bar{x}_1 \sim x_2 & \text{if and only if} & R_0 = R_3 \\
\text{(iv)} & f{:}x_1 \sim \bar{x}_2 & \text{if and only if} & R_0 = R_3
\end{array}
\right\}
\qquad (7.19)
$$

Das and Sheng's technique depends on the repeated expansion of the logic function with respect to two variables at a time, and on comparison of residues. They pointed out that as many as $n(n-1)/2$ permutations may be needed to detect partial symmetry, and $(n-1)$ permutations to detect total symmetry. This is a tedious job, unless a computer is used, which is what Das and Sheng did.

However, a casual designer with no access to such a program will probably resort to hand computation, especially for small logic functions of less than six or seven variables and, in such a case, a knowledge of the pattern technique described in this section will prove very useful. This may be done by treating the variables under test as row variables, and other variables as column variables; one variable will still be used as a residue variable. Then it is a matter of testing if any exchange of sub-matrices will produce identical rows as described in Equation (7.19). If so, the logic function is symmetric with respect to these two row variables, thus eliminating the need for many of the tedious expansions. Conversely, if a function is known to be symmetric with respect to two variables, say, x_1 and x_2, then using these variables in the row level will result in two similar rows.

It might be helpful to mention that if a function $f(x_1, x_2, \ldots, x_n)$ is totally or partially symmetric, then its complement $\bar{f}(x_1, \ldots, x_n)$ is also totally or partially symmetric. This means that what is a good expansion and implementation for a particular function is also good for its complement.

In general, the ULM implementation of a complemented logic function (symmetric or not) is the same as that of the function itself, except that the residues are complemented. This is shown by the following n-variable function:

$$
\begin{aligned}
f(x_1, \ldots, x_n) \ = \ & \bar{x}_1\bar{x}_2\, R(0, 0, x_3, \ldots, x_n) \ + \\
& \bar{x}_1 x_2\, R(0, 1, x_3, \ldots, x_n) \ + \\
& x_1\bar{x}_2\, R(1, 0, x_3, \ldots, x_n) \ + \\
& x_1 x_2\, R(1, 1, x_3, \ldots, x_n)
\end{aligned}
$$

$$\bar{f}(x_1, \ldots, x_n) = [x_1 + x_2 + \bar{R}(0, 0, x_3, \ldots, x_n)] \ [x_1 + \bar{x}_2 + \bar{R}(0, 1, x_3, \ldots, x_n)]$$
$$[\bar{x}_1 + x_2 + \bar{R}(1, 0, x_3, \ldots, x_n)] \ [\bar{x}_1 + \bar{x}_2 + \bar{R}(1, 1, x_3, \ldots, x_n)]$$

$$= \bar{x}_1\bar{x}_2 \ \bar{R}(0, 0, x_3, \ldots, x_n) +$$
$$\bar{x}_1x_2 \ \bar{R}(0, 1, x_3, \ldots, x_n) +$$
$$x_1\bar{x}_2 \ \bar{R}(1, 0, x_3, \ldots, x_n) +$$
$$x_1x_2 \ \bar{R}(1, 1, x_3, \ldots, x_n)$$

$$= \bar{x}_1\bar{x}_2\bar{R}_0 + \bar{x}_1x_2\bar{R}_1 + x_1\bar{x}_2\bar{R}_2 + x_1x_2\bar{R}_3$$

To illustrate this, consider Example (7.7).

Example 7.7
Carry out the ULM implementation of the expression:

$$f = \bar{x}_1x_2 + x_1\bar{x}_2\bar{x}_3 + x_1\bar{x}_3x_4$$
$$= \bar{x}_1 \ [\bar{x}_2\bar{x}_3(0) + \bar{x}_2x_3(0) + x_2\bar{x}_3(1) + x_2x_3(1)] +$$
$$x_1 \ [\bar{x}_2\bar{x}_3(1) + \bar{x}_2x_3(0) + x_2\bar{x}_3(x_4) + x_2x_3(0)]$$

Then the ULM implementation for this expression is given in Fig. 7.20(a) using $x_1, x_2,$ x_3, as the control variables for the first and second levels respectively.

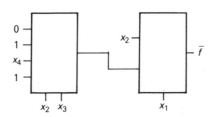

Fig. 7.20(b) ULM implementation of Example (7.7)

$$\bar{f} = \bar{x}_1\bar{x}_2 + x_1x_3 + x_1x_2\bar{x}_4$$
$$\bar{f} = \bar{x}_1 \ [\bar{x}_2\bar{x}_3(1) + \bar{x}_2x_3(1) + x_2\bar{x}_3(0) + x_2x_3(0)] +$$
$$x_1 \ [\bar{x}_2\bar{x}_3(0) + \bar{x}_2x_3(1) + x_2\bar{x}_3(\bar{x}_4) + x_2x_3(1)]$$

The ULM implementation of $\bar{f}$ is given in Fig. 7.20(b):

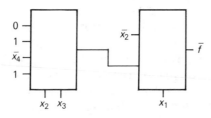

Fig. 7.20(b) ULM implementation of $\bar{f}$, where f is given in Example (7.7)

Furthermore, since the dual of a logic function is found by complementing the function as above, then complementing all literals, the dual will have the same circuit structure as the original function except the front inputs are permuted and the logical constants are complemented.

From the above, the dual of the logical expression f is given by the following:

$$fd(x_1, ..., n_n) = x_1x_2\, \bar{R}(0, 0, \bar{x}_3, ..., \bar{x}_n)\, +$$
$$x_1\bar{x}_2\, \bar{R}(0, 1, \bar{x}_3, ..., \bar{x}_n)\, +$$
$$\bar{x}_1x_2\, \bar{R}(1, 0, \bar{x}_3, ..., \bar{x}_n)\, +$$
$$\bar{x}_1\bar{x}_2\, \bar{R}(1, 1, \bar{x}_3, ..., \bar{x}_n)$$

For Example (7.7) we have:

$$fd = x_1x_2 + \bar{x}_1\bar{x}_3 + \bar{x}_1\bar{x}_2x_4$$
$$fd = \bar{x}_1\, [\bar{x}_2\bar{x}_3(1) + \bar{x}_2x_3(x_4) + x_2\bar{x}_3(1) + x_2x_3(0)]\, +$$
$$x_1\, [\bar{x}_2\bar{x}_3(0) + \bar{x}_2x_3(0) + x_2\bar{x}_3(1) + x_2x_3(1)]$$

The ULM implementation of fd is given in Fig. 7.20(c).

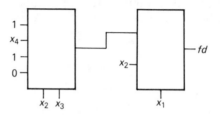

Fig. 7.20(c)   **ULM implementation of fd, where f is given in Example (7.7)**

It is therefore possible to say that, if a good choice of control variables is found for a logic function, the same choice may be used for the complement or the dual of the function.

7.4.6 A Computer Program for Level Minimization

Level Minimization Versus Exhaustive Search

Example (7.6) serves to illustrate how the residue interchange pattern becomes difficult to detect as the number of variables increases. This example is just about as complex as one might expect to solve conveniently using the manual pattern recognition technique. For functions having more than about eight variables, computer aid is desirable in expanding the function and selecting a suitable implementation. To use a completely exhaustive technique in permuting the control and residue variables is a formidable task, even for a computer, and requires excessive computation time. In the approach proposed in this section, the control variables are selected one level at a time, starting with level 1 and working down. The control variables, once chosen, are considered to be fixed and therefore do not appear in selections at any subsequent levels.

This is essentially the same basic procedure as the manual pattern recognition technique previously described except that, instead of interchanging patterns, we use a computer to generate the residue matrices and compare for identical rows for each permutation of control and residue variables.

For two-level networks the search is exhaustive, but for larger multi-level networks, it is not so, as will be seen. The method does not therefore guarantee an optimum solution, but the chances of obtaining a better one by exhaustive enumeration are, as will be shown later, very small indeed. The main advantage of the method is that, for more than two levels, the number of permutations of control variables is only a fraction of the number required using complete enumeration, which gives a corresponding reduction in computation time and storage.

To show the reduction in computation which can be achieved for any n-variable logic function, let $Q = n - 1$ be the total number of control variables. Let us neglect the permutations of control variables within a given level (since they have no effect on the complexity of that level); then if the number of control variables per level is the same, the number of permutations using complete enumeration is given by:

$$Pe = n\left[\binom{Q}{c}\binom{Q-c}{c}\cdots\binom{Q-(l-2)c}{c}\right]$$

$$= n\cdot\prod_{i=0}^{l-2}\binom{Q-ic}{c} \tag{7.20}$$

If P_j is the number of permutations in level j, then

$$Pe = n(P_1 \cdot P_2 \ldots P_{l-1}) \tag{7.21}$$

where $P_1 = \dfrac{Q!}{(Q-c)!c!}$

.

.

.

$$P_{l-1} = \dfrac{[Q-(l-2)c]!}{[Q-(l-1)c]!c!}$$

giving $Pe = n\cdot\dfrac{Q!}{[c!]^l}$ $\tag{7.22}$

If the number of control variables is not the same at all levels (this is true if $(n-1)/c$ is not an integer), the number of permutations may be found for each level individually, and the total is calculated from Equation (7.21). Substituting numerical values will clearly show the rapid increase in Pe as shown in Table 7.10.

Using the technique described in Section 7.4.4 which is to optimize one level at a time, starting from the first level up to the $l-1$ level, the permutations can be greatly reduced. We note here that for a four-level network using three-variable modules, a branch containing $1 + 4 + 16$ modules is saved by finding two similar rows in the first-level expansion, compared with $1 + 4$ modules at the second level, and 1 module at the third level for the same finding. It is therefore reasonable to adopt this technique, especially if the number of permutations required is so dramatically reduced as shown in Table 7.10. This technique uses only a fraction of the computation time that would be needed for an exhaustive search.

Variables	Exhaustive search	Level minimization	
n	Pe	P_{max}	P_{min}
5	30	30	30
7	630	210	111
9	22680	1134	273
11	1247400	5610	544
13	97297200	26598	952

Table 7.10   Total number of permutations of control and residue variables for ULM trees using three-variable modules

Using the level-by-level minimization technique, the minimum number of permutations is given by:

$$P_{min} = nP_1 + P_2 + ... + P_{l-1} \tag{7.23}$$

This represents the case when it is possible to make a definite choice of control variables at each level, i.e. a choice which gives a saving of modules in that it gives two or more equivalent rows in the residue matrix. The computer generates n-residue matrices, one for each of the n-variables used as the residue variable in turn. Each of these matrices is rearranged for different permutations of control variables at the first level, and the matrix having the largest number of identical rows is chosen as optimum. The first-level control variables, and the residue variables which give this matrix are then taken as fixed. This will involve nP_1 permutations to make the choice.

If more than one matrix satisfies this criterion, the optimum matrix is taken as the one having the most constant rows (independent of the residue variables). If again, the score is the same, the best choice is taken as the matrix having most logical constants. It was indicated earlier that for a particular choice of residue variable, the permutations of control variables can only change the positions of the elements within the residue matrix. Therefore the number of logical constants depends only on the choice of residue variable and is independent of the choice of control variables.

If no saving can be made at the first level, no control or residue variables are fixed at this stage. The next set of permutations will then be taken over both first and second levels simultaneously. This will increase the number of permutations P to:

$$P = nP_1 + nP_{12} + P_3 + ... + P_{l-1}$$

$$\text{where } P_{12} = \frac{Q!}{(Q-2c)! \, 2c!}$$

This procedure is adopted at all subsequent levels if no savings can be made. The worst case occurs if no minimization is possible, or only possible at level $l-1$. This results in the maximum number of permutations P_{max} given by:

$$P_{max} = n(P_1 + P_{12} + P_{123} + \dots + P_{12\dots l-1}) \tag{7.24}$$

where $P_{1 \dots lk} = \dfrac{Q!}{(Q - l_k c)! \, l_k c!}$

Table 7.10 shows that even the maximum number of permutations is very much less than the number required for an exhaustive search.

Justification for Level Minimization

In level minimization, the computer examines one level at a time. If after trying all possibilities at the first level it is found that at least one branch can be saved, the residue variable and the control variables for the first level which give the maximum saving are fixed. The computer then moves to the second level, and so on. Since any fixed variables no longer appear in any further permutations, there is a possibility of missing a better solution at the second or subsequent levels. The probability of this happening is so small, however, that it is not worth the effort and additional computation time required in employing an exhaustive search.

To show this, we consider Example 7.6, which consists of three levels. One branch was saved at the first level, which effectively saved five modules. In a worst-case design, we assume that the permutations of control variables between the second and third levels produce no further saving. In this case, we might say that for an exhaustive search to produce a better result, at least six modules must be saved at the second level. This means that in the basic 16×4 second-level residue matrix, six rows should be redundant.

Each element of the matrix can take one of the four possible values $0, 1, x_7$, or $\bar{x}_7$, which are assumed to have equal probability. If row i is the same as row j, and $i \neq j$, then $R_{ik} = R_{jk}$, for $k = 0, \dots, 3$. The probability of a particular row arrangement is given by:

$$\left[4 \left(\frac{1}{4} \right)^2 \right]^4 = \left(\frac{1}{4} \right)^4$$

One way of saving six modules is for seven rows to be the same. Then:

(i) The probability that a particular row arrangement occurs somewhere exactly seven times, but no other arrangement is repeated, is given by

$$^{16}C_7 \left[\left(\frac{1}{4} \right)^4 \right]^7 \cdot \frac{4^4 - 1}{4^4} \cdot \frac{4^4 - 2}{4^4} \cdot \dots \cdot \frac{4^4 - 9}{4^4}$$

The total number of such arrangements is 4^4; hence the probability $P(7)$ of getting seven identical rows with no other row repeated is

$$P(7) = (4^4) \, ^{16}C_7 \left[\left(\frac{1}{4} \right)^4 \right]^7 \cdot \frac{4^4 - 1}{4^4} \cdot \frac{4^4 - 2}{4^4} \cdot \dots \cdot \frac{4^4 - 9}{4^4}$$

$$= (4^4) \left(\frac{16!}{7! \, 9!} \right) \left[\left(\frac{1}{4} \right)^4 \right]^7 \cdot \frac{4^4 - 1}{4^4} \cdot \frac{4^4 - 2}{4^4} \cdot \dots \cdot \frac{4^4 - 9}{4^4}$$

It is also possible to save six modules by finding six similar rows and another two similar rows. Then:

(ii)   The probability $P(6, 2)$ of six rows having one specified arrangement, two rows having a second specified arrangement, and no other rows being repeated, is

$$P(6, 2) = (4^4)(4^4 - 1)\left(\frac{16!}{6!\ 2!\ 8!}\right)\left(\frac{1}{4^4}\right)^6\left(\frac{1}{4^4 - 1}\right)^2 \cdot \frac{4^4 - 2}{4^4} \cdot$$

$$\cdot \frac{4^4 - 3}{4^4} \cdot \ \ldots\ \cdot \frac{4^4 - 9}{4^4}$$

(iii)   The probability $P(5, 2, 2)$ of three specified row arrangements appearing with multiplicities 5, 2, and 2, all other rows having multiplicity 1 is

$$P(5, 2, 2) = (4^4)(4^4 - 1)(4^4 - 2)\left(\frac{16!}{5!\ 2!\ 2!\ 7!}\right)\left(\frac{1}{4^4}\right)^5\left(\frac{1}{4^4 - 1}\right)^2 \cdot$$

$$\left(\frac{1}{4^4 - 2}\right)^2 \cdot \ \ldots\ \cdot \frac{4^4 - 3}{4^4} \cdot \ \ldots\ \cdot \frac{4^4 - 9}{4^4}$$

Similarly, we can work out the other arrangements $P(5, 3)$, $P(4, 4)$, ..., $P(2, 2, 2, 2, 2, 2)$. The probability $P(T)$ of being able to save six modules at the second level is then the sum of these probabilities, and is given by:

$$P(T) \simeq 0.5 \times 10^{-4}$$

In conclusion, we say that, if no saving is possible at an early level, one automatically moves to the succeeding level but, if a saving is possible, advantage should be made of this with the knowledge that it will most probably lead to an optimum or near optimum solution. What makes this more true is that in the forego-ing analysis we assumed a worst-case design situation in which no further saving after the first level was possible. In practice, other modules may be saved at succeeding levels, as in the second level with Example (7.6). This is not surprising if we know that the probability of finding one redundant row out of the remaining twelve rows left is 0.204, which is rather high.

How to Use the Program
A digital computer program written in 1900 Fortran was tested using the level-by-level minimization technique, and found very satisfactory. The program can tackle any number of logic functions, and each function can have any number of variables and minterms. The computer will carry out the expansions for all permuta-tions of control and residue variables as described earlier, compare these and choose the optimum residue matrix. The output is printed in a form suitable for direct implementation.

In Fig. 7.21 is a general flow chart for the program. The program listing is given in Appendix 7.4.

The data are given in the following form:

(a)   The number of logic functions is given first.

(b)   For each logic function the following specifications are given: (i) number of minterms; (ii) number of variables; (iii) number of controls used in the first level; (iv) number of controls of the module; and (v) integer (1) if all in-

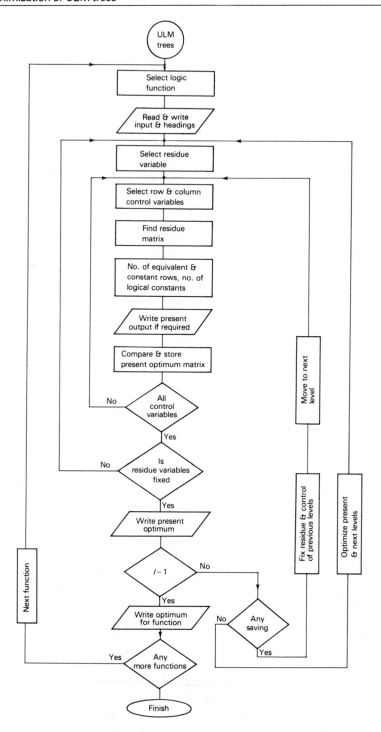

Fig. 7.21   A simplified flow chart for the control variable selection procedure

termediate data is to be printed, or (0) if only the optimal matrix for each level is to be printed.

(c)   Finally, the binary values of minterms are given.

For the two examples used in this chapter the input data would be:

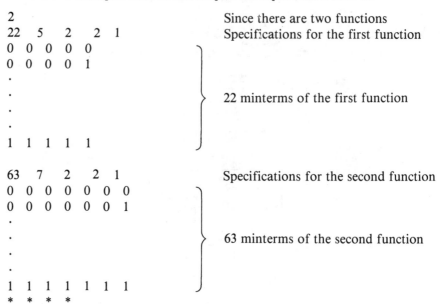

If all intermediate data are required, the computer print-out for each logic function will be of the following form:

(i)   A print-out of the logic function.

(ii)   A print-out of all the expansions required, indicating for each expansion the number of logical constants, the row and column control variables, and the residue matrix.

(iii)   A print-out of the optimum residue matrix in a form suitable for direct implementation.

For Example (7.5), the program indicated that variables x_1 and x_4 should be used in the first level, variables x_2 and x_3 in the second level, and that the residue matrix is as given in the last matrix of Example (7.5). Similarly, the computer's solution for Example (7.6) is in agreement with the solution arrived at by pattern recognition.

The two examples in this chapter took a total time of 55 s and less than 20 K of storage, using the ICL 1904A computer. Many other problems were run, and the program was found to be very efficient and fast.

The computation time is proportional to the number of permutations, which in turn is proportional to the number of variables. The increase in computation time for functions of a larger number of variables is therefore expected to be proportional to the increase in permutations as shown in Table 7.10. the program should prove very useful for regular users of ULMs particularly when large logic functions are involved.

APPENDIX 7.1   COMPUTER PROGRAM FOR COLUMN PARTITIONING

```
              MASTER PARTITIONING
              INTEGER A(500,30),C(500),G,D,P,L(50)
              READ(1,102)IW
      102     FORMAT(I0)
              IDD=0
      3       READ(1,100)N1,N2,IG,IH
      100     FORMAT(4I0)
              DO 1 J1=1,N1
      1       READ(1,101)(A(J1,J2),J2=1,N2)
      101     FORMAT(30I0)
              WRITE(2,203)
      203     FORMAT(///)
              WRITE(2,200)
      200     FORMAT(1H1 ,30X,17HORIGINAL FUNCTION/)
              DO 2 J1=1,N1
      2       WRITE(2,201)J1,(A(J1,J2),J2=1,N2)
      201     FORMAT(1H1 ,5X,I3,5X,30(5X,I1)/)
              WRITE(2,202)
      202     FORMAT(1H1 ,30X,19HPOSSIBLE PARTITIONS/)
              IY=0
              DO 40 J=IG,IH
              IX=0
              DO 4 I=1,J
      4       L(I)=I
      6       D=0
              DO 20 M=1,N1
              IP=0
              DO 30 IR=1,J
              P=A(M,L(IR))*(2**(J-IR))
      30      IP=IP+P
              IF(D.EQ.0) GO TO 21
              DO 22 K=1,D
              IF(IP.EQ.C(K)) GO TO 20
      22      CONTINUE
      21      D=D+1
              C(D)=IP
      20      CONTINUE
              G=2**(J-1)
              IF(D.LE.G) GO TO 25
              GO TO 27
```

```
25      IY=1
        IF(IX.NE.0) GO TO 26
        WRITE(2,204)
204     FORMAT(8X,'D',10X,'PARTITION AND RECODE
       1'THE FOLLOWING COLUMNS'/)
        IX=1
26      WRITE(2,205)D,(L(II),II=1,J)
205     FORMAT(1H1 ,5X,I3,15X,20(5X,I2)/)
27      IF(L(J).EQ.N2) GO TO 31
        L(J)=L(J)+1
        GO TO 6
31      IE=J-1
32      ID=N2+(IE-J)
        IF(L(IE).EQ.ID) GO TO 35
        L(IE)=L(IE)+1
        DO 33 K=IE+1,J
33      L(K)=L(K-1)+1
        GO TO 6
35      IE=IE-1
        IF(IE.EQ.0) GO TO 40
        GO TO 32
40      CONTINUE
        IF(IY.EQ.1) GO TO 50
        WRITE(2,206)
206     FORMAT(1H1 ,30X,20HNO USEFUL PARTITIONS/)
50      CONTINUE
        IDD=IDD+1
        IF(IDD.LT.IW) GO TO 3
        STOP
        END
        FINISH
   ****
```

APPENDIX 7.2   EXAMPLE (7.2) ENTERED ON A SIX-VARIABLE DECOMPOSITION CHART

uvwxyz

```
 0  1  2  3  4  5  6  7  8  9 10 11 12 13 14 15 16 17 18 19 20 21 22 23 24 25 26 27 28 29 30 31
32 33 34 35 36 37 38 39 40 41 42 43 44 45 46 47 48 49 50 51 52 53 54 55 56 57 58 59 60 61 62 63
```

u — vwxyz

```
 0  1  2  3  4  5  6  7  8  9 10 11 12 13 14 15 16 17 18 19 20 21 22 23 24 25 26 27 28 29 30 31
32 33 34 35 36 37 38 39 40 41 42 43 44 45 46 47 48 49 50 51 52 53 54 55 56 57 58 59 60 61 62 63
```

v — uwxyz

```
 0  1  2  3  4  5  6  7  8  9 10 11 12 13 14 15 32 33 34 35 36 37 38 39 40 41 42 43 44 45 46 47
16 17 18 19 20 21 22 23 24 25 26 27 28 29 30 31 48 49 50 51 52 53 54 55 56 57 58 59 60 61 62 63
```

w — uvxyz

```
 0  1  2  3  4  5  6  7 16 17 18 19 20 21 22 23 32 33 34 35 36 37 38 39 48 49 50 51 52 53 54 55
 8  9 10 11 12 13 14 15 24 25 26 27 28 29 30 31 40 41 42 43 44 45 46 47 56 57 58 59 60 61 62 63
```

x — uvwyz

```
 0  1  2  3  8  9 10 11 16 17 18 19 24 25 26 27 32 33 34 35 40 41 42 43 48 49 50 51 56 57 58 59
 4  5  6  7 12 13 14 15 20 21 22 23 28 29 30 31 36 37 38 39 44 45 46 47 52 53 54 55 60 61 62 63
```

y — uvwxz

```
 0  1  4  5  8  9 12 13 16 17 20 21 24 25 28 29 32 33 36 37 40 41 44 45 48 49 52 53 56 57 60 61
 2  3  6  7 10 11 14 15 18 19 22 23 26 27 30 31 34 35 38 39 42 43 46 47 50 51 54 55 58 59 62 63
```

z — uvwxy

```
 0  2  4  6  8 10 12 14 16 18 20 22 24 26 28 30 32 34 36 38 40 42 44 46 48 50 52 54 56 58 60 62
 1  3  5  7  9 11 13 15 17 19 21 23 25 27 29 31 33 35 37 39 41 43 45 47 49 51 53 55 57 59 61 63
```

(a)

uv — wxyz

```
 0  1  2  3  4  5  6  7  8  9 10 11 12 13 14 15
16 17 18 19 20 21 22 23 24 25 26 27 28 29 30 31
32 33 34 35 36 37 38 39 40 41 42 43 44 45 46 47
48 49 50 51 52 53 54 55 56 57 58 59 60 61 62 63
```

uw — vxyz

```
 0  1  2  3  4  5  6  7 16 17 18 19 20 21 22 23
 8  9 10 11 12 13 14 15 24 25 26 27 28 29 30 31
32 33 34 35 36 37 38 39 48 49 50 51 52 53 54 55
40 41 42 43 44 45 46 47 56 57 58 59 60 61 62 63
```

ux — vwyz

```
 0  1  2  3  8  9 10 11 16 17 18 19 24 25 26 27
 4  5  6  7 12 13 14 15 20 21 22 23 28 29 30 31
32 33 34 35 40 41 42 43 48 49 50 51 56 57 58 59
36 37 38 39 44 45 46 47 52 53 54 55 60 61 62 63
```

uy — vwxz

```
 0  1  4  5  8  9 12 13 16 17 20 21 24 25 28 29
 2  3  6  7 10 11 14 15 18 19 22 23 26 27 30 31
32 33 36 37 40 41 44 45 48 49 52 53 56 57 60 61
34 35 38 39 42 43 46 47 50 51 54 55 58 59 62 63
```

uz — vwxy

```
 0  2  4  6  8 10 12 14 16 18 20 22 24 26 28 30
 1  3  5  7  9 11 13 15 17 19 21 23 25 27 29 31
32 34 36 38 40 42 44 46 48 50 52 54 56 58 60 62
33 35 37 39 41 43 45 47 49 51 53 55 57 59 61 63
```

vw — uxyz

```
 0  1  2  3  4  5  6  7 32 33 34 35 36 37 38 39
 8  9 10 11 12 13 14 15 40 41 42 43 44 45 46 47
16 17 18 19 20 21 22 23 48 49 50 51 52 53 54 55
24 25 26 27 28 29 30 31 56 57 58 59 60 61 62 63
```

vx — uwyz

```
 0  1  2  3  8  9 10 11 32 33 34 35 40 41 42 43
 4  5  6  7 12 13 14 15 36 37 38 39 44 45 46 47
16 17 18 19 24 25 26 27 48 49 50 51 56 57 58 59
20 21 22 23 28 29 30 31 52 53 54 55 60 61 62 63
```

vy — uwxz

```
 0  1  4  5  8  9 12 13 32 33 36 37 40 41 44 45
 2  3  6  7 10 11 14 15 34 35 38 39 42 43 46 47
16 17 20 21 24 25 28 29 48 49 52 53 56 57 60 61
18 19 22 23 26 27 30 31 50 51 54 55 58 59 62 63
```

(b)

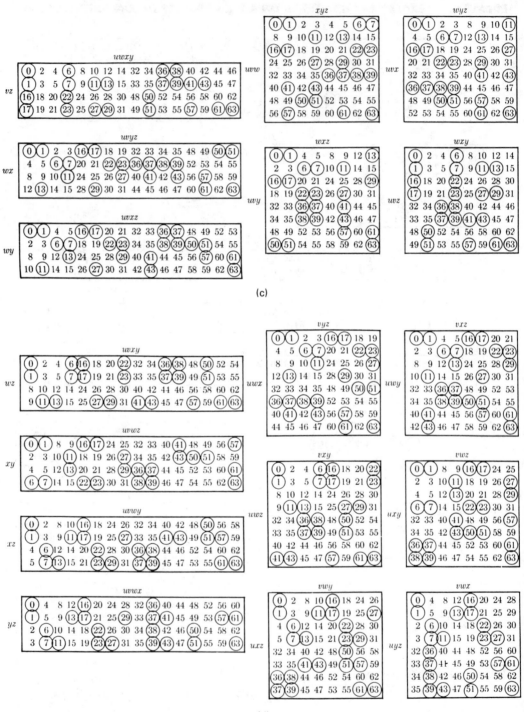

(c)

(d)

The charts are reproduced from reference 20

APPENDIX 7.3   COMPUTER PROGRAM FOR MINIMIZING EXCLUSIVE LOGIC FUNCTIONS

```
10 REM DIM X[M+1,N],S[N,N],P[N],Q[N],Z[M],H[M],G[M],
D[N,?],R[N,N],E[N],L[N]
20 DIM X[1,1],S[1,1],P[1],Q[1],Z[1],H[1],G[1],D[1,1],
R[1,1],E[1],L[1]
30 REM
40 REM                                              Reading Data
50 READ N,M
60 PRINT "      FUNCTIONS = "N,"      WORDS = "M
70 PRINT "     ******************          *******
** "
80 PRINT
90 PRINT " THE DATA"
100 PRINT " ********"
110 PRINT
120 PRINT "FUNCTIONS            MINTERMS"
130 PRINT "***********      ************"
140 FOR K = 1 TO N
150 READ D[K,1]
160 PRINT " F"K"      ";
170 D[K,1] = D[K,1] + 1
180 FOR K2 = 2 TO D[K,1]
190 READ D[K,K2]
200 PRINT D[K,K2];
210 F = (K2 - 1)/8
220 IF F£INTF THEN 250
230 PRINT
240 PRINT "              ";
250 NEXT K2
260 PRINT
270 NEXT K
280 REM                          Finding S-Table
290 PRINT
300 PRINT
310 PRINT "     S   TABLE"
320 PRINT "    *************"
330 PRINT
340 S[1,1] = 1
350 FOR K = 2 TO N
360 F = 1
370 FOR K2 = 1 TO K - 1
380 FOR K3 = 2 TO D[K,1]
390 FOR K4 = 2 TO D[K2,1]
400 IF D[K,K3] = D[K2,K4] THEN 450
410 NEXT K4
420 NEXT K3
430 F = F + 1
440 S[K,F] = K2
450 NEXT K2
460 S[K,1] = F
470 NEXT K
480 FOR K = 2 TO N
490 PRINT " S"K" = ";
500 IF S[K,1] = 1 THEN 580
```

```
510 FOR K2 = 2 TO S[K,1]
520 PRINT S[K,K2];
530 F = (K2 − 1)/8
540 IF F£INTF THEN 570
550 PRINT
560 PRINT "                    ";
570 NEXT K2
580 PRINT
590 NEXT K
600 REM                                    Finding E-List
610 V = 0
620 FOR K = 1 TO N
630 V2 = V
640 IF S[K,1] > 1 THEN 690
650 V = V + 1
660 X[V,2] = K
670 X[V,1] = 2
680 GOTO 1450
690 MAT P = ZER
700 MAT Q = ZER
710 FOR K2 = 2 TO S[K,1]
720 F = S[K,K2]
730 Q[F] = 1
740 NEXT K2
750 FOR K2 = 1 TO V2
760 Z[K2] = 0
770 H[K2] = 0
780 FOR K3 = 2 TO X[K2,1]
790 F = X[K2,K3]
800 H[K2] = H[K2] + Q[F]
810 NEXT K3
820 IF H[K2] > 1 THEN 840
830 Z[K2] = 1
840 IF H[K2] < X[K2,1] − 1 THEN 930
850 FOR K3 = 2 TO X[K2,1]
860 F = X[K2,K3]
870 P[F] = 1
880 NEXT K3
890 X[K2,1] = X[K2,1] + 1
900 F = X[K2,1]
910 X[K2,F] = K
920 Z[K2] = 1
930 NEXT K2
940 U = 0
950 U = U + 1
960 IF U < V2 THEN 990
970 IF Z[U] > 0 THEN 1370
980 GOTO 1050
990 IF Z[U] > 0 THEN 950
1000 FOR K2 = U + 1 TO V2
```

```
1010 IF Z[K2] >0 THEN 1040
1020 IF H[U] > = H[K2] THEN 1040
1030 U = K2
1040 NEXT K2
1050 Z[U] = 1
1060 FOR K2 = 2 TO X[U,1]
1070 F = X[U,K2]
1080 IF Q[F] >P[F] THEN 1240
1090 NEXT K2
1100 FOR K2 = 1 TO V
1110 F = X[K2,1]
1120 IF X[K2,F] <K THEN 1230
1130 FOR K3 = 2 TO X[U,1]
1140 F = X[U,K3]
1150 IF Q[F] = 0 THEN 1210
1160 FOR K4 = 2 TO X[K2,1] − 1
1170 IF F = X[K2,K4] THEN 1210
1180 IF F<X[K2,K4] THEN 1230
1190 NEXT K4
1200 GOTO 1230
1210 NEXT K3
1220 GOTO 940
1230 NEXT K2
1240 V = V + 1
1250 F = 1
1260 FOR K2 = 2 TO X[U,1]
1270 W = X[U,K2]
1280 IF Q[W] = 0 THEN 1320
1290 F = F + 1
1300 X[V,F] = W
1310 P[W] = 1
1320 NEXT K2
1330 F = F + 1
1340 X[V,F] = K
1350 X[V,1] = F
1360 GOTO 940
1370 FOR K2 = 2 TO S[K,1]
1380 F = S[K,K2]
1390 IF P[F] >0 THEN 1440
1400 V = V + 1
1410 X[V,2] = F
1420 X[V,3] = K
1430 X[V,1] = 3
1440 NEXT K2
1450 NEXT K
1460 PRINT
1470 PRINT " THE POSSIBLE GROUPS"
1480 PRINT " *********************"
1490 PRINT
1500 FOR K = 1 TO V
```

```
1510 PRINT K"          ";
1520 FOR K2 = 2 TO X[K,1]
1530 PRINT X[K,K2];
1540 F = (K2 - 1)/8
1550 IF F£INTF THEN 1580
1560 PRINT
1570 PRINT "                ";
1580 NEXT K2
1590 PRINT
1600 NEXT K
1610 REM                    Choosing min. no. of groups covering all functions
1620 FOR K = 1 TO N
1630 Q[K] = 1
1640 P[K] = 0
1650 NEXT K
1660 MAT Z = ZER
1670 FOR K = 1 TO V
1680 FOR K2 = 2 TO X[K,1]
1690 F = X[K,K2]
1700 P[F] = P]F] + 1
1710 NEXT K2
1720 NEXT K
1730 U = 0
1740 FOR K = 1 TO V
1750 FOR K2 = 2 TO X[K,1]
1760 F = X[K,K2]
1770 IF P[F] = 1 THEN 1800
1780 NEXT K2
1790 GOTO 1880
1800 U = U + 1
1810 FOR K2 = 2 TO X[K,1]
1820 F = X[K,K2]
1830 Q[F] = 0
1840 S[U,K2] = F
1850 NEXT K2
1860 S[U,1] = X[K,1]
1870 Z[K] = 1
1880 NEXT K
1890 F = 0
1900 FOR K = 1 TO N
1910 F = F + Q[K]
1920 NEXT K
1930 IF F = 0 THEN 2330
1940 IF U < 2 THEN 2020
1950 MAT·P = ZER
1960 FOR K = 1 TO U
1970 FOR K2 = 2 TO S[K,1]
1980 F = S[K,K2]
1990 P[F] = P[F] + 1
·2000 NEXT K2
```

```
2010 NEXT K
2020 FOR K = 1 TO V
2030 G[K] = 0
2040 H[K] = 0
2050 IF Z[K] > 0 THEN 2110
2060 FOR K2 = 2 TO X[K,1]
2070 F = X[K,K2]
2080 G[K] = G[K] + Q[F]
2090 H[K] = H[K] + P[F]
2100 NEXT K2
2110 NEXT K
2120 W = 0
2130 W = W + 1
2140 IF G[W] = 0 THEN 2130
2150 IF W = V THEN 2240
2160 FOR K = W + 1 TO V
2170 IF G[W] > G[K] THEN 2230
2180 IF G[W] < G[K] THEN 2220
2190 IF X[W,1] > X[K,1] THEN 2230
2200 IF X[W,1] < X[K,1] THEN 2220
2210 IF H[W] < = H[K] THEN 2230
2220 W = K
2230 NEXT K
2240 U = U + 1
2250 FOR K = 2 TO X[W,1]
2260 F = X[W,K]
2270 Q[F] = 0
2280 S[U,K] = F
2290 NEXT K
2300 S[U,1] = X[W,1]
2310 Z[W] = 1
2320 GOTO 1890
2330 REM
2340 REM                                    Reshaping the chosen groups
2350 MAT E = ZER
2360 FOR K = 1 TO U
2370 F = 0
2380 FOR K2 = 2 TO S[K,1]
2390 W = S[K,K2]
2400 F = F + D[W,1] − 1
2410 NEXT K2
2420 IF F < M THEN 2440
2430 E[K] = 1
2440 NEXT K
2450 MAT Z = ZER
2460 W = 0
2470 W = W + 1
2480 FOR K = 1 TO N
2490 P[K] = 0
2500 Q[K] = N
```

```
2510 NEXT K
2520 FOR K = W TO U
2530 FOR K2 = 2 TO S[K,1]
2540 F = S[K,K2]
2550 P[F] = P[F] + 1
2560 NEXT K2
2570 NEXT K
2580 FOR K = W TO U
2590 G[K] = 0
2600 H[K] = 0
2610 FOR K2 = 2 TO S[K,1]
2620 F = S[K,K2]
2630 IF P[F] > 1 THEN 2650
2640 G[K] = G[K] + 1
2650 H[K] = H[K] + P[F]
2660 NEXT K2
2670 NEXT K
2680 FOR K = W TO U − 1
2690 F = K
2700 FOR K2 = K + 1 TO U
2710 IF G[F] < G[K2] THEN 2770
2720 IF G[F] > G[K2] THEN 2760
2730 IF S[F,1] < S[K2,1] THEN 2770
2740 IF S[F,1] > S[K2,1] THEN 2760
2750 IF H[F] > = H[K2] THEN 2770
2760 F = K2
2770 NEXT K2
2780 IF F = K THEN 3000
2790 FOR K2 = 2 TO S[F,1]
2800 L[K2] = S[F,K2]
2810 NEXT K2
2820 FOR K2 = 2 TO S[K,1]
2830 S[F,K2] = S[K,K2]
2840 NEXT K2
2850 FOR K2 = 2 TO S[F,1]
2860 S[K,K2] = L[K2]
2870 NEXT K2
2880 L1 = G[F]
2890 L2 = H[F]
2900 L3 = S[F,1]
2910 L4 = E[F]
2920 G[F] = G[K]
2930 H[F] = H[K]
2940 S[F,1] = S[K,1]
2950 E[F] = E[K]
2960 G[K] = L1
2970 H[K] = L2
2980 S[K,1] = L3
2990 E[K] = L4
3000 NEXT K
```

```
3010 FOR K = W TO U
3020 L[K] = INT(LOG(G[K])/LOG(2) + 1)
3030 F = INT(LOG(S[K,1] − 1)/LOG(2) + 1)
3040 IF L[K] < F AND (S[K,1] − 1 > 2^L[K] OR E[K] = 0) THEN 3260
3050 IF K = W THEN 3210
3060 FOR K2 = 2 TO S[W,1]
3070 L[K2] = S[W,K2]
3080 NEXT K2
3090 FOR K2 = 2 TO S[K,1]
3100 S[W,K2] = S[K,K2]
3110 NEXT K2
3120 FOR K2 = 2 TO S[W,1]
3130 S[K,K2] = L[K2]
3140 NEXT K2
3150 L1 = S[W,1]
3160 L2 = E[W]
3170 S[W,1] = S[K,1]
3180 E[W] = E[K]
3190 S[K,1] = L1
3200 E[K] = L2
3210 FOR K2 = 2 TO S[W,1]
3220 F = S[W,K2]
3230 Z[F] = 1
3240 NEXT K2
3250 GOTO 3730
3260 NEXT K
3270 E[W] = 0
3280 FOR K = W + 1 TO U
3290 V = S[K,1] − 2^L[K]
3300 FOR K2 = 2 TO S[K,1]
3310 F = S[K,K2]
3320 IF Q[F] < = V THEN 3340
3330 Q[F] = V
3340 NEXT K2
3350 NEXT K
3360 V = 2^L[W]
3370 V2 = 1
3380 FOR K = 2 TO S[W,1]
3390 F = S[W,K]
3400 IF P[F] > 1 THEN 3450
3410 V2 = V2 + 1
3420 S[W,K] = S[W,V2]
3430 S[W,V2] = F
3440 Z[F] = 1
3450 NEXT K
3460 GOTO 3590
3470 V2 = V2 + 1
3480 K2 = V2
3490 W2 = S[W,K2]
3500 FOR K = V2 + 1 TO S[W,1]
```

```
3510 F = S[W,K]
3520 IF Q[W2] < = Q[F] THEN 3550
3530 K2 = K
3540 W2 = F
3550 NEXT K
3560 S[W,K2] = S[W,V2]
3570 S[W,V2] = W2
3580 Z[W2] = 1
3590 IF V2<V THEN 3470
3600 S[W,1] = V
3610 IF S[W,1] <3 THEN 3730
3620 FOR K = 2 TO S[W,1] − 1
3630 K3 = K
3640 F = S[W,K3]
3650 FOR K2 = K + 1 TO S[W,1]
3660 IF S[W,K3] < = S[W,K2] THEN 3690
3670 K3 = K2
3680 F = S[W,K3]
3690 NEXT K2
3700 S[W,K3] = S[W,K]
3710 S[W,K] = F
3720 NEXT K
3730 FOR K = W + 1 TO U
3740 V = 1
3750 FOR K2 = 2 TO S[K,1]
3760 F = S[K,K2]
3770 IF Z[F] = 1 THEN 3800
3780 V = V + 1
3790 S[K,V] = F
3800 NEXT K2
3810 IF S[K,1] = V then 3840
3820 E[K] = 0
3830 S[K,1] = V
3840 NEXT K
3850 IF W<U − 1 THEN 2470
3860 V = U
3870 FOR K = 1 TO V
3880 L1 = S[K,1]
3890 F = LOG(L1 − 1)/LOG2
3900 IF F>INTF OR L1 = 2 OR E[K] = 1 THEN 3950
3910 U = U + 1
3920 S[U,1] = 2
3930 S[U,2] = S[K,L1]
3940 S[K,1] = L1 − 1
3950 NEXT K
3960 FOR K = 1 TO U − 1
3970 K3 = K
3980 FOR K2 = K + 1 TO U
3990 IF S[K3,1] > S[K2,1] OR (S[K3,1] = S[K2,1] AND E[K3] > = E[K2])
THEN 4010
4000 K3 = K2
```

```
4010 NEXT K2
4020 IF K = K3 THEN 4150
4030 FOR K2 = 1 TO S[K3,1]
4040 X[1,K2] = S[K3,K2]
4050 NEXT K2
4060 FOR K2 = 1    S[K,1]
4070 S[K3,K2] = S[K,K2]
4080 NEXT K2
4090 FOR K2 = 1 TO X[1,1]
4100 S[K,K2] = X[1,K2]
4110 NEXT K2
4120 L1 = E[K]
4130 E[K] = E[K3]
4140 E[K3] = L1
4150 NEXT K
4160 PRINT
4170 PRINT " THE CHOSEN GROUPS"
4180 PRINT " *****************"
4190 PRINT
4200 FOR K = 1 TO U
4210 PRINT K"   ";
4220 FOR K2 = 2 TO S[K,1]
4230 PRINT S[K,K2];
4240 F = (K2 − 1)/8
4250 IF F£INTF THEN 4280
4260 PRINT
4270 PRINT "   £;   ";
4280 NEXT K2
4290 PRINT
4300 NEXT K
4310 REM      Choosing the code and finding final output
4320 FOR K = 1 TO U
4330 P[K] = INT(LOG(S[K,1] − 1 − E[K])/LOG(2)) + 1
4340 NEXT K
4350 B = 0
4360 FOR K = 1 TO U
4370 B2 = B + 1
4380 B = B + P[K]
4390 IF E[K] = 1 THEN 4450
4400 S[K,1] = S[K,1] + 1
4410 FOR K2 = S[K,1] TO 3 STEP − 1
4420 S[K,K2] = S[K,K2 − 1]
4430 NEXT K2
4440 S[K,2] = 0
4450 FOR K2 = 1 TO S[K,1] − 1
4460 F = K2 − 1
4470 FOR K3 = B TO B2 STEP − 1
4480 R[K3,K2] = F − 2*INT(F/2)
4490 F = INT(F/2)
4500 NEXT K3
4510 NEXT K2
```

```
4520 FOR K2 = B2 TO B
4530 F = 1
4540 FOR K3 = 1 TO S[K,1] − 1
4550 IF R[K2,K3] = 0 THEN 4610
4560 W = S[K,K3 + 1]
4570 FOR K4 = 2 TO D[W,1]
4580 F = F + 1
4590 X[F,K2] = D[W,K4]
4600 NEXT K4
4610 NEXT K3
4620 X[1,K2] = F
4630 NEXT K2
4640 NEXT K
4650 PRINT
4660 PRINT "    THE CODE "
4670 PRINT "    ********* "
4680 B = 0
4690 FOR K = 1 TO U
4700 B2 = B + 1
4710 B = B + P[K]
4720 PRINT
4730 PRINT "GROUP NO;" K, "ELEMENTS = " S[K,1] − 1, "DEGREE = " P[K]
4740 PRINT "*************"
4750 PRINT "      ";
4760 FOR K2 = B2 TO B
4770 PRINT "R" K2;
4780 NEXT K2
4790 PRINT
4800 FOR K2 = 2 TO S[K, − ]
4810 IF S[K,K2] = 0 THEN 4840
4820 PRINT "F" S[K,K2] "   ";
4830 GOTO 4850
4840 PRINT "NON     ";
4850 FOR K3 = B2 TO B
4860 PRINT R[K3,K2 − 1] "   ";
4870 NEXT K3
4880 PRINT
4890 NEXT K2
4900 NEXT K
4910 FOR K = 1 TO B
4920 FOR K2 = 2 TO X[1,K] − 1
4930 K4 = K2
4940 F = X[K4,K]
4950 FOR K3 = K2 + 1 TO X[1,K]
4960 IF F < = X[K3,K] THEN 4990
4970 K4 = K3
4980 F = X[K4,K]
4990 NEXT K3
5000 X[K4,K] = X[K2,K]
5010 X[K2,K] = F
```

```
5020 NEXT K2
5030 NEXT K
5040 REM      Write the final output
5050 PRINT
5060 PRINT "   THE FINAL OUTPUT"
5070 PRINT "   ****************"
5080 PRINT
5090 PRINT "FUNCTIONS      MINTERMS"
5100 PRINT "*********      ********"
5110 FOR K = 1 TO B
5120 PRINT "  R"K;
5130 FOR K2 = 2 TO X[1,K]
5140 PRINT X[K2,K];
5150 F = (K2 - 1)/8
5160 IF F£INTF THEN 5190
5170 PRINT
5180 PRINT "      ";
5190 NEXT K2
5200 PRINT
5210 NEXT K
5220 PRINT
5230 PRINT
5240 READ I
5250 GOSUB I OF 5590, 5270
5260 END
5270 REM      Reshaping the output for UL Ms
5280 PRINT "FOR U.L.M."
5290 PRINT "*********"
5300 FOR K = 1 TO B
5310 PRINT
5320 PRINT
5330 PRINT
5340 N1 = X[1,K] - 1
5350 N2 = INT(LOG(X[X[1,K],K])/LOG(2)) + 1
5360 PRINT "R"K;"  WORDS = "N1;"  NO. OF INPUTS = "N2
5370 PRINT "***"
5380 PRINT
5390 PRINT "WORD DECIMAL    BINARY"
5400 PRINT "            ";
5410 FOR K2 = 1 TO N2
5420 PRINT "X"K2;
5430 NEXT K2
5440 PRINT
5450 FOR K2 = 2 TO X[1,K]
5460 F = X[K2,K]
5470 FOR K3 = N2 TO 1 STEP -1
5480 Q[K3] = F - 2*INT(F/2)
5490 F = INT(F/2)
5500 NEXT K3
5510 PRINT K2 - 1;X[K2,K];
```

```
5520 FOR K3 = 1 TO N2
5530 PRINT " "Q[K3];
5540 NEXT K3
5550 PRINT
5560 NEXT K2
5570 NEXT K
5580 RETURN
5590 REM      Reshaping the output for ROMs
5600 PRINT "FOR ROMS "
5610 PRINT "******** "
5620 FOR K = 1 TO B
5630 F = X[1,K]
5640 W = X[F,K]
5650 FOR K2 = M + 1 TO 2 STEP - 1
5660 IF W = K2 - 2 THEN 5690
5670 X[K2,K] = 0
5680 GOTO 5740
5690 X[K2,K] = 1
5700 F = F - 1
5710 W = X[F,K]
5720 IF F > 1 THEN 5740
5730 W = - 1
5740 NEXT K2
5750 NEXT K
5760 PRINT
5770 PRINT "        ";
5780 FOR K = 1 TO B
5790 PRINT " R"K;
5800 NEXT K
5810 PRINT
5820 PRINT "WORD"
5830 FOR K = 2 TO M + 1
5840 PRINT K - 2;
5850 FOR K2 = 1 TO B
5860 PRINT "    "X[K,K2];
5870 NEXT K2
5880 PRINT
5890 NEXT K
5900 RETURN
```

APPENDIX 7.4   COMPUTER PROGRAM FOR MINIMIZING ULM TREES

```
        MASTER ULM TREES
C       THIS PROGRAM MINIMIZES ULM TREES
        INTEGER C1,P1,P2,P,G,D,L(30),C(200),A(300,20),B(64,64),CM,
       1L1(20),L2(20),L1M(20),L2M(20),CC,LM(20),BN(20),BM(64,64)
C       GIVE THE NO. OF FUNCTIONS
        READ(1,100)IW
100     FORMAT(IO)
        IDD=0
C       GIVE PARAMETERS FOR PRESENT FUNCTION AND MINTERMS
1       READ(1,102) N1,N2,C1,CC,IPR
102     FORMAT(5IO)
        DO 2 J1=1,N1
2       READ(1,101)(A(J1,J2),J2=1,N2)
101     FORMAT(30IO)
        WRITE(2,203)
203     FORMAT(//)
        IF(IPR.EQ.0) GO TO 49
        WRITE(2,200)
200     FORMAT(3X,5HNO,OF,18X,15HMINTERMS OF THE/)
        WRITE(2,207)
207     FORMAT(2X,8HMINTERMS,16X,14HLOGIC FUNCTION/)
        DO 3 J1=1,N1
3       WRITE(2,201)J1,(A(J1,J2),J2=1,N2)
201     FORMAT(1H1 ,5X,I3,5X,30(5X,I1)/)
49      WRITE(2,203)
        WRITE(2,202)
202     FORMAT(2X,'3 AND 2 IN THE RESIDUE MATRIX REFER TO THE',
       1' RESIDUE VARIABLE AND ITS COMPLEMENT RESPECTIVELY'/)
        IF(IPR.EQ.0) GO TO 98
        WRITE(2,203)
        WRITE(2,208)
208     FORMAT(2X,5HNO.OF,6X,7HRESIDUE,8X,7HCONTROL,9X,7HRESIDUE/)
        WRITE(2,204)
204     FORMAT(2X,9HCONSTANTS,2X,8HVARIABLE,6X,9HVARIABLES,8X,6HMATRIX/)
98      CONTINUE
C       SELECT VARIABLES AND RESIDUE MATRIX
        J=N2-1
```

```
          CM=J-CC
5         JK,IC=0
          DO 4 I=1,J
4         L(I)=I
6         WRITE(2,203)
          DO 60 IV=1,N2
          DO 63 IZ=1,J
          IF(IV.EQ.L(IZ)) GO TO 60
63        CONTINUE
          GO TO 64
60        CONTINUE
64        CONTINUE
          DO 90 I3=1,C1
90        L1(I3)=L(I3)
7         D,I1=0
          IA=0
          DO 92 IX=1,J
          DO 93 IU=1,C1
          IF(L(IX).EQ.L1(IU)) GO TO 92
93        CONTINUE
          IA=IA+1
          L2(IA)=L(IX)
92        CONTINUE
          DO 20 M=1,N1
          IP=0
          IP1,IP2=1
          DO 30 IR1=1,C1
          P1=A(M,L1(IR1))*(2**(C1-IR1))
30        IP1=IP1+P1
          DO 39 IR2=1,IA
          P2=A(M,L2(IR2))*(2**(IA-IR2))
39        IP2=IP2+P2
          DO 33 IR= 1,J
          P=A(M,L(IR))*(2**(J-IR))
38        IP=IP+P
          IF(D.EQ.0) GO TO 21
          DO 22 K=1,D
          IF(IP.EQ.C(K)) GO TO 19
22        CONTINUE
```

```
21      D=D+1
        C(D)=IP
61      IF(A(M,IV).EQ.1) GO TO 62
        B(IP1,IP2)=2
        GO TO 65
62      B(IP1,IP2)=3
65      CONTINUE
        GO TO 20
19      I1=I1+1
        B(IP1,IP2)=1
20      CONTINUE
        G=2**J
        I2=G-D
        IY=I1+I2
        J3=2**C1
        J4=2**(J-C1)
        IF(IPR.EQ.0) GO TO 175
C       PRINT THE PRESENT SOLUTION IF REQUIRED
        WRITE(2,205) IY,IV,(L1(II),II=1,C1)
205     FORMAT(2X,I3,8X,I2,9X,20(2X,I2)/)
        WRITE(2,209)(L2(II),II=1,IA)
209     FORMAT(24X,20(2X,I2)/)
        DO 70 J1=1,J3
70      WRITE(2,206)(B(J1,J2),J2=1,J4)
206     FORMAT(1H1 ,40X,70(3X,I1)/)
C       TEST RESIDUE MATRIX AND STORE AS PRESENT OPTIMAL WITH PARAMETERS
175     IB1=0
        DO 133 J1=1,J3-1
        DO 133 J0=J1+1,J3
        IF(IB1.EQ.0) GO TO 136
        DO 136 JJ=1,IB1
        IF(J1.EQ.BN(JJ)) GO TO 133
        IF(J0.EQ.BN(JJ)) GO TO 135
136     CONTINUE
        DO 132 J2=1,J4
        IF(B(J1,J2).NE.B(J0,J2))GO TO 135
132     CONTINUE
        IB1=IB1+1
```

```
          BN(IB1)=JO
135    CONTINUE
133    CONTINUE
          IY1=0
          DO 141 J1=1,J3
          DO 140 J2=1,J4
          IF(B(J1,J2)-1) 140,140,141
140    CONTINUE
          IY1=IY1+1
141    CONTINUE
C      COMPARE PRESENT SOLUTION WITH PRESENT OPTIMAL
          IF(JK.EQ.0) GO TO 150
          IF(IB1.GT.IBM) GO TO 150
          IF(IB1.LT.IBM) GO TO 160
          IF(IY1.GT.IRC) GO TO 150
          IF(IY1.LT.IRC) GO TO 160
          IF(IY.GT.IYM) GO TO 150
          GO TO 160
150    DO 130 J1=1,J3
          DO 131 J2=1,J4
131    BM(J1,J2)=B(J1,J2)
130    CONTINUE
          IBM=IB1
          IYM=IY
          IRC=IY1
          IVM=IV
          DO 170 JJ=1,C1
170    L1M(JJ)=L1(JJ)
          DO 171 JJ=1,IA
171    L2M(JJ)=L2(JJ)
          DO 172 JJ=1,J
172    LM(JJ)=L(JJ)
          JK=JK+1
160    CONTINUE
C      PRESENT SOLUTION BECOMES OPTIMAL OR ERASED
          DO 83 KK=1,J3
          DO 82 JJ=1,J4
```

```
82      B(KK,JJ)=0
83      CONTINUE
C       TEST OTHER POSSIBLE SOLUTIONS IF ANY
84      IF(L1(C1).EQ.N2) GO TO 91
        L1(C1)=L1(C1)+1
        IF(IXX.EQ.0) GO TO 179
        DO 177 JJ=1,IC
177     IF(L1(C1).EQ.L1(JJ)) GO TO 84
179     CONTINUE
        IF(L1(C1).NE.IV) GO TO 7
        L1(C1)=L1(C1)+1
        IF(IXX.EQ.0) GO TO 189
        DO 187 JJ=1,IC
187     IF(L1(C1).EQ.L1(JJ)) GO TO 84
189     CONTINUE
        IF(L1(C1).LE.N2) GO TO 7
        L1(C1)=L1(C1)-2
91      IE1=C1-1
        IF(IE1.EQ.IC) GO TO 27
95      ID1=N2+IE1-C1
162     IF(L1(IE1).EQ.ID1) GO TO 96
        L1(IE1)+L1(IE1)+1
        IF(IXX.EQ.0) GO TO 186
        DO 185 JJ=1,IC
185     IF(L1(IE1).EQ.L1(JJ)) GO TO 162
186     CONTINUE
        IF(L1(IE1).NE.IV) GO TO 99
        L1(IE1)=L1(IE1)+1
        IF(IXX.EQ.0) GO TO 196
        DO 195 JJ=1,IC
195     IF(L1(IE1).EQ.L1(JJ)) GO TO 162
196     CONTINUE
        IF(L1(IE1).LE.ID1) GO TO 99
        GO TO 96
99      ID2=ID1
        DO 97 K1=IE1+1,C1
163     L1(K1)=D1(K1-1)+1
        ID2=ID2+1
```

```
            IF(IXX.EQ.0) GO TO 173
188         DO 174 JJ=1,IC
174         IF(L1(K1).EQ.L1(JJ)) GO TO 178
            GO TO 173
178         L1(K1)=L1(K1)+1
            GO TO 188
173         CONTINUE
164         IF(L1(K1).GT.ID2) GO TO 96
            IF(L1(K1).NE.IV) GO TO 97
            L1(K1)=L1(K1)+1
            IF(IXX.EQ.0) GO TO 164
            GO TO 188
97          CONTINUE
            GO TO 7
96          IE1=IE1-1
            IF(IE1.EQ.IC) GO TO 27
            GO TO 95
27          IF(IXX.EQ.1) GO TO 40
            IF(L(J).EQ.N2) GO TO 31
            L(J)=L(J)+1
            GO TO 6
31          IE=J-1
32          ID=N2+(IE-J)
            IF(L(IE).EQ.ID) GO TO 35
            L(IE)=L(IE)+1
            DO 33 K=IE+1,J
33          L(K)=L(K-1)+1
            GO TO 6
35          IE=IE-1
            IF(IE.EQ.0) GO TO 40
            GO TO 32
40          WRITE(2,203)
C           PRINT OPTIMAL SOLUTION
            WRITE(2,210)
210         FORMAT(1H1 ,20X,22HOPTIMUM RESIDUE MATRIX/)
            WRITE(2,208)
            WRITE(2,204)
            WRITE(2,205) IYM,IVM,(L1M(II),II=1,C1)
            WRITE(2,209)(I2M(II),II=1,IA)
```

```
         DO 161 J1=1,J3
161      WRITE(2,206)(BM(J1,J2),J2=1,J4)
         IF(C1.EQ.CM) GO TO 190
         IF(IBM.GT.0) GO TO 199
         WRITE(2,203)
         WRITE(2,215)
215      FORMAT(1H1 ,20X,23HNO SAVING AT THIS STAGE/)
C        IF NO SAVING  GO BACK AND OPTIMIZE PRESENT AND NEXT LEVELS.
C        ELSE FIX RESIDUE VARIABLE AND CONTROLS OF PRESENT LEVEL
C        THEN MOVE TO NEXT LEVEL
         IXX=0
         C1=C1+CC
         GO TO 5
199      IXX=1
         IC=C1
         C1=C1+CC
         JK=0
         IV=IVM
         DO 191 JJ=1,IC
191      L1(JJ)=L1M(JJ)
         DO 192 JJ=IC+1,C1
192      L1(JJ)=L2M(JJ-CC)
         DO 193 JJ=1,J
193      L(JJ)=LM(JJ)
         GO TO 7
190      CONTINUE
         IDD=IDD+1
C        TEST IF ALL FUNCTIONS ARE SOLVED
         IF(IDD.EQ.IW) GO TO 85
         DO 80 KK=1,J
80       L(KK)=0
         DO 180 KK=1,J3
         DO 181 JJ=1,J4
181      BM(KK,JJ)=0
180      CONTINUE
         GO TO 1
85       CONTINUE
         STOP
         END
         FINISH
```

8

Partitioning of Sequential Circuits

8.1 THE STATE ASSIGNMENT

A major step in the design of a finite-state sequential machine is the state assignment. If binary variables are used to represent the internal states of a machine having n distinct states, s state variables are required for the assignment, as described in Chapter 4, and we have

$$s = \lceil \log_2(n) \rceil \qquad (8.1)$$

It was also indicated in Chapter 4 that arbitrary assignments may be uneconomical in terms of hardware, and that techniques are available for finding economical state assignments.

For a state table of a sequential machine having n rows and s state variables, there are $A(n)$ different assignments, where $A(n)$ is given $\langle 77 \rangle$ by Equation (8.2).

$$A(n) = \frac{(2^s - 1)!}{(2^s - n)! s!} \qquad (8.2)$$

Table 8.1 shows that the number of possible assignments increases very rapidly, thus making it almost impossible to try all possible assignments for large machines.

Humphry $\langle 50 \rangle$ suggested simple rules for the state assignment, but they do not necessarily lead to the best solution. Briefly the rules are:

No. of states n	No. of different assignments
2	1
3	3
4	3
5	140
6	420
7	840
8	840
9	10 810 800
.	.
.	.
.	.
16	$\simeq 5.10^{10}$
17	$\simeq 10^{20}$

Table 8.1 Number of different assignments for sequential machines

(*a*) If two or more states have the same next state, they should be made adjacent in the assignment.

(*b*) It is advisable to give adjacent coding to states, if they are both next states of a present state. If there is a discrepancy between the two rules, the first one should take precedence.

Armstrong ⟨7, 8⟩ described a fast programmed algorithm which was similar to that of Humphry. The program looked for assignments that result in simplified two-level AND-OR, or OR-AND, form for the next state variables, when expressed as Boolean functions of the present internal state variables, and input variables. The method was intended as a fast algorithm for large machines, to produce good assignments in most cases, but not necessarily the best.

Another useful technique was presented by Dolotta and McCluskey ⟨28⟩, and is known to give good results, but not always optimum ones. The choice of assignment is determined by scoring functions which could be altered to suit different situations as described by the user. The method requires large amounts of work particularly for hand computation, but has the advantage that it has been programmed. The model assumes that the combinational network uses only a sum of products, or a product of sums, and two-state logic; the cost is defined as the number of gates and diodes needed for the combinational part.

A technique which is more systematic and results in economical implementation is due to Hartmanis and Stearns ⟨42, 43, 101⟩. The technique is briefly outlined in this chapter and used in Chapters 9 and 10. Hartmanis's state assignment depends on the existence of what are called partitions with the substitution property (s.p. partitions). These are also called closed partitions. It was found that a state assignment based on such partitions results in a subset of the state variables that is independent of the remaining state variables, thus avoiding the necessity of testing all possible assignments in order to arrive at an optimum implementation. The reduced dependency between the variables results in simpler logical expressions. The sequential circuit is effectively decomposed into smaller and simpler blocks.

Before outlining the theory, some basic definitions are necessary:

(i)   A set S is a collection of distinct objects, which are called elements, or members, hence one writes: $a \in A$ to say that a is an element of the set A. A set which has no elements is called a *null* or *empty* set.

(ii) Two sets A and B are said to be equal if they contain the same elements $A = B$.

(iii) $A \subseteq B$ This indicates that A is a subset of B, which happens when every element of A is also an element of B. If B contains at least one element which is not in A, then A is a proper subset of B, written as $A \subset B$.

(iv) Two sets A and B are *disjoint*, if they have no common elements.

(v) $A \cup B$ is the union of A and B, and is the set containing all elements that are elements of either A or B or both; symbolically this is written as

$$A \cup B = \{a \mid a \in A \text{ or } a \in B\}.$$

(vi) $A \cap B$ is the intersection of A and B, and is the set containing elements that are members of both A and B, symbolically.

$$A \cap B = \{a \mid a \in A \text{ and } a \in B\}$$

(vii)   A partition π on a set S is a collection of disjoint subsets of S, such that their entire union is S. The subsets of a partition π are called its *blocks*.

(viii)   A partition on the states of a sequential machine is called a closed partition or s.p. partition with respect to the machine, if for any two states S_i and S_j belonging to the same block of π and any input x, the states $S_i x$ and $S_j x$ are also in a common block of π, where $S_i x$ denotes the x-successor of S_i.

Hartmanis showed that if s binary variables $y_1, y_2, \ldots, y_s$ are assigned to the states of a sequential machine with n states, and if the first k variables, $1 \leqslant k \leqslant s$, designating the next states can be computed from the input and first k-variables of the old states then there exists a partition π with the substitution property for the machine, which identifies all the states that agree in the first k variables.

Therefore, instead of an arbitrary assignment resulting in next-state variables that might depend on all present states as well as input variables, assignments based on s.p. partitions will result in next-state variables that are independent of some of the present state variables. This is, in fact, a decomposition of the original machine into smaller submachines that are simpler and easier to implement, but still give the same overall performance as the original machine.

8.2   CLOSED PARTITIONS

If we are to base our assignments on closed (or s.p.) partitions, we need first of all to know some of their properties, and know how to compute them.

A partition π_1 on S is said to be smaller than or equal to π_2 on S, i.e. $\pi_1 \leqslant \pi_2$, if and only if each block of π_1 is contained in a block of π_2. Thus the smallest partition is the one which contains a single element in each block, while the largest partition has only one block containing all the elements. These two partitions are trivial and called the 0 and I partitions respectively.

$$\text{If} \quad \pi_1 = \overline{ab}, \overline{cd}, \overline{efgh}$$
$$\pi_2 = \overline{ac}, \overline{bd}, \overline{e}, \overline{fgh}$$
$$\pi_3 = \overline{abcd}, \overline{efgh}$$

Then $\pi_3 > \pi_2$

$\pi_3 > \pi_1$

But $\pi_2 \not\geqslant \pi_1$ and $\pi_1 \not\geqslant \pi_2$

Therefore π_1 and π_2 are not comparable

In such a case, we define the least upper bound (lub). For π_1 and π_2 the lub is π_4, such that if $\pi_2 \leqslant \pi_4$ and $\pi_1 \leqslant \pi_4$ and if $\pi_2 \leqslant \pi_5$ and $\pi_1 \leqslant \pi_5$ then $\pi_4 \leqslant \pi_5$. The lub of π_1 and π_2 is given by $\pi_1 + \pi_2$.

Similarly the greatest lower bound (glb) is given by $\pi_1 \cdot \pi_2$.

From above

$$\pi_1 + \pi_2 = \overline{abcd}, \overline{efgh}$$
$$\pi_1 \cdot \pi_2 = \overline{a}, \overline{b}, \overline{c}, \overline{d}, \overline{e}, \overline{fgh}$$

It can be seen that the blocks of $\pi_1 \cdot \pi_2$ are obtained by intersecting the blocks of π_1 and π_2. Thus any two elements are contained in the same block of $\pi_1 \cdot \pi_2$, if and only if they are contained in the same block in π_1 and π_2.

As for $\pi_1 + \pi_2$, let B be a block of π_1; then in the partition $\pi_1 + \pi_2$, the block which contains B is the set union of all blocks of π_1 and π_2 that are chain-connected to B, in the family of subsets consisting of the blocks of π_1 and π_2. Further, if π_1 and π_2 are s.p. partitions, then $\pi_1 \cdot \pi_2$ and $\pi_1 + \pi_2$ are also s.p. partitions. A computer program for computing all s.p. partitions is described next.

8.3   A COMPUTER PROGRAM FOR S.P. PARTITIONS

The main tools in the decomposition theory are the s.p. partitions. It is not always good enough to find just a couple of partitions and start decomposing the machine. This effectively means that before we can claim a good assignment, or an economical implementation, one really needs to have on hand a list of all useful s.p. partitions, and preferably a sketch of the partition lattice so as to be able to make a reasonable assignment. The task of finding all s.p. partitions is probably what deters many people from using this powerful technique, since it is a very lengthy and tedious process especially for large-state tables. It was therefore felt that a digital computer program would be very useful.

To find an s.p. partition, we start by identifying any two distinct states. A list is then made of pairs induced by all inputs. Then the pairs induced by these new pairs are found, and so on, until no new pairs are generated. The subsets of the partition are formed by taking the partition union of the pairs obtained. We define a partition union of two subsets as the union of the two subsets when there is at least one element in common. Pairs formed by the same states are ignored. Comparing partitions and taking their union is continued until no new change occurs. Any missing states are then added to form a partition. The partition obtained is thus the family of subsets that are disjoint and whose union is the entire set. It is closed since it satisfies the substitution property. If this is repeated for all possible pairs of states, a list of all basic closed partitions is formed.

To illustrate this consider the state table in Table 8.2.

Present state	Next state	
	I_1	I_2
1	3	5
2	2	5
3	1	6
4	5	1
5	4	2
6	6	3

Table 8.2   A state table

Start with state pair (1 2). The implied pairs are (2 3) under I_1 and (5 5) (or just 5) under I_2. The implied pair (2 3) results in (1 2) under I_1 (already been considered) and (5 6) which is new and must be considered, and so on. This results in the following:

$$
\begin{aligned}
1\,2 &\rightarrow & 2\,3 \quad 5 \\
2\,3 &\rightarrow & 1\,2 \quad 5\,6 \\
5\,6 &\rightarrow & 4\,6 \quad 2\,3 \\
4\,6 &\rightarrow & 5\,6 \quad 1\,3 \\
1\,3 &\rightarrow & 1\,3 \quad 5\,6
\end{aligned}
$$

Since there are no new pairs to be considered, the partition union of all pairs on the left is taken. (1 2), (2 3), and (1 3) results in subset 1 2 3, and (5 6) and (4 6) result in subset 4 5 6.

Hence the table has a partition $\pi_1 = 1\,2\,3, 4\,5\,6$. The partitions are usually written as follows $\overline{1\,2\,3, 4\,5\,6}$.

The procedure is then repeated for all other pairs, namely (1 3), (1 4), (1 5), (1 6), (2 3), (2 4), ..., and (5 6), a total of 15 iterations. Most of these partitions, however, are likely to be trivial.

The program then computes the input consistent partitions and the state independent partitions (if any), and prints them after the basic partitions. These partitions are defined in Section 8.4.

For any n-state machine, one has to try all the $n(n-1)/2$ distinct pairs. If π_1, π_2, ..., π_t distinct non-trivial partitions are generated, then one has to add these partitions, if they can be added, that is $\pi_1 + \pi_2, \pi_1 + \pi_3, ..., \pi_{t-1} + \pi_t$. Next add the new partitions just generated after elimininating repetition, and so on until no new non-trivial closed partitions are found. The sum of two partitions is found by comparing and taking partition union of the subsets from both partitions. This means that the summation process will commence if the last process has produced at least two new non-trivial partitions.

Present state	Next state	
	$\bar{x}$	x
1	3	4
2	4	3
3	6	6
4	8	6
5	1	2
6	5	7
7	2	2
8	7	5

Table 8.3   State table for Example (8.1)

Consider the relatively small machine in Table 8.3. This has up to $(8 \times 7)/2 = 28$ basic partitions, giving rise to a possibility of $(28 \times 27)/2 = 338$ sums, then a sum of sums, etc. This means that a very large store is required, especially for large tables, to keep all the partitions, together with enough information on the number and size

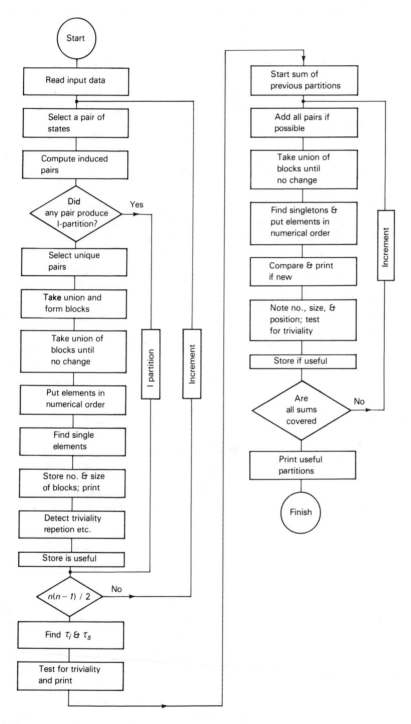

Fig. 8.1   A simplified flow chart for the partition generation program

of subsets in each partition in order to add and compare in the process of looking for new partitions. Thus it becomes necessary to make short cuts, like predicting I-partitions for any induced pair that has already resulted in a trivial partition. Such a partition will be eliminated after printing 'I-partition'. Repeated partitions are also eliminated and some unwanted intermediate results are erased to save storage. Results computed are stored in the matrices declared in the first statement after putting the elements of each block in numerical order. The program listing is not given here but is available in the program library of the Institution of Electrical Engineers, or, on request, from the author. The flow chart is shown in Fig. 8.1.

The computer will print all the basic partitions, the state and input partitions, plus any new partitions obtained from the summation process. This could add up to a large number of partitions, most of which are not useful. The partitions are examined in order to reduce the list of partition candidates.

After examining a partition, a note is printed to identify non-useful partitions, such as an I-partition, a trivial partition, a repeated partition (if the partition is the same as a previously computed partition), or state variables increased (if $s_D > s$ but the individual components are smaller than the original). This last note is a warning to the designer that he should use this partition only if he has no choice, and only in special cases such as parallel decomposition as will be explained later in this chapter. The designer may therefore skip all the labelled partitions. After testing over 20 different machine tables of varying size, it was found that this process leaves only the good partitions, and these are normally very few. The useful partitions are selected and stored. They are then printed separately and can be examined for a possible decomposition.

The program therefore consists of two major sections.

(i) Generating all the basic closed partitions, plus the state and input partitions and testing them for triviality, etc.
(ii) Generating larger closed partitions by adding pairs of already computed partitions, testing them for triviality, etc.

The program will also collect all the useful partitions and print them separately. The criterion for deciding whether a partition is useful will be discussed in due course.

8.4   DECOMPOSITION OF SYNCHRONOUS SEQUENTIAL SYSTEMS

Having computed all the s.p. partitions, we briefly list the main results that help us use these partitions to decompose a given machine into serial, parallel or composite structures that will result in more economical implementations.

Let $\#(\pi)$ be the number of distinct blocks in a partition π.

8.4.1 Serial Decomposition

The existence of a partition τ, and a closed partition π on the set of states of a sequential machine, such that $\pi \cdot \tau = \pi(0)$, is a sufficient condition for the machine to be decomposed into two components operating in series. The first component (predecessor) consists of $\lceil \log_2 \# (\pi) \rceil$ state variables to distinguish the blocks of π. These variables are independent of the remaining $\lceil \log_2 \# (\tau) \rceil$ variables, assigned to distinguish between the blocks of τ. The state variables of the successor may depend on each other, as well as the state variables of the predecessor.

We note that since $\pi \cdot \tau = \pi(0)$, the state variables distinguishing the blocks of τ also distinguish the states within the blocks of π. The lattice and circuit structure are shown in Fig. 8.2.

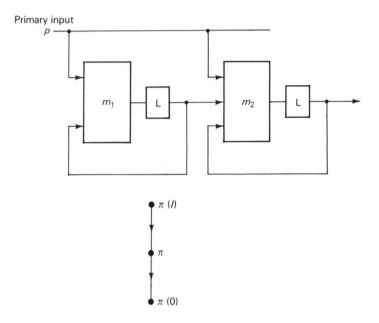

Fig. 8.2   Block diagram and π-lattice of a machine decomposed into two components in series

Some machines may be decomposed into more than two components in a cascaded chain. If for example there are m- closed partitions such that $\pi_1 > \pi_2 > \ldots > \pi_m$, we may select another set of partitions $\tau_1, \tau_2, \ldots, \tau_{m-1}$ such that for any value of i in the range $1 \leqslant i \leqslant m-1$, we have:

$$\pi_i \cdot \tau_i = \pi_{i+1}$$
and
$$\pi_1 \cdot \tau_1 \cdot \tau_2 \ldots \tau_{m-1} = \pi(0)$$

The first component m_1 has $\lceil \log_2 \# (\pi_1) \rceil$ state variables assigned to distinguish the blocks of π_1. The first component is independent of all the successor components. The second component m_2 consists of $\lceil \log_2 \# (\tau_1) \rceil$ state variables, and since $\pi_1 \cdot \tau_1 = \pi_2$, the m_2 component depends on its predecessor m_1 but is independent of its successors, and so on.

8.4.2 Parallel Decomposition

The existence of two non-trivial closed partitions π_1 and π_2 on the states of a machine, such that $\pi_1 \cdot \pi_2 = \pi(0)$ is a sufficient condition for it to be decomposed into two independent components operating in parallel. This decomposition requires a minimum number of state variables if

$$\lceil \log_2 \# (\pi_1) \rceil + \lceil \log_2 \# (\pi_2) \rceil = \lceil \log_2 (n) \rceil,$$

where n is the number of states of the machine.

One should observe that if the product of the partitions $= \pi(0)$, we are assured that the new components do cover the original system and give a complete description of it. We also hint that if $\pi_1 + \pi_2 \neq \pi(I)$, there is a redundancy which may be factored out, as will be shown later.

The lattice and circuit structure of this basic decomposition is shown in Fig. 8.3.

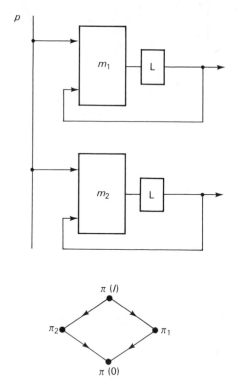

Fig. 8.3   Block diagram and π-lattice of a machine decomposed into two independent components in parallel

If there exist m closed partitions $\pi_1, \pi_2, \ldots, \pi_m$ such that $\pi_1 \cdot \pi_2 \ldots \pi_m = \pi(0)$, the machine may be decomposed into m-components operating in parallel independently of each other.

8.4.3 Composite Decomposition

There are cases where more complex decompositions are possible. In Fig. 8.4 are two common cases where the parallel components have a predecessor Fig. 8.4(*a*) or a successor Fig. 8.4(*b*). One of these cases will be demonstrated by means of an example that follows. It is also possible to have both, a successor and a predecessor, as shown in Fig. 8.4(*c*).

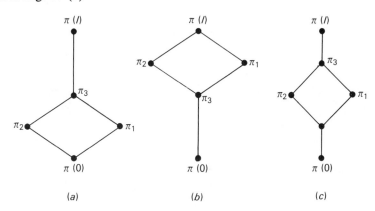

(*a*) (*b*) (*c*)

Fig. 8.4   Lattice diagrams for composite structures

The technique of finding a common submachine which may be factored out to serve as predecessor for two or more machines, and other properties of cascade decomposition, are further discussed by Kohavi and Smith ⟨58, 96⟩.

As well as simplifications resulting from reduced dependence between the state variables, it is possible in some cases to simplify the output expressions ⟨59⟩, and/or reduce the dependence on primary inputs. Some of these cases are briefly discussed.

8.4.4 Reduced Dependence Partitions

(i)   A partition τ_o is called an *output-consistent partition*, if for every block of τ_o and every input, all the states in a block of τ_o have the same output.

Such a partition, if available, ensures an assignment in which the output depends on at the most the primary inputs and variables assigned to the blocks of τ_o. This does not of course guarantee a simpler overall system, since it does not guarantee simpler next-state functions. One should therefore look for a closed partition π, such that $\pi \cdot \tau_o = \pi(0)$ and

$$\lceil \log_2 \# (\pi) \rceil + \lceil \log_2 \# (\tau_o) \rceil \nsupseteq \lceil \log_2 (n) \rceil$$

Otherwise one might end up using more state variables which might offset the saving due to the reduced dependence of the output.

(ii)   A partition τ_s is called a *state-independent partition* if for every primary input x_i and all secondaries $S_1, S_2, \ldots, S_n$, all the next states $S_1 x_i, \ldots, S_n x_i$ are in the same block of τ_s, i.e. the blocks are along columns of the state table.

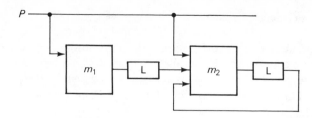

Fig. 8.5   A state-independent predecessor m_1 has no feedback loop

The existence of such a partition means that it is possible to have a submachine whose state variables are assigned to the blocks of τ_s, which is dependent only on the primary inputs, as shown in Fig. 8.5.

(iii)   A partition τ_i is called *input-consistent*, if for every state S_i and all inputs x_1, $x_2, \ldots, x_p$, all the next states $S_i x_1$, $S_i x_2$, $\ldots$, $S_i x_p$ are in the same block of τ_i.

This means that the $\lceil \log_2 \# (\tau_i) \rceil$ state variables distinguishing the blocks of τ_i are independent of the input.

It is also known that if the machine possesses a closed partition $\pi \geqslant \tau_i$, the state variables assigned to the blocks of π are independent of the input and other state variables, as will be shown in a following example.

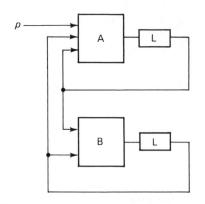

Fig. 8.6(a)   Submachine B is independent of primary input

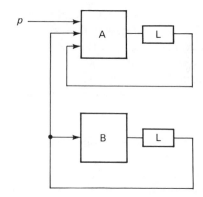

Fig. 8.6(b)   Submachine B is autonomous

In Fig. 8.6, we distinguish two cases. In Fig. 8.6(*a*) the submachine B is *input-independent*, i.e. the present primary inputs have no effect on the immediate next state, but they could still affect the next states at some time in the future. This is because the submachine B may still depend on secondaries of the other component (or components) that are not *input-independent*. Fig. 8.6(*b*) shows the case where the primary inputs have no effect at all on the *input-independent* submachine B, which is called *autonomous*.

There are cases where the table does not possess an input-consistent partition, but has a suitable cover*. In such a case, it is possible to employ state splitting to force a partition; the table is then modified accordingly (this also applies to state-independent partitions). Consider the flow table below which has a cover $\overline{12}$, $\overline{13}$. If state 1 is split into $1a$ and $1b$, then we obtain the partition $\tau_i = \overline{1a\,2}$, $\overline{1b\,3}$, without increasing the state variables necessary to encode the states.

	Input 0	1
1	2	1
2	3	1
3	1	3

Original table

	Input 0	1
1a	2	1a
1b	2	1a
2	3	1b
3	1b	3

Modified table

One could also have partial input independence, where a submachine is independent of one or more primary inputs, but not all. The computer program described earlier computes the input and state independent partitions.

8.4.5 Partition Pairs

Another important concept, is the partition pair assignment. A partition pair (τ, τ') on the states of a sequential machine Mc is an ordered pair of partitions, such that if S_i and S_j are in the same block of τ, then for every input x_k, $x_k S_i$ and $x_k S_j$ are in the same block of τ'. Assignments based on partition pairs results in cross-dependence as will be shown shortly.

If $\tau = \tau'$, then τ is a closed partition, since it will have its own successor blocks. This is because τ' consists of all the successor blocks of τ.

Example (8.1)
Consider the machine given in Table 8.3, which has eight states and two input columns. This machine will be decomposed in more than one way to illustrate the theory already outlined.

This machine has three s.p. partitions. The π-lattice is shown in Fig. 8.7, and the three non-trivial closed partitions are:

$$\tau_i = \pi_1 = \overline{12}, \overline{34}, \overline{57}, \overline{68}$$
$$\pi_2 = \overline{18}, \overline{37}, \overline{45}, \overline{26}$$
$$\pi_3 = \overline{1268}, \overline{3457}$$

(i)   We observe that this machine has an input consistent partition $\tau_i = \pi_1$. If we choose to make use of this finding, the state variables assigned to distinguish between

---

* A cover is a collection of subsets, whose union is S, such that no subset is included in another subset in the collection.

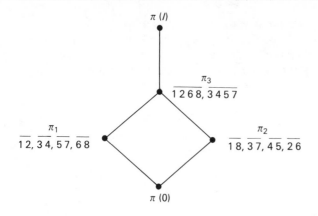

$y_1\,y_2\,y_3$	$Y_1\quad Y_2\quad Y_3$	
	$\bar{x}$	x
0 0 0	1 0 0	1 0 1
0 0 1	1 0 1	1 0 0
1 0 0	1 1 0	1 1 0
1 0 1	1 1 1	1 1 0
0 1 0	0 0 0	0 0 1
1 1 0	0 1 0	0 1 1
0 1 1	0 0 1	0 0 1
1 1 1	0 1 1	0 1 0

Fig. 8.7   Partition lattice for Example (8.1)

Table 8.4   Input-consistent state assignment for Example 8.1

the blocks of τ_i will be independent of the input variable x. These variables are y_1 and y_2, as shown in Table 8.4. Only two variables are needed since $\lceil \log_2 \,\#\, (\tau_i) \rceil = 2$. We need another partition τ, such that $\pi_1 \cdot \tau = \pi(0)$. One such partition is $\tau = \overline{1356}$, $\overline{2478}$, since $(\overline{1356,\ 2478}) \cdot (\overline{12,\ 34,\ 57,\ 68}) = \pi(0)$. A third variable y_3 is assigned to distinguish between the blocks of τ. Note that in Table 8.4, y_3 assigns code 0 to states 1,3,5 and 6 and assigns code 1 to states 2,4,7 and 8.

From Table 8.4, the following is obtained:

$$
\begin{aligned}
Y_1 &= \bar{y}_2 \\
Y_2 &= y_1 \\
Y_3 &= \bar{x}y_3 + x(\bar{y}_1\bar{y}_3 + \bar{y}_1 y_2 + y_2 \bar{y}_3)
\end{aligned}
\tag{8.3}
$$

For such an assignment we would normally expect a two-component structure, the first (the predecessor) consisting of two state variables y_1 and y_2 that are assigned to a closed partition which is also an input-consistent partition. The state variables of the predecessor should therefore be independent of the third variable y_3 and of the input variable x but may depend on each other. The state variable of the successor component is y_3, and since this is assigned to a non-closed partition, it may have no independencies.

From Equation 8.3, we notice that y_1 and y_2 have a cross-dependence which is simpler than we expected. This is due to an unexpected partition pair assignment.

Note that y_1 as given in Table 8.4 distinguishes the blocks of

$$
\tau_{y1} = \overline{1257},\ \overline{3468}
$$

And y_2 distinguishes the blocks of

$$
\tau_{y2} = \overline{1234},\ \overline{5678}
$$

Where $\tau_{y1} \cdot \tau_{y2} = \pi_1$

$\tau_{y1},\ \tau_{y2}$ is a partition pair, which is the reason why y_1 depends only on y_2 and y_2 depends only on y_1, as shown in Fig. 8.8.

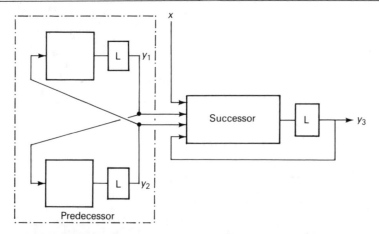

**Fig. 8.8   Circuit structure for the assignment in Table 8.4.
The predecessor is independent of the primary input**

(ii)   Looking again at the input-consistent partition π_1, we note that π_3 is obtainable by the pair-wise addition of the blocks of π_1. Partition π_3 therefore describes a component of the input-independent machine, and is a larger input-consistent partition.

In Table 8.5 is another assignment in which y_1 and y_2 distinguish the blocks of π_1. The state variables y_1 and y_2 are independent of the input and of other state variables. The state variable y_1 distinguishes the blocks of π_3, which makes it a submachine of the input-independent component. Since π_3 is a closed partition, y_1 is independent of all other state variables (see Fig. 8.9). From Table 8.5 we obtain the following:

$$Y_1 = \bar{y}_1$$
$$Y_2 = \bar{y}_1 y_2 + y_1 \bar{y}_2 \tag{8.4}$$
$$Y_3 = \bar{y}_1 \bar{y}_3 + \bar{x} y_1 y_3 + x \bar{y}_1 + x \bar{y}_3$$

Input-independence can have some interesting properties that may be exploited as will be shown in Chapter 9.

$y_1\, y_2\, y_3$	$Y_1 \quad Y_2 \quad Y_3$	
	$\bar{x}$	x
1 1 0	0 0 0	0 0 1
1 1 1	0 0 1	0 0 0
0 0 0	1 0 1	1 0 1
0 0 1	1 0 0	1 0 1
0 1 1	1 1 0	1 1 1
1 0 1	0 1 1	0 1 0
0 1 0	1 1 1	1 1 1
1 0 0	0 1 0	0 1 1

**Table 8.5   Assignment based on partitions
π_1 and π_3 for Example (8.1)**

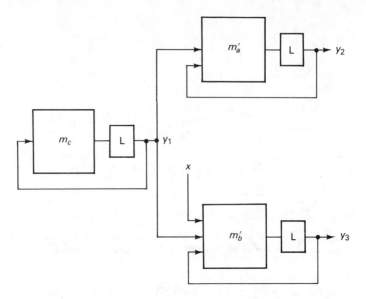

Fig. 8.9 Circuit structure for the assignment in Table 8.5
The components m_c and m_a' are input-independent

(iii) The same machine in Example (8.1) can be decomposed as a serial connection of three components. This may be seen from the lattice diagram (Fig. 8.7) part of which is reproduced here for convenience:

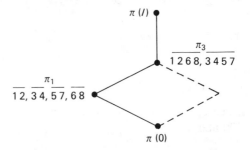

The first component m_1 based on π_3 consists of one state variable y_1 to distinguish the blocks of π_3 which will be independent of all successor components. The second component m_2 is identified by τ_2 in such a way that

$$\tau_2 \cdot \pi_3 = \pi_1$$

$$\tau_2 = \overline{1234,\ 5678}$$

Therefore m_2 consists of a single-state variable y_2 to distinguish the blocks of τ_2 which may depend on y_1 of its predecessor, but is independent of its successor m_3.

Since y_1, y_2 also distinguish the blocks of π_1 which is equal to the input consistent partition τ_i, y_1 and y_2 will be independent of the input.

The state variable y_3 of the tail component m_3 is assigned to the blocks of τ_3 such

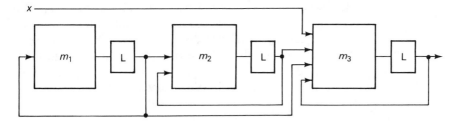

Fig. 8.10   Circuit structure for the assignment in Table 8.6
The components m_1 and m_2 are input-independent

$y_1 y_2 y_3$	Y_1 Y_2 Y_3	
	$\bar{x}$	x
0 0 0	1 0 0	1 0 1
0 0 1	1 0 1	1 0 0
1 0 0	0 1 0	0 1 0
1 0 1	0 1 1	0 1 0
1 1 0	0 0 0	0 0 1
0 1 0	1 1 0	1 1 1
1 1 1	0 0 1	0 0 1
0 1 1	1 1 1	1 1 0

Table 8.6   Serial-decomposition state assignment for Example (8.1)

$y_1 y_2 y_3 y_4$	Y_1 Y_2 Y_3 Y_4	
	$\bar{x}$	x
0 0 0 0	0 1 0 1	0 1 1 0
0 0 1 1	0 1 1 0	0 1 0 1
0 1 0 1	1 1 1 1	1 1 1 1
0 1 1 0	1 1 0 0	1 1 1 1
1 0 1 0	0 0 0 0	0 0 1 1
1 1 1 1	1 0 1 0	1 0 0 1
1 0 0 1	0 0 1 1	0 0 1 1
1 1 0 0	1 0 0 1	1 0 1 0

Table 8.7   Parallel-decomposition state assignment for Example (8.1)

that $\tau_3 \cdot \pi_1 = \pi(0)$, and may depend on all state and input variables. Then $\tau_3 = \overline{1356}$, $\overline{2478}$.

This state assignment is shown in Table 8.6, the circuit structure is shown in Fig. 8.10 and the next state variables are given in Equation (8.5). We have

$$Y_1 = \bar{y}_1$$
$$Y_2 = \bar{y}_1 y_2 + y_1 \bar{y}_2 \qquad\qquad (8.5)$$
$$Y_3 = \bar{x} y_3 + x y_1 y_2 + x \bar{y}_1 \bar{y}_3$$

(iv)   It is also possible to decompose the machine into two components in parallel independently of each other. One component m_a has y_1 and y_2 to distinguish the blocks of π_1. These two variables are independent of y_3 and y_4 in the second component m_b based on π_2; y_3 and y_4 are also independent of y_1 and y_2. Since $\pi_1 = \tau_i$, m_a is also input-independent. This is possible since:

$$(\overline{12},\, \overline{34},\, \overline{57},\, \overline{68}) \cdot (\overline{18},\, \overline{37},\, \overline{45},\, \overline{26}) = \overline{1},\, \overline{2},\, \overline{3},\, \overline{4},\, \overline{5},\, \overline{6},\, \overline{7},\, \overline{8}$$

i.e. $\pi_1 \times \pi_2$ $\qquad\qquad\qquad = \pi(0)$

From the state assignment in Table 8.7, Equations (8.6) are derived for the next state variables of the two components.

Parallel decomposition is normally the most economical, but that is only true if $\pi_1 + \pi_2 = \pi(I)$. In this case $\pi_1 + \pi_2 = \pi_3$, which is smaller than $\pi(I)$.

From this, and by inspecting the lattice diagram, we see that the reason for needing four state variables, instead of the three variables required by the original machine, is that there is a predecessor which is common to both m_a and m_b and which should be factored out if an economical realization is desired. We have

$$Y_1 = y_2$$

$$Y_2 = \bar{y}_1 \tag{8.6}$$

$$Y_3 = \bar{x}\,y_4 + x\,\bar{y}_4 + \bar{y}_3 y_4$$

$$Y_4 = \bar{x}\,\bar{y}_3 + x\,y_3 + \bar{y}_3 y_4$$

Present state	Next state $\bar{x}$	x
$a = 12$	b	b
$b = 34$	d	d
$c = 57$	a	a
$d = 68$	c	c

State table for m_a component

Present state	Next state $\bar{x}$	x
$e = 18$	f	g
$f = 37$	h	h
$g = 45$	e	h
$h = 26$	g	f

State table for m_b component

$$\pi_a = \overline{ad,\ bc} \qquad\qquad \pi_b = \overline{eh,\ fg}$$

$$\tau_a = \overline{ab,\ cd} \qquad\qquad \tau_b = \overline{ef,\ gh}$$

$$\pi_a \cdot \tau_a = \pi(0) \qquad\qquad \pi_b \cdot \tau_b = \pi(0)$$

It is obvious that both parallel components can be serially decomposed, and the state tables of the decomposed components are shown below:

$$m_a$$

	$\bar{x}$	x
$ad = P_1$	Q_1	Q_1
$bc = Q_1$	P_1	P_1

$$Y_5 = \bar{y}_5$$

	$y_5 x$ 00	01	10	11
$ab = \alpha_1$	α_1	α_1	β_1	β_1
$cd = \beta_1$	β_1	β_1	α_1	α_1

The second component may be simplified to the following:

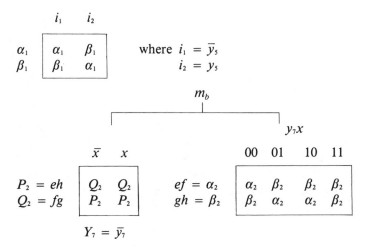

$$\begin{array}{c|cc} & i_1 & i_2 \\ \hline \alpha_1 & \alpha_1 & \beta_1 \\ \beta_1 & \beta_1 & \alpha_1 \end{array}$$

where $i_1 = \bar{y}_5$

$\quad\quad\; i_2 = y_5$

$$m_b$$

$$\begin{array}{c|cc} & \bar{x} & x \\ \hline P_2 = eh & Q_2 & Q_2 \\ Q_2 = fg & P_2 & P_2 \end{array}$$

$$\begin{array}{c} ef = \alpha_2 \\ gh = \beta_2 \end{array}$$

$$\begin{array}{c} y_7 x \\ \begin{array}{cccc} 00 & 01 & 10 & 11 \\ \hline \alpha_2 & \beta_2 & \beta_2 & \beta_2 \\ \beta_2 & \alpha_2 & \alpha_2 & \beta_2 \end{array} \end{array}$$

$$Y_7 = \bar{y}_7$$

The second component can be simplified as well to the following:

$$\begin{array}{c|ccc} & i_3 & i_4 & i_5 \\ \hline \alpha_2 & \alpha_2 & \beta_2 & \beta_2 \\ \beta_2 & \beta_2 & \alpha_2 & \beta_2 \end{array}$$

where $i_3 = \bar{y}_7\bar{x}$

$\quad\quad\; i_4 = \bar{y}_7 x + y_7 \bar{x}$

$\quad\quad\; i_5 = y_7 x$

We observe that the first components in both cases above are of the same structure, and may be factored out.

This is achieved now by decomposing both m_a and m_b serially and factoring the predecessor. The assignment of m_a and m_b is shown in Tables 8.8(a) and 8.8(b).

	Y_5	Y_6
$y_5\, y_6$	$\bar{x}$	x
0 0	1 0	1 0
1 0	0 1	0 1
1 1	0 0	0 0
0 1	1 1	1 1

Table 8.8(a) Assignment for component m_a in Example (8.1)

	Y_7	Y_8
$y_7\, y_8$	$\bar{x}$	x
0 0	1 0	1 1
1 0	0 1	0 1
1 1	0 0	0 1
0 1	1 1	1 0

Table 8.8(b) Assignment for component m_b in Example (8.1)

From these tables the following is obtained for m_a:

$$Y_5 = \bar{y}_5$$

$$Y_6 = y_5\bar{y}_6 + \bar{y}_5 y_6$$

$$\left.\begin{array}{c}\\ \\\end{array}\right\} \quad (8.7)$$

Also the following is obtained for m_b:

$$Y_7 = \bar{y}_7$$

$$Y_8 = \bar{x}\,\bar{y}_7 y_8 + y_7\bar{y}_8 + xy_7 + x\bar{y}_8$$

$$\left.\begin{array}{c}\\ \\\end{array}\right\} \quad (8.8)$$

By factoring the predecessor and re-labelling the variable, expressions similar to those in Equation (8.4) are obtained.

Since m_a is input-independent, the predecessor and m_a' are also input-independent. Further both m_a' and m_b' are dependent on their common predecessor m_c. This composite decomposition structure was shown in Fig. 8.9.

$y_1\,y_2\,y_3$	$Y_1\quad Y_2\quad Y_3$	
	$\bar{x}$	x
0 0 0	0 1 0	0 1 1
0 0 1	0 1 1	0 1 0
0 1 0	1 0 1	1 0 1
0 1 1	1 1 1	1 0 1
1 0 0	0 0 0	0 0 1
1 0 1	1 0 0	1 1 0
1 1 0	0 0 1	0 0 1
1 1 1	1 1 0	1 0 0

Table 8.9   An arbitary state assignment for Example 8.1

The reduced dependency amongst the state variables is obvious in all cases, if compared with an arbitrary assignment which might result in each state variable being dependent on all input and state variables.

(v)   Having seen the various degrees of reduced dependency, we consider what would happen if the assignment was arbitrary, as shown in Table 8.9.

From Table 8.9 the following next-state expressions are derived:

$$Y_1 = \bar{y}_1 y_2 + y_1 y_3 + y_2 y_3$$

$$Y_2 = \bar{y}_1 y_2 + \bar{x} y_2 y_3 + \bar{x}\,\bar{y}_1 y_3 + x\,\bar{y}_2 y_3$$

$$Y_3 = y_2\bar{y}_3 + \bar{y}_1 y_2 + x\,\bar{y}_3 + \bar{x}\,\bar{y}_1 y_3$$

$$\left.\begin{array}{c}\\ \\ \\\end{array}\right\} \quad (8.9)$$

One can see that each state variable depends on all other state variables. In fact the only independence is that Y_1 does not depend on the input variable x. This is because the code for Y_1 happened to partition the states into $\overline{1234}\ \overline{5678}$, which is an input-independent partition that may be derived from τ_i, which is π_1 in Fig. 8.7.

Implementations of Example (8.1) using ROMs will be given in Chapter 11.

The effect of decomposition and reduced dependence on discrete gates is very well known and accepted; its effect on MSI and LSI components will be discussed in Chapter 9.

8.5   STATE SPLITTING ⟨44, 57, 60⟩

In practice, not all sequential systems possess s.p. partitions and consequently cannot be decomposed. In such cases, it is possible to carry out what is known as state splitting. State splitting is the opposite of state reduction, which is the stage normally carried out before the state assignment.

In the state reduction process, two equivalent states are merged, and the entries of the state table are modified accordingly. Two states are considered equivalent if their next states and outputs are identical for all possible inputs. This could be useful if it results in a reduction in the number of state variables required for assignment. The process should be carried out with care, since it could disrupt the systematic information flow in the unreduced machine and spoil both the chance of obtaining s.p. partitions, and the possibility of decomposition.

In state splitting, a state is split into more than one, and the state table is modified accordingly. This process will increase the number of states and might increase the number of state variables. This point will be discussed further in Chapter 9. The state splitting process will be outlined now starting with some definitions.

(i)   The implication graph is a directed graph with vertices corresponding to subsets of the set of states of the machine. The subsets consist of the states to be identified in the state table, or the states that are implied by previously identified subsets of the states. The arc labelled x_i represents the transition from one subset of states to its x_i-successor.

(ii)   A closed implication graph is a sub-graph of the implication graph having the following properties:

(a) For any vertex in the sub-graph all outgoing arcs and their terminating vertices belong to the sub-graph.

(b) All states are represented at least once.

The procedure for generating an augmented machine Mc' possessing at least one closed partition from the machine Mc is summarized as follows:

(a) Draw the implication graph, and choose a sub-graph having the least number of vertices. If any state S_i is present in more than one vertex, we label S_i in the first vertex S_i', and the second S_i'' and so on.

(b) For every split state, split the corresponding states in the original state table, and modify the entries of the new table.

The sub-graph produces what is called a *cover* on the machine. After splitting, all states are now distinguishable, and the vertices are now the blocks of a closed partition of the augmented machine Mc', which is equivalent to the original machine Mc.

9

Partition-based Design for
Synchronous Sequential Circuits

This chapter discusses the effect of decomposition on the hardware requirement of sequential circuits implemented with ROMs, Shift registers, and ULM multiplexers. In situations where more than one decomposition is possible, Lagrange's method of undetermined multipliers is used to optimize the design.

9.1 SHIFT REGISTER IMPLEMENTATION OF SEQUENTIAL CIRCUITS
⟨25, 26, 63, 75, 104⟩

9.1.1 Feedback Shift Register (FSR) Realization

Sequential machines (SM) containing input-consistent partitions are of particular interest in that the input-independent submachine can in some cases be further decomposed into smaller components. This gives rise to a serial internal structure and a serial flow of information which renders itself liable to feedback shift register implementation as shown in Fig. 9.1.

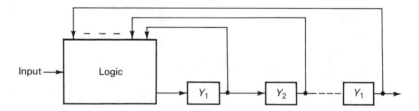

Fig. 9.1 A General FSR implementation

In terms of input consistent partitions, if larger partitions can be found by the pair-wise addition of the blocks of τ_i as was defined in Chapter 8, the larger partitions imply that a submachine within the input independent component can be obtained. The partition τ_i found by collecting all the states in blocks according to the rows is the smallest input-consistent partition. Next states within the input-independent component described by these larger partitions are also input-independent. For example if

$\tau_i = \overline{1\,2}\ \ \overline{3\,4}\ \ \overline{5\,6}\ \ \overline{7\,8}$, then the larger partitions are:

$\tau_{i_1} = \overline{1\,2\,3\,4}\ \ \overline{5\,6\,7\,8}$

$\tau_{i_2} = \overline{1\,2\,5\,6}\ \ \overline{3\,4\,7\,8}$

$\tau_{i_3} = \overline{1\,2\,7\,8}\ \ \overline{3\,4\,5\,6}$

The independence tree, also called the *composition tree*, can be used to indicate whether a sequential machine is FSR realizable, as will be shown in Example (9.1).

The regularity, repeatability and availability of shift registers (SR) as MSI devices make their use for sequential machines very desirable. A SR is a serial connection of *l* unit delays interconnected in such a way that at the occurrence of a shift signal, the contents of the *i*th delay is shifted into the $(i+1)$ delay, as described in Chapter 4.

Example (9.1)

Table 9.1 and the state diagram in Fig. 9.2 represent a sequential system with the input consistent partition τ_i given below. Investigate a shift register implementation for the system.

$$\tau_i = \overline{1\,2}\,,\overline{3\,4}\,,\overline{5\,6}\,,\overline{7\,8}$$

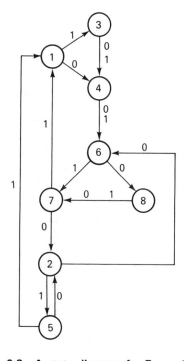

Assignment			Present	Next	
Y_1	Y_2	Y_3	state	state	
				Input	
				0	1
0	0	1	1	4	3
1	0	1	2	6	5
0	0	0	3	4	4
1	0	0	4	6	6
0	1	0	5	2	1
1	1	0	6	8	7
0	1	1	7	2	1
1	1	1	8	7	7

**Table 9.1   Flow table and assignment
for Example (9.1)**

Fig. 9.2   A state diagram for Example (9.1)

SOLUTION   When the independence tree is drawn, τ_i is the partition at level 1. The partition at the next level is formed by taking the successors of the blocks of τ_i regardless of the input conditions.

For the flow table in this example the independence tree in Fig. 9.3 is drawn.

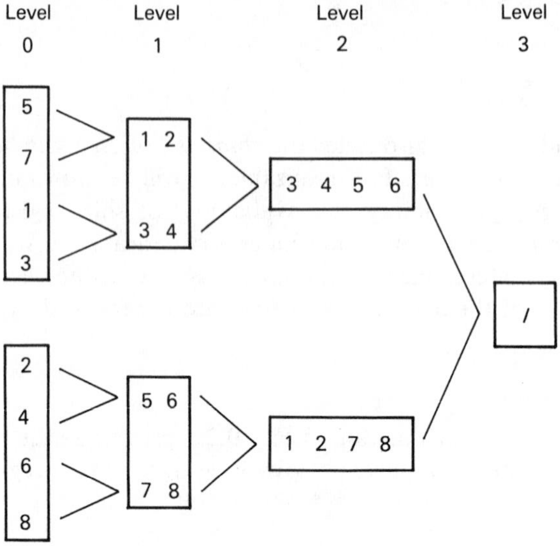

Fig. 9.3   Independence tree for Example (9.1)

A three-level independence tree indicates a possibility of a single three-bit shift register, preceded by some combinational logic as shown in Fig. 9.1. To determine the nature of this combinational logic, we assign the three state variables Y_1, Y_2 and Y_3 to the partition in the tree as follows:

Y_1 to distinguish the blocks of $\overline{1\,3\,5\,7}$ $\overline{2\,4\,6\,8}$

Y_2 to distinguish the blocks of $\overline{1\,2\,3\,4}$ $\overline{5\,6\,7\,8}$

Y_3 to distinguish the blocks of $\overline{1\,2\,7\,8}$ $\overline{3\,4\,5\,6}$

We note that the product of the three partitions is $\pi(0)$, and that Y_1 distinguishes the blocks of a partition that is not input-consistent; Y_1 is therefore dependent upon the primary input.

The assignment shown in Table 9.1 results in the circuit diagram of Fig. 9.4, which implements the following next-state expressions:

$$Y_1 = \bar{x}(\bar{y}_1 + \bar{y}_2 + \bar{y}_3) + \bar{y}_2\bar{y}_3$$

$$Y_2 = y_1$$

$$Y_3 = y_2$$

In the independence tree of Fig. 9.3, each node was addressed by no more than two nodes at the previous level, so that a single three-bit (three-level) shift register can be used. If however, the largest node had 2^K predecessors, then K shift registers in parallel are required.

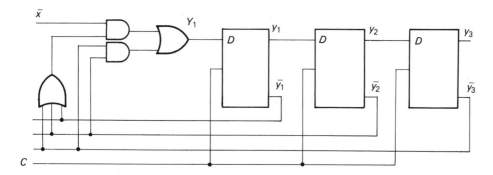

Fig. 9.4   Circuit diagram of Example (9.1)

9.1.2 Feedback-Free Shift Register Realizations

We noted in Chapter 8 that a state-independent submachine is dependent upon the primary input only and can be implemented without feedback, though feedback may be present in the successor. The smallest state-independent partition τ_s is computed by putting into a single block, all the next states contained within one column. This partition describes the largest state-independent component. This component can in some cases be broken down into smaller components based on larger partitions that are formed by the pair-wise addition of the blocks of τ_s.

If a predecessor is independent of the states of its successors then we have a serial flow of information without feedback, which is a case where shift registers may be used.

Looking back at Fig. 8.5 we can see that m_2 regards the primary input and the output of m_1 as inputs. In some cases the successor m_2 contains another submachine that is dependent upon the new primary inputs only. If the process of isolating loop-free predecessors can be continued all the way through, a loop-free structure will result which lends itself to shift register realization without feedback.

In parallel with Section 9.1.1, the successor tree, (ambiguity tree), can be used to see if it is possible to specify individual states at the lth level; that is to see if every state has one or more unique homing sequences of finite length. If so the machine can be implemented using a l-bit shift register as a finite input memory machine.

A sequential machine has a finite input memory of degree l, if the knowledge of the last l but not $l-1$ inputs uniquely determine its present state. Such a machine can be realized as a feedback-free circuit by storing the last l inputs in a shift register followed by combinational logic.

Consider Table 9.2 and its successor tree in Fig. 9.5. τ_s is a state independent partition; the two blocks are the elements of the two columns in the table. The partitions at the second and third levels are formed by collecting the next states of the blocks of the previous partitions along individual columns. The partition at any level defines the predecessor up to that level.

The tree is a good indication of the number of levels required and whether a loop-free structure is possible. Had the tree contained a self-loop at the second level,

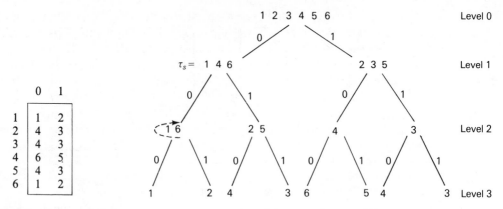

Table 9.2   A flow table

	0	1
1	1	2
2	4	3
3	4	3
4	6	5
5	4	3
6	1	2

Fig. 9.5   A successor tree

say, we know then that the successor must have a feedback. This situation could arise if the next state in row 6 column 0 is 6 instead of 1; then the successor of (1 6) would be (1 6) also, which results in a self-loop as shown dotted in Fig. 9.5.

Martin described the memory span test, which is identical to the successor tree to determine if a sequential machine (SM) has finite input memory ⟨75⟩. As shown in Fig. 9.6 one first forms the set containing all the states (U). The successor sets under each input are then formed $(U)I_j$. Repeated entries and DON'T CARE in $(U)I_j$ are eliminated. The resulting sets $(U)I_1$, $(U)I_2$, ... describe the set of all states that the machine could be in after the occurrence of inputs I_1, I_2, ... regardless of starting

	0	1
(U) 1, 2, 3, 4, 5, 6	1 4 6	2 3 5
$(U)^1$　　　1　4　6	1 6	2 5
2　3　5	4	3
$(U)^2$　　　　1　6	1	2
2　5	4	2
4	6	5
3	4	3
$(U)^3$　　Singletons		

Fig. 9.6   Memory span test for the machine in Table 9.2

states. This is called $(U)^1$. If all the sets are singletons* or DON'T CARE, then the SM has finite input memory of degree one, and the length of the memory-span test is one. If any of the sets $(U)I_1$, $(U)I_2$, ... is not a singleton or DON'T CARE, then form the successor sets of $(U)^1$ under each input; these are the $(U)^2$ sets. This is continued until the first time that all the sets of the lth iteration $(U)^l$ are either singleton or DON'T CARE, or until every set in $(U)^l$ is present in $(U)^{l+1}$. If the first alternative occurs, the SM has input memory of degree l. Otherwise the SM has infinite memory.

If a shift register is available, it seems reasonable to examine the successor and/or independence tree of the machine first, to see if the machine is suitable for shift register realization. It is also shown how easy it is to test if a table contains an input or state-independent partitions. If this test proves negative, one might proceed to compute the s.p. partitions, examine them and start to decompose the system wherever necessary and possible, then implement the sub-systems using the available devices like ULMs, ROMs, PLAs, or a mixture of various modules, if it is thought advantageous.

9.2   ULM IMPLEMENTATION OF SEQUENTIAL CIRCUITS

9.2.1 Introduction

In Chapter 3, ULMs were used for combinational logic. Here we introduce ULMs, in conjunction with delays or latches for sequential operations, and examine the relevance of the s.p. partition assignments; and the effect of decomposition, either with or without state splitting, on the total number of modules required.

Next state variables Y and output variables Z of a sequential machine may be written in terms of present-state variables y, and input variables x as follows:

$$Y_1 = f_1 (y_1, \ldots, y_s, x_1, \ldots, x_p)$$
.
.
.
$$Y_s = f_s (y_1, \ldots, y_s, x_1, \ldots, x_p)$$

$$Z_1 = g_1 (y_1, \ldots, y_s, x_1, \ldots, x_p)$$ (9.1)
.
.
.
$$Z_m = g_m (y_1, \ldots, y_s, x_1, \ldots, x_p)$$

where
s = number of state variables.
p = number of primary input variables.
m = number of output variables.

* A set having a single element.

These expressions can be expanded using Shannon's expansion technique as was shown in Equation (3.5), and ULM-multiplexers are then used with a latch in the feedback path to complete the sequential circuits.

Table 3.1 is still valid for the number of modules required for a next state or output expression if we let $v = s + p$. Equations (3.6) to (3.11) may still be used to compute the number of levels and the number of modules required for implementation. It is also reasonable to expect that if the assignment of a sequential machine is based on closed partitions, and the system is decomposed into two or more components operating in series or parallel, then a reduced dependency between the state variables will be obtained, and consequently fewer modules and possibly fewer levels will be required. This will result in a simpler and more economical circuit structure. We next look at the effect of decomposition, with particular reference to the effect on the number of ULMs for a particular sequential machine.

9.2.2 Effect of Decomposition

Example (9.2)

Machine M_1 – a system having useful closed partitions.

As an example of the effect decomposition can have in reducing the number of modules in a ULM realization of a sequential machine, consider the machine M_1 specified by the state table given in Table 9.3.

Present state	Next state Inputs $x_1 x_2$				Z
	00	01	11	10	
1	1	3	4	6	1
2	3	2	6	5	0
3	1	2	6	4	0
4	5	6	2	3	0
5	5	4	3	2	0
6	4	6	2	1	0

Table 9.3 State table for machine M_1

a	b	c
000	000	000
010	010	011
101	011	010
110	111	110
111	100	100
001	110	111

Table 9.4 Showing three possible assignments for machine M_1

This machine has the following closed partitions:

$$\begin{cases} \pi_1 = \overline{123}, \overline{456} \\ \pi_2 = \overline{15}, \overline{26}, \overline{34} \end{cases} \tag{9.2}$$

Three possible assignments of state variables are shown in Table 9.4. Assignment (*a*) is an arbitrary assignment and gives rise to the following next-state functions, set out in full expanded form to illustrate the correspondence between this form of the function and its implementation.

$$Y_1 = \bar{x}_1\bar{x}_2 [\bar{y}_1\bar{y}_2(y_3) + \bar{y}_1 y_2(\bar{y}_3) + y_1\bar{y}_2(0) + y_1 y_2(1)]$$
$$+ \bar{x}_1 x_2 [\bar{y}_1\bar{y}_2(\bar{y}_3) + \bar{y}_1 y_2(0) + y_1\bar{y}_2(0) + y_1 y_2(y_3)]$$
$$+ x_1\bar{x}_2 [\bar{y}_1\bar{y}_2(0) + \bar{y}_1 y_2(\bar{y}_3) + y_1\bar{y}_2(y_3) + y_1 y_2(\bar{y}_3)]$$
$$+ x_1 x_2 [\bar{y}_1\bar{y}_2(\bar{y}_3) + \bar{y}_1 y_2(0) + y_1\bar{y}_2(0) + y_1 y_2(y_3)]$$

$$Y_2 = \bar{x}_1\bar{x}_2 [\bar{y}_1\bar{y}_2(y_3) + \bar{y}_1 y_2(0) + y_1\bar{y}_2(0) + y_1 y_2(1)]$$
$$+ \bar{x}_1 x_2 [\bar{y}_1\bar{y}_2(0) + \bar{y}_1 y_2(\bar{y}_3) + y_1\bar{y}_2(y_3) + y_1 y_2(y_3)]$$
$$+ x_1\bar{x}_2 [\bar{y}_1\bar{y}_2(0) + \bar{y}_1 y_2(\bar{y}_3) + y_1\bar{y}_2(y_3) + y_1 y_2(y_3)]$$
$$+ x_1 x_2 [\bar{y}_1\bar{y}_2(1) + \bar{y}_1 y_2(0) + y_1\bar{y}_2(0) + y_1 y_2(\bar{y}_3)]$$

$$\tag{9.3}$$

$$Y_3 = \bar{x}_1\bar{x}_2 [\bar{y}_1\bar{y}_2(0) + \bar{y}_1 y_2(\bar{y}_3) + y_1\bar{y}_2(0) + y_1 y_2(1)]$$
$$+ \bar{x}_1 x_2 [\bar{y}_1\bar{y}_2(1) + \bar{y}_1 y_2(0) + y_1\bar{y}_2(0) + y_1 y_2(\bar{y}_3)]$$
$$+ x_1\bar{x}_2 [\bar{y}_1\bar{y}_2(\bar{y}_3) + \bar{y}_1 y_2(\bar{y}_3) + y_1\bar{y}_2(0) + y_1 y_2(\bar{y}_3)]$$
$$+ x_1 x_2 [\bar{y}_1\bar{y}_2(0) + \bar{y}_1 y_2(\bar{y}_3) + y_1\bar{y}_2(y_3) + y_1 y_2(y_3)]$$

$$Z = \bar{y}_1\bar{y}_2\bar{y}_3$$

Using three-variable modules, the output function Z requires one module. Since all the next-state variables are functions of x_1, x_2, y_1, y_2, y_3, then from Equation (3.6), each requires $(5-1)/2 = 2$ levels. From Equation (3.7) the three variables require a maximum Mo of $3(1-4^2)/(1-4) = 15$ modules. This assignment, therefore needs a maximum total of 16 modules.

Assignment (b) is based on the two closed partitions given in Equation (9.2), and results in the following:

$$Y_1 = \bar{x}_1 y_1 + x_1\bar{y}_1$$
$$Y_2 = \bar{y}_2 [\bar{x}_1\bar{x}_2(0) + \bar{x}_1 x_2(1) + x_1\bar{x}_2(1) + x_1 x_2(1)] +$$
$$y_2 [\bar{x}_1\bar{x}_2(\bar{y}_3) + \bar{x}_1 x_2(1) + x_1\bar{x}_2(y_3) + x_1 x_2(1)]$$
$$Y_3 = \bar{y}_2 [\bar{x}_1\bar{x}_2(0) + \bar{x}_1 x_2(1) + x_1\bar{x}_2(0) + x_1 x_2(1)] +$$
$$y_2 [\bar{x}_1\bar{x}_2(\bar{y}_3) + \bar{x}_1 x_2(0) + x_1\bar{x}_2(y_3) + x_1 x_2(0)]$$

$$\tag{9.4}$$

$$Z = \bar{y}_1\bar{y}_2\bar{y}_3$$

This assignment requires a maximum of eight modules.

Finally, assignment (c) which is based on partition pairs results in the following:

$$Y_1 = \bar{x}_1 y_1 + x_1\bar{y}_1$$
$$Y_2 = \bar{x}_1\bar{x}_2(y_3) + \bar{x}_1 x_2(1) + x_1\bar{x}_2(\bar{y}_3) + x_1 x_2(1)$$
$$Y_3 = \bar{x}_1\bar{x}_2(0) + \bar{x}_1 x_2(y_2) + x_1\bar{x}_2(\bar{y}_2) + x_1 x_2(y_2)$$

$$\tag{9.5}$$

$$Z = \bar{y}_1\bar{y}_2\bar{y}_3$$

This assignment requires a maximum of four modules.

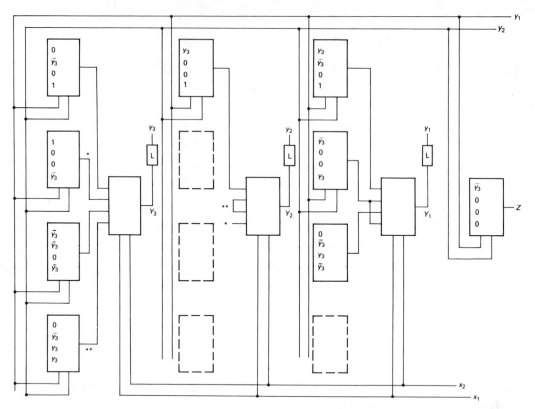

Fig. 9.7(a)   ULM implementation of Example (9.2), assignment (a)

The implementations based on these three assignments, using three-variable multiplexers, are shown in Fig. 9.7 (where L = latch). Due to redundancies, assignment (*a*) requires 12 modules instead of the maximum of 16 predicted above [Fig. 9.7(a)]. The result of applying decomposition is to reduce the total number of modules to 7 out of a maximum of 8, as shown in Fig. 9.7(b). Further reduction to 4 modules is achieved with the partition-pair assignment, as shown in Fig. 9.7(c).

Example (9.2) indicates how ULM-modules are used for sequential systems, and gives an indication of the size of possible saving if the system is decomposable.

Table 9.5 shows the maximum number of modules that might be required to implement the next states of a machine hypothetically assumed to have two primary input variables, and no input independence, neither before nor after decomposition; i.e. all components depend on all primary inputs. The machine is then decomposed into two components operating in series (Fig. 9.8, ii), and two equal or nearly equal components operating in parallel (Fig. 9.8, iii). The size of the components given in Table 9.5 is the most favorable. The table and the graph show that a large reduction in modules is possible after decomposition.

If it is possible to decompose a system into more than two components then more saving is expected; also, more saving may be achieved if some of the components are independent of some or all the primary inputs.

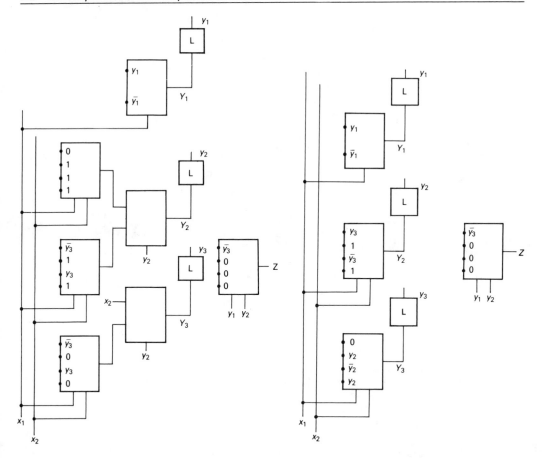

Fig. 9.7(b) ULM implementation of example (9.2), assignment (b)

Fig. 9.7(c) ULM implementation of Example (9.2), assignment (c)

No. of state variables s	Original No. of modules	Parallel Decomposition No. of modules	Serial Decomposition	
			No. of modules	State variable distribution
2	6	2	4	1 + 1
3	15	7	11	2 + 1
4	44	12	26	3 + 1
5	105	21	57	3 + 2
6	258	30	130	4 + 2
7	595	59	275	5 + 2
8	1368	88	600	6 + 2

Table 9.5 Maximum number of three-variable ULM modules for a sequential system having two primary input variables, both before and after decomposition

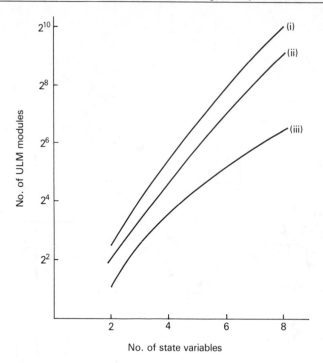

Fig. 9.8   Maximum No. of ULMs required for a sequential system having $p = 2$
(i)   Before decomposition
(ii) Two submachines in series
(iii) Two submachines in parallel

It is necessary to point out that when the next state and output expressions are implemented using ULMs, whether before or after decomposition, modules may be saved by carefully choosing the control variables. The technique illustrated in Chapter 7 is therefore still useful.

9.2.3 State Splitting

In real life, not all systems are decomposable, and in many cases, state splitting may be employed to arrive at a decomposable system. If the state splitting does not result in any increase in the number of state variables, a saving in modules is almost certain. However, if the number of state variables increases, saving is not assured.

For a state machine having s state variables, and p primary input variables, let:

Mo = No. of modules for the original system.
Mo_D = No. of modules after decomposition.
l_i = No. of levels for state variable Y_i.
c_{wi} = No. of control inputs for variables Y_i, in level w.
d_i = No. of state and input variables of which the variable Y_i is
            independent.

Then Equations (3.9) and (3.10) may be modified to calculate the total number of modules required for the system at hand as follows:

$$Mo = \sum_{i=1}^{s} (1 + 2^{cli} + 2^{cli + c2i} + \ldots + 2^{cli + \cdots + cl - 1i}) \tag{9.6}$$

or

$$Mo = \sum_{i=1}^{s} \left[1 + 2^{cli} \left(\frac{1 - I^{li-1}}{1 - I} \right) \right] \tag{9.7}$$

where

$$l_i = \left\lceil \frac{s_i + p_i - d_i - 1}{c} \right\rceil$$

After state splitting, the state variables may increase to a value s_s (suffix s indicates state splitting). The total modules required for the new decomposed machine Mo_D is given by:

$$Mo_D = \sum_{i=1}^{s_s} \left[1 + 2^{cli} \left(\frac{1 - I^{lsi-1}}{1 - I} \right) \right] \tag{9.8}$$

where

$$l_{si} = \left\lceil \frac{s_{si} + p_i - d_{si} - 1}{c} \right\rceil$$

The difference in the number of state variables, $s_s - s$, at the point where Equations (9.7) and (9.8) are equal, gives an indication of the maximum allowable redundancy that is useful.

It should be noted that even though $s_s > s$ due to state splitting, $d_{si} > > d_i$ due to reduced dependencies between the state variables after decomposing the augmented machine; hence the possibility of saving modules still applies. One should remember, however, that there are other factors influencing the final count of modules required. These are:

(i)   The dependency of the state variables on each other and on the input variables.
(ii)   The distribution of the state variables among the components.
(iii) Whether we have serial, parallel or composite decomposition.
(iv)   The amount of added redundancy to obtain a decomposable system.

To give a general idea, Figs 9.9 and 9.10 are drawn for an arbitrarily chosen sequential machine having five state variables and four input variables. Due to state splitting, the state variables are increased to six, and the machine is decomposed into:

(a) Two components operating in parallel (Fig. 9.10)
(b) Two components operating in series (Fig. 9.9)

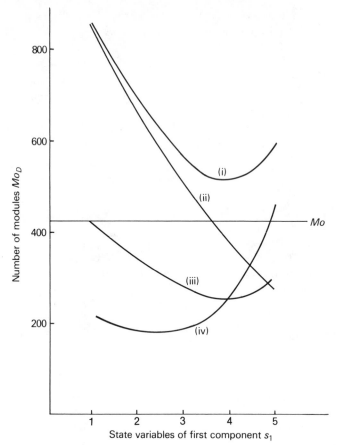

Fig. 9.9   Modules required for two components in series
(i) $d_1 = d_2 = 0$
(ii) $d_1 = 2, d_2 = 0$
(iii) $d_1 = d_2 = 1$
(iv) $d_1 = 0, d_2 = 2$

These graphs show the variation of the total number of modules Mo_D with the number of state variables in one component s_1, for various degrees of redundancy. It should be pointed out that Mo_D is defined only for integer values of s; nevertheless, the graphs are drawn continuous for convenience. The graphs show that the number of modules depends on the distribution of the state variables in both components, i.e. the number of state variables (s_1 and s_2), in the two components and on the in-dependencies d. It can be seen that saving is very much more likely with parallel decomposition; then nearly all points are below the Mo line, where Mo is the total number of modules required by the original non-decomposed, non-augmented system.

State splitting may also be applied to the smaller components to obtain further decomposition, and saving can be obtained, especially if the splitting does not result in further increase in state variables.

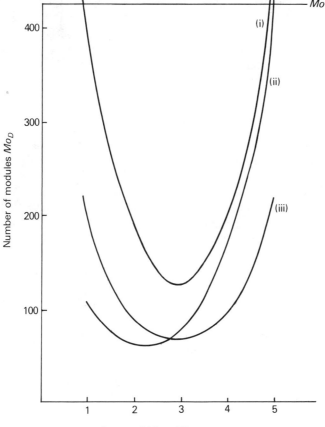

Fig. 9.10   Modules required for two components in parallel
(i) $d_1 = d_2 = 0$
(ii) $d_1 = 0, d_2 = 2$
(iii) $d_1 = d_2 = 1$

It is a difficult problem to specify limits on the allowed increase in state variables, which might be called the maximum allowable redundancy. This is because the number of modules depends on many factors as already mentioned, particularly the independencies d (for example if there is less dependency on input variables after decomposition), and the distribution of state variables, which will be studied further in the next section. However, it is possible to give the following generalization.

(a) Parallel Decomposition into two Equal or Nearly Equal Components
(i)   Assume a worst-case design in which all next-state variables in a unit depend on all present-state variables in that unit, as well as all input variables.
(ii)   The same kind of modules are used before and after decomposition. Let $N = Mo - 1$ be the number of modules in all but the first level, required for any next state variable.

The limits to state splitting may be summarized by the following statement:

In a worst-case design of a sequential system having s state variables, decomposable after augmentation into at least two equal or nearly equal components in parallel, it is possible to add up to $(s-2)$ state variables due to state splitting, and still achieve saving in modules, if and only if

$$s \geqslant 3 \quad \text{and} \quad N > s-2$$

To show this is true, consider the following:

If $s < 3$, then s may be 2. State splitting will increase s to at least 3. If this is decomposed, then at best the two components will have 1 and 2 state variables each. One of these components is as large as the original. Similarly for $s = 1$. Hence there is no saving of modules if $s < 3$.

If more than $(s-2)$ state variables can be added, then at least $(s-1)$ state variables may be added. The new system will have $(s+s-1)$ state variables. If this is decomposed, then the two parallel components will have s and $(s-1)$ state variables respectively. One of the components is as large as the original. Hence no saving is possible if more than $(s-2)$ variables are added.

The maximum number of modules for the original system is given by $Mo = s(1+N)$. If the state variables increase by $s-2$, the new system will have $(s+s-2)$ variables. After decomposition into two parallel components, each will have $(s-1)$ variables.

Since each component has one variable less than the original, each variable needs a maximum of $(1+N/2)$ modules ($N/2$ only if $c_1 = 1$ originally). That this is true may be seen from Equation (9.7), which shows that when the number of control variables in the first level is reduced by one, the number of modules in succeeding levels is reduced by half. However, if the first level has only one control variable to start with, that module will be redundant, thus reducing the number of modules from $1+N$ to $N/2$.

For both components, the maximum number of modules is given by:

$$Mo_D = 2(s-1)(1+N/2)$$
$$= s(1+N) + (s-2) - N$$

If $N > s-2$
Then $Mo_D < Mo$

Therefore it is possible to add up to $(s-2)$ variables and still achieve saving if $N > s-2$.

This condition is easily met in practice; for example when

$$s = 3, N \geqslant 2, Mo_D < M_o.$$

As s increases, N increases faster, making saving more certain.

In the foregoing it was assumed that it is best to decompose into two equal components, but the choice depends on d as defined earlier in this section. This is illustrated in Fig. 9.10. Further, if the system can be decomposed into more than two components, or if the dependency of the state variables on each other and on the input variables is reduced, the conditions may then be relaxed, and more saving may be expected.

(b) Serial Decomposition into Two Components

For serial decomposition, module saving depends on the number of state and input variables, on the type of modules, and on the relative size of the components. Normally it is more economical to have the large components first, as predecessor, but as can be seen from Fig. 9.9, this depends largely on the independencies. In general, state splitting that leads to any increase in state variables with serial decomposition is only advisable for large systems having more than about eight variables. Even then, saving is not large unless d is large, as shown in Fig. 9.9, or unless it is possible to break the system into more than two components.

(c) Composite Decomposition

As well as serial and parallel decomposition, it is possible to have composite decomposition as was discussed in Chapter 8. There are obviously many possible situations, one of which is shown in Figs 9.11 and 9.12. These graphs are similar to

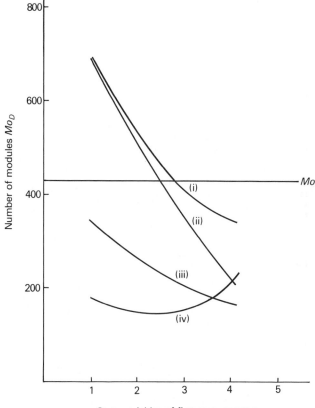

Fig. 9.11   **Modules requires for a composite structure of three components;**
    $s_2 = 1$ **(fixed)**
(i) $d_1 = d_2 = d_3 = 0$
(ii) $d_1 = 2, d_2 = d_3 = 0$
(iii) $d_1 = d_2 = d_3 = 1$
(iv) $d_1 = d_2 = 0, d_3 = 2$

those in Figs 9.9 and 9.10, but now we have two components in parallel having s_1 and s_2 state variables followed by a successor having s_3 state variables.

In Fig. 9.11 it is assumed that one of the two parallel components is small and fixed. The graphs are therefore expected to look like those for serial decomposition in Fig. 9.9. Fig. 9.12 shows the case where the successor is small and fixed. This graph resembles that for parallel decomposition.

One should note here that, in all these graphs, the decomposed system has one state variable more than the original system. A reasonably good saving in modules is still possible as can be seen by comparing with the original requirement indicated by the *Mo* line.

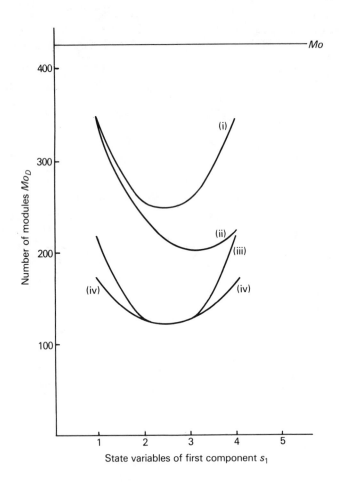

Fig. 9.12 Modules required for a composite structure of three components;
$s_3 = 1$ **(successor is fixed)**
(i) $d_1 = d_2 = d_3 = 0$
(ii) $d_1 = 2, d_2 = d_3 = 0$
(iii) $d_1 = d_2 = 0, d_3 = 2$
(iv) $d_1 = d_2 = d_3 = 1$

9.3   ROM IMPLEMENTATION OF SEQUENTIAL CIRCUITS

Read-only-memory devices may be used with clocked latches or a register of appropriate size in the feedback path loop to implement synchronous sequential machines — as shown in Fig. 9.13.

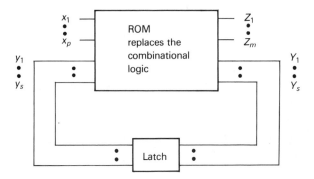

Fig. 9.13   Read-only-memory realization of sequential systems

The next states given in the state table are stored in the ROM and are addressed by the total states consisting of present states and primary inputs. The ROM shown in Fig. 9.13 has 2^{p+s} words, addressed by the 2^{p+s} address lines corresponding to the outputs of the ROMs decoder, which has $(p+s)$ inputs. The system shown in Fig. 9.13 has $(s+m)$ outputs, which means that every word has $(s+m)$ bits.

$$\text{Total memory } M = (s+m)\, 2^{s+p} \text{ bits} \tag{9.9}$$

In Section 9.2, ULMs only were considered. The techniques discussed however, like decomposition and state splitting, are general and would apply to other modules. In this section the use of ROMs (or similar memory devices), and the effect of the decomposition technique will be briefly outlined.

When we examine the next states, we find the total memory of the system is given by:

$$M = s \cdot 2^{s+p} \tag{9.10}$$

If this is decomposed into k equal components in parallel, the memory required is given by:

$$M_D = s \cdot 2^{(s/k)+p} \tag{9.11}$$

This indicates that even for small values of k, say 2, large savings in memory are possible.

In general, if the original system shown in Fig. 9.13 is decomposed into j components operating in series or parallel as shown in Fig. 9.14, the total memory required for the decomposed system, M_D in the parallel case, is given by:

$$M_D = \sum_{i=1}^{j} (m+s_s-\beta_i-\gamma_i)2^{(p+s_s-\alpha_i-\beta_i)} \tag{9.12}$$

(a) Original system

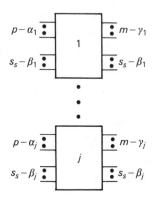

(b) System with *j* sub-units in parallel

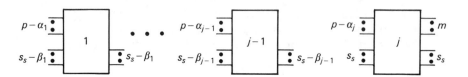

(c) System with *j* sub-units in series

Fig. 9.14   ROM implementation of a sequential system

where α_i is the number of primary inputs that sub-unit i is independent of.

β_i is the number of state variables that sub-unit i is independent of.

γ_i is the number of output variables that sub-unit i is independent of.

the subscript s indicates state splitting.

For j sub-units in series, the first $(j-1)$ components have $\gamma_i = m$, $i = 1, \ldots, j-1$.

The jth unit has $\gamma_j = 0$, and $\beta_j = 0$.

Then, if the original machine can be decomposed without state splitting or if state splitting does not cause any increase in the number of state variables, $s_s = s$.

Substituting numerical values in Equations (9.9–9.12) soon shows that a large saving in storage is achieved if decomposition is possible. This concept however can be misleading in some cases. For example, if it is possible to reduce the storage requirement from 250 bits to, say, 130 bits, that may look impressive but, since ROM

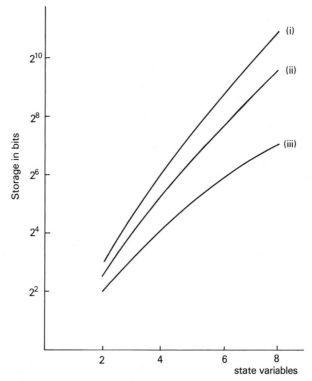

Fig. 9.15   Storage requirement for a sequential system using ROMs
(i) Before decomposition
(ii) Two submachines in series
(iii) Two submachines in parallel

modules are only available in standard sizes, the designers might still have to use a 256-bit module. Fig. 9.15 gives an idea about the sort of saving for serial and parallel decompositions.

9.4   OPTIMUM DISTRIBUTION OF STATE VARIABLES

So far, we have learned how the memory and/or module requirements are influenced by the type of decomposition, and the distribution of the state variables between the various components. Further, the type of decomposition and the number of state variables required by each component are directly dependent upon the partitions. It is therefore useful to try to study these aspects in some detail. This is done by optimizing the expressions for the number of modules, or storage locations, with respect to the state variables of the decomposed system, and by relating these state variables to the algebraic properties of the partitions. The understanding gained from this should enable us to make a good choice of partitions for our decomposition when a choice is possible.

9.4.1 ULM Modules

It was mentioned earlier that the number of modules required for a particular decomposed system depends on the relative size of the components and on the number of independencies d. Here, Lagrange's method of undetermined multipliers is used to obtain relationships that give some indication on how to decompose a system if a choice of decomposition structure exists.

Equations (9.7) and (9.8) relate the total number of modules Mo to the number of state variables s. Both Mo and s are integers, and so it is necessary to assume that there is a continuous function that coincides with Mo for all integral values of s, as shown in Figs 9.9 to 9.12. To simplify further development it will be assumed that for any sub-group of state variables s_i, in sub-system M_i, all the state variables have on average the same independency d_i. By considering a large number of machines, it has been found that this assumption is not far from reality. Mo_D, the total number of modules for a decomposed system, can be minimized by partially differentiating and substituting in Equation (9.14) below, under the constraint:

$$\phi\,(s_1, s_2, \ldots, s_j) = 0 \tag{9.13}$$

Where s_i is the number of state variables in sub-system i, $i = 1, 2, \ldots, j$.

Using Lagrange's method, we have:

$$\frac{\delta Mo_D}{\delta s_1} + \lambda\,\frac{\delta \phi}{\delta s_1} = 0$$

$$\begin{array}{c} \cdot \\ \cdot \\ \cdot \end{array} \tag{9.14}$$

$$\frac{\delta Mo_D}{\delta s_j} + \lambda\,\frac{\delta \phi}{\delta s_j} = 0$$

where λ is the Lagrange multiplier.

Since $s - (s_1 + s_2 + \ldots + s_j) = 0 = \phi$

$$\frac{\delta \phi}{\delta s_i} = 1 \qquad \text{for } i = 1, \ldots, q, \ldots, k, \ldots, j \qquad \text{where } j > k > q \tag{9.15}$$

(i) Parallel Decomposition into j Components

$$Mo_D \doteq s_1 \left[1 + 2^{c_{11}} \left(\frac{1 - I^{(s_1 + p - d_1 - 1 - c)/c}}{1 - I}\right)\right] + \ldots$$
$$+ s_j \left[1 + 2^{c_{1j}} \left(\frac{1 - I^{(s_j + p - d_j - 1 - c)/c}}{1 - I}\right)\right] \tag{9.16}$$

where c_{1_i} refers to the number of control variables used in the first level in component i.

Differentiating with respect to the state variables of any two components s_q and s_k, substituting Equation (9.15) in Equation (9.14) and equating, we get:

$$I^{(s_q+p-d_q-1-c)/c}\,[2^{c1_q}]\left[1 + \frac{s_q \ln (I)}{c}\right] - 2^{c1_q}$$

$$= I^{(s_k+p-d_k-1-c)/c}\,[2^{c1_k}]\left[1 + \frac{s_k \ln (I)}{c}\right] - 2^{c1_k} \tag{9.17}$$

This gives the relationship between the qth and kth components for a minimum number of modules; e.g. for two components in parallel, $q = 1$, $k = 2$.

The symmetry of Equation (9.17) confirms that, with equal independencies, the components should be equal or nearly equal for minimal implementations.

If $c_{1_q} = c_{1_k}$, Equation (9.17) simplifies to the following:

$$I^{(s_q-d_q)/c}\left(1 + \frac{s_q \ln (I)}{c}\right) = I^{(s_k-d_k)/c}\left(1 + \frac{s_k \ln (I)}{c}\right) \tag{9.18}$$

Numerical values show that this equation is best satisfied at minimum values of Mo_D as in Fig. 9.10.

(ii) Serial Decomposition into j Components

Here any state variable in a component may depend on all input variables and all state variables of its predecessors, as well as its own state variables.

Then $Mo_D = s_1\left[1 + 2^{c1_1}\left(\dfrac{1 - I^{(s_1+p-d_1-1-c)/c}}{1 - I}\right)\right] + \ldots$

$$+ s_j\left[1 + 2^{c1_j}\left(\frac{1 - I^{(s_1+\ldots+s_q+\ldots+s_k+\ldots+s_j+p-d_j-1-c)/c}}{1 - I}\right)\right]$$

Differentiating with respect to s_q and s_k, and substituting as in the parallel case above we get:

$$s_q\left[\frac{-2^{c1_q}}{c(1 - I)}\ln (I)\, I^{(s_1+\ldots+s_q+p-d_q-1-c)/c}\right] + \left[1 + 2^{c1_q}\left(\frac{1 - I^{(s_1+\ldots+s_q+p-d_q-1-c)/c}}{1 - I}\right)\right]$$

$$+ s_{q+1}\left[\frac{-2^{c1_{q+1}}}{c(1 - I)}\ln (I)\, I^{(s_1+\ldots+s_q+s_{q+1}+p-d_{q+1}-1-c)/c}\right] + \ldots$$

$$+ s_{k-1}\left[\frac{-2^{c1_{k-1}}}{c(1 - I)}\ln (I)\, I^{(s_1+\ldots+s_q+\ldots+s_{k-1}+p-d_{k-1}-1-c)/c}\right] =$$

$$1 + 2^{c1_k}\left(\frac{1 - I^{(s_1+\ldots+s_q+\ldots+s_k+p-d_k-1-c)/c}}{1 - I}\right) \tag{9.19}$$

This is for any two components q and k where $k > q$, in other words q is the predecessor of k or the predecessor of one of k's predecessors.

For two components in series, Equation (9.19) reduces to the following:

$2^{c_{12}} I^{(s_1 + s_2 + p - d_2 - 1 - c)/c}$

$$= I^{(s_1 + p - d_1 - 1 - c)/c} \left(1 + \frac{s_1 \ln (I)}{c}\right) 2^{c_{11}} + 2^{c_{12}} - 2^{c_{11}} \tag{9.20}$$

If $c_{11} = c_{12}$ then:

$$I^{(s_2 - d_2 + d_1)/c} = 1 + \frac{s_1 \ln (I)}{c} \tag{9.21}$$

Substituting numerical values, it can be seen that these equations are best satisfied when Mo_D is a minimum as in Fig. 9.9. These equations are intended as a guide to the most economical implementation for systems having many closed partitions and hence more than one way of decomposition.

(iii) Composite Decomposition
Here we assume the case given in Fig. 9.16. The total number of modules is given in the equation below:

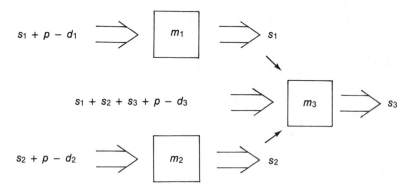

Fig. 9.16 Composite structure of a sequential machine

$$Mo_D = s_1 \left[1 + 2^{c_{11}} \left(\frac{1 - I^{(s_1 + p - d_1 - 1 - c)/c}}{1 - I}\right)\right] +$$

$$s_2 \left[1 + 2^{c_{12}} \left(\frac{1 - I^{(s_2 + p - d_2 - 1 - c)/c}}{1 - I}\right)\right] +$$

$$s_3 \left[1 + 2^{c_{13}} \left(\frac{1 - I^{(s_1 + s_2 + s_3 + p - d_3 - 1 - c)/c}}{1 - I}\right)\right]$$

If M_3 is assumed small, we have a case similar to parallel decomposition, as shown in Fig. 9.12. For a realistic but easy way to manipulate the case, we assume m_2 is small compared to m_1, that is $s_1 > s_2$. This case is shown in Fig. 9.11. Then Mo_D may be differentiated with respect to s_1 and s_3, and substituting as before we obtain:

$$I^{(s_1+p-d_1-1-c)/c} \left[2^{c_{11}} \right] \left[1 + \frac{s_1 \ln (I)}{c} \right] - 2^{c_{11}} =$$

$$I^{(s_1+s_2+s_3+p-d_3-1-c)/c} \left[2^{c_{13}} \right] - 2^{c_{13}} \tag{9.22}$$

If $c_{11} = c_{13}$, that is an equal number of control variables in the first level, then the relationship between s_1 and s_3 for minimum module requirement is given by:

$$I^{(s_2+s_3-d_3+d_1)/c} = 1 + \frac{s_1 \ln (I)}{c} \tag{9.23}$$

The relationships developed by using Lagrange's method may be used in making a good choice of partitions for an economical assignment, especially in automated design. In other words, if a general computer program is developed, and if a choice between various assignments is to be made, then the choice that satisfies these relationships will result in the most economical implementation.

As a simple illustration, we consider a system having six state variables, and two input variables, and we look at the two simple cases of two components operating in parallel and in series. Throughout we assume a maximum dependency, i.e. $d = 0$.

(i) Two Components in Parallel
Assume we have a choice between three decompositions. In the first both components are equal and have three state variables each.

From Equation (9.18)

$$I^{(s_1-d_1)/c} \left(1 + \frac{s_1 \ln (I)}{c} \right) = I^{(s_2-d_2)/c} \left(1 + \frac{s_2 \ln (I)}{c} \right)$$

Substituting numerical values, we get

$$4^{3/2} \left(1 + \frac{3 \ln (4)}{2} \right) = 4^{3/2} \left(1 + \frac{3 \ln (4)}{2} \right)$$

The two sides of the equation are equal, and we conclude that this is a good choice. In fact the maximum number of modules required is 30 three-variable modules. The ratio of the L.H.S. and R.H.S. is $1:1$.

In the second choice, one component has four state variables, while the other has two state variables. Substituting in the same equation we get:

$$\text{L.H.S.} = 4^{4/2} \left(1 + \frac{4 \ln (4)}{2} \right)$$

$$\text{R.H.S.} = 4^{2/2} \left(1 + \frac{2 \ln (4)}{2} \right)$$

The ratio of the L.H.S. and R.H.S. is $6.3:1$.

In the third choice, one component has five state variables while the other has only one state variable. Substituting numerical values, we get a ratio of L.H.S. and R.H.S. of $42:1$.

This choice is therefore not as good as the first two choices. The maximum number of modules required is 106 compared with 50 and 30 in the second and first choices respectively.

(ii) Two Components in Series

Now we consider the case where the choice is between two serial decompositions. The first has a four-state variable predecessor, and a two-state variable successor. From Equation (9.21), we have:

$$I^{(s_2 - d_2 + d_1)/c} = 1 + \frac{s_1 \ln (I)}{c}$$

Substituting numerical values, we get:

L.H.S. = 4
R.H.S. = 3.8

The ratio of the L.H.S. and the R.H.S. is 1.05.: 1. We leave this for the time being, and look at the second choice in which the first component has two state variables. Again from Equation (9.21) we get:

L.H.S. = 16
R.H.S. = 2.4

The ratio of the L.H.S. and the R.H.S. is 6.7 : 1.

It is clear that the first choice satisfies Equation (9.21) better than the second choice, and would therefore result in a more economical realization.

Checking this, we find that the first choice requires a maximum of 130 modules, as compared to a maximum of 178 modules required by the second choice.

9.4.2 ROM Modules

To be compatible with the sections on ULMs, let $d_i = a_i + \beta_i$, i.e. the state variables of the ith component may be independent of d_i state and input variables. The memory required for the state variables of j sub-units in series may be written as:

$$M_D = 2^p \ [s_1 2^{s_1 - d_1} + s_2 2^{s_1 + s_2 - d_2} + \ldots + s_j 2^{s_1 + \ldots + sj - dj}] \tag{9.24}$$

where s_i is the number of state variables in sub-system i, and $i = 1, 2, \ldots, j$
For j components in parallel, we write

$$M_D = 2^p \ [s_1 2^{s_1 - d_1} + s_2 2^{s_2 - d_2} + \ldots + s_j 2^{sj - dj}] \tag{9.25}$$

Lagrange's method of undetermined multipliers may be used, as was the case with ULMs, to find the relationships between the state variables in various components for minimum storage conditions. This is done by differentiating expressions (9.24) and (9.25) and substituting in Equation (9.14) under the constraint given in Equation (9.13).

For any two components q and k in series, where $k > q$, we have

$$1 + s_q \ln(2) = 2^{sk - dk + qk} \tag{9.26}$$

When there are only two components in series, $q = 1$, $k = 2$, and then

$$2^{s_2 - d_2 + d_1} = 1 + s_1 \ln(2) \tag{9.27}$$

For two components q and k in parallel, we have

$$2^{sq-dq} [1 + s_q \ln(2)] = 2^{sk-dk} [1 + s_k \ln(2)] \tag{9.28}$$

When there are only two components in parallel, $q = 1$, $k = 2$, and then

$$2^{s_1-d_1} [1 + s_1 \ln(2)] = 2^{s_2-d_2} [1 + s_2 \ln(2)] \tag{9.29}$$

These expressions are of the same form as Equations (9.21) and (9.18) respectively. Further, if memory requirement is plotted against the state variable distribution, graphs similar to those of Figs 9.9 to 9.12 for ULMs will be obtained. The limiting case for the number of components in the non-trivial decomposition is when $j = s$. In most cases encountered, j is in the region 2–4. This is quite acceptable if we remember that we are using MSI devices and not valves or even gates.

9.4.3 Design Procedure

(*a*)   Compute all closed partitions.

(*b*)   Compute the input-consistent and state-independent partitions if any.

(*c*)   For every unique partition π_i other than $\pi(0)$ and $\pi(I)$, compute the following

$$\#(\pi_i), b(\pi_i), \lceil \log_2 \#(\pi_i) \rceil, \lceil \log_2 b(\pi_i) \rceil$$

and $\lceil \log_2 \#(\pi_i) \rceil + \lceil \log_2 b(\pi_i) \rceil$

where $b(\pi_i) = $ Number of states in the largest block of π_i.

(*d*)   Eliminate repeated, redundant and non-useful partitions.

(*e*)   Sketch the lattice of the remaining partitions.

(*f*)   Depending upon the nature of the problem, design requirement, and available modules, we choose partition candidates. These are normally seen from the lattice structure, and we keep in mind the following two points:

(i)   For parallel decomposition, the Lagrange multiplier expressions and other discussions in Chapters 8 and 9 reveal that we are interested in uniform components, and that $s_D \gg s$.

(ii)   For serial decomposition, the relationship between any two successive components is worked out using the Lagrange multiplier expressions given in Equations 9.21 and 9.26; again $s_D \gg s$.

(*g*)   The lattice should tell us if the system has a common predecessor, successor or both.

(*h*)   Take advantage of any independencies by using τ_i and τ_s when possible as was shown in Example (8.1).

10

Partition-based Design for Asynchronous Sequential Circuits

10.1 INTRODUCTION

Synchronous sequential machines have been conveniently chosen to illustrate the decomposition of sequential switching systems. The convenience stems from the fact that the state assignment for synchronous circuits is relatively straightforward.

Unlike synchronous machines, asynchronous machines have no clock pulses to regulate their operation; hence there is a problem of ensuring that the system functions, as specified, independently of the variations in transmission delays within the circuit. One good thing about asynchronous circuits is that full advantage can be taken of the device speed, since the circuit does not have to wait for the arrival of clock pulses between transitions. However the penalty is that transient conditions must be taken into account, and several state variables are allowed to change simultaneously (a *race*), only if the resulting state does not depend on the order in which the variables change (a *non-critical race*). Races can be taken care of by restricting the state assignment in such a manner that there are no state transitions involving critical races.

Table 10.1 is a typical flow table, where each row represents an internal state, and each column represents an input state. Next internal states constitute the entries of the table. A next state is said to be stable if, for a given input state, the next state is the same as the present state. Stable states are denoted by a circled entry on the flow table, as explained in Chapter 5.

$$x_1 x_2$$

	00	01	11	10
1	①	①	4	2
2	②	3	②	②
3	③	③	4	–
4	④	–	④	5
5	⑤	1	–	⑤

Table 10.1 A flow table for an asynchronous sequential machine [Example (10.1)]

414

In synchronous sequential systems, any state assignment which has a unique coding for every internal state may yield a correct circuit. Due to the possibility of races and hazards among the circuit variables, this is not a sufficient condition for asynchronous systems in which the assignment must ensure that no critical races exist among the state variables. As was the case with synchronous circuits, the circuit complexity is dependent upon the binary state code chosen. Most work done in this field is concerned with finding assignments using the minimum number of state variables that are also free of critical races.

Hartmanis's decomposition technique will be used to decompose the system when possible as was the case with synchronous systems, only now the components will be assigned with minimal variable unicode single transition time (USTT). This way, we hope that we can minimize circuit complexity, save hardware and yet preserve speed.

Section 10.2 gives a brief review illustrating the basic theory for the Liu-Tracy USTT assignment and Tan's adaptation of the decomposition techniques to asynchronous systems.

Some references covering the subject-matter of this chapter are ⟨49, 54, 71, 105, 106, 107, 111⟩.

10.2   UNICODE SINGLE TRANSITION TIME ASSIGNMENT

If G and H are two disjoint subsets of the flow-table rows then the unordered pair (G, H) is called a *partial state dichotomy*. For a pair of transitions $i \rightarrow j$, $k \rightarrow m$, where i and j are each different from k and m, let $G = \{i, j\}$ and $H = \{k, m\}$; then the dichotomy is (G, H), usually written as (ij, km). If one of the transitions is degenerate, say $k = m$, the dichotomy is (ij, k). The dichotomy would be (i, k) if both transitions were degenerate. A state variable y_i is said to cover a dichotomy (G, H) if $y_i = 0$ for every state in H, and $y_i = 1$ for every state in G, or vice versa. A row assignment for a single output change (SOC) flow table is a valid USTT assignment if and only if (iff), for every pair of transitions $i \rightarrow j$ and $k \rightarrow m$ appearing in the same column such that $j \neq m$, the dichotomy (ij, km) is covered by at least one y_i of the assignment.

Once a set of dichotomies is derived, we assign the state variable so as to cover every member of the set. Techniques are available for computing the smallest set of dichotomies and hence minimizing the number of state variables. The technique is not reproduced here but it is helpful to note that if G^* contains G and if H^* contains H, then the dichotomy (G^*, H^*) covers the dichotomy (G, H), since the state variable covering (G^*, H^*) must also cover (G, H), then (G, H) may be ignored. This implies that if $i \rightarrow j$ is a transition in some column, then a $j \rightarrow j$ transition in that column need not be considered since any dichotomy covering the former must also cover the latter.

Furthermore if one of the dichotomies is (ij, km), then the dichotomy (i, km) generated in another column may be discarded. This will become clear in a later example.

The Liu-Tracy method for minimal y_i USTT assignment for SOC flow tables may be summarized as follows:

(i)   Find the transitions for each column including transitions such as $i \rightarrow i$ in columns in which i is a stable state, and is not in the destination of any other transition. States with unspecified next-state entries may be ignored.

(ii) Generate a dichotomy (ij, km) for every pair of transitions $i \rightarrow j$ and $k \rightarrow m$, where $j \neq m$. Repeat for all columns.

(iii) Find a set of covering dichotomies.

(iv) Form a flow matrix associating a covering state variable with each of the covering dichotomies.

10.3   DECOMPOSITION OF ASYNCHRONOUS SEQUENTIAL SYSTEMS

In Chapters 8 and 9, it was shown how synchronous circuits can be decomposed into smaller sub-units operating in series or parallel. In the following sections the same technique is adapted to asynchronous circuits. The effect of decomposing a given system into two components, operating in series or parallel, on the hardware requirements of a system when implemented with LSI/MSI devices is investigated.

10.3.1 Parallel Decomposition

A machine can be decomposed into two components operating in parallel independently of each other iff there exists two s.p. partitions π_1 and π_2, such that $\pi_1 \cdot \pi_2 = \pi(0)$. This case is similar to that of synchronous circuits, only now the submachine must be free of critical races, which is possible if the components are given STT assignments. This effectively means that the number of state variables required by a submachine M_i may exceed $\lceil \log_2 \# \pi_i \rceil$, where π_i defines M_i.

Example (10.1)

Investigate the possibility of decomposing the machine in Table 10.1.

SOLUTION   This machine has the following s.p. partitions:

$$\pi_1 = \overline{1\,3}, \overline{2\,4\,5}$$
$$\pi_2 = \overline{1}, \overline{2}, \overline{3\,4}, \overline{5}$$
$$\pi_1 \cdot \pi_2 = \pi(0)$$

Two submachines M_1 and M_2 corresponding to π_1 and π_2 are shown in Table 10.2(a) and 10.2(b). M_1 and M_2 operate in parallel independently of each other. Both are given USTT assignment as shown, resulting in the next-state expressions given below.

It is clear that reduced dependence due to decomposition is possible. The overall saving however, is likely to be less than with synchronous machines, because of the need to eliminate critical races.

x_1x_2

	y_2	y_3	y_4		
	0	0	0	c	= (1)
	1	1	0	d	= (2)
	1	0	1	e	= (34)
	0	1	1	f	= (5)

Table 10.2(a)   A flow table for M_1

x_1x_2

y_1		00	01	11	10
0	a = (13)	ⓐ	ⓐ	b	b
1	b = (245)	ⓑ	a	ⓑ	ⓑ

Table 10.2(a)   A flow table for M_1 of Example (10.1) with assignment

x_1x_2

	00	01	11	10
	ⓒ	ⓒ	e	d
	ⓓ	e	ⓓ	ⓓ
	ⓔ	ⓔ	ⓔ	f
	ⓕ	c	–	ⓕ

Table 10.2(b)   A flow table for M_2 of Example (10.1) with assignment

From Table 10.2(a), the following next-state expression is obtained:

$$Y_1 = \bar{x}_1\bar{x}_2(y_1) + \bar{x}_1x_2(0) + x_1\bar{x}_2(1) - x_1x_2(1)$$

This expression is independent of the state variables of M_2. Similarly the next state variables of M_2 are derived from Table 10.2(b). These are independent of y_1, and are given by:

$$
\begin{aligned}
Y_2 = \ & \bar{x}_1\bar{x}_2\bar{y}_2\bar{y}_3(0) + \\
& \bar{x}_1\bar{x}_2\bar{y}_2y_3(0) + \\
& \bar{x}_1\bar{x}_2y_2\bar{y}_3(y_4) + \\
& \bar{x}_1\bar{x}_2y_2y_3(\bar{y}_4) + \\
& \bar{x}_1x_2\bar{y}_2\bar{y}_3(0) + \\
& \bar{x}_1x_2\bar{y}_2y_3(0) + \\
& \bar{x}_1x_2y_2\bar{y}_3(y_4) + \\
& \bar{x}_1x_2y_2y_3(y_4) + \\
& x_1\bar{x}_2\bar{y}_2\bar{y}_3(y_4) + \\
& x_1\bar{x}_2\bar{y}_2y_3(0) + \\
& x_1\bar{x}_2y_2\bar{y}_3(0) + \\
& x_1\bar{x}_2y_2y_3(y_4) + \\
& x_1x_2\bar{y}_2\bar{y}_3(y_4) + \\
& x_1x_2\bar{y}_2y_3(0) + \\
& x_1x_2y_2\bar{y}_3(y_4) + \\
& x_1x_2y_2y_3(y_4) \quad .
\end{aligned}
$$

$$
\begin{aligned}
Y_3 = \ & \bar{x}_1\bar{x}_2\bar{y}_2\bar{y}_3(0) + \\
& \bar{x}_1\bar{x}_2\bar{y}_2y_3(y_4) + \\
& \bar{x}_1\bar{x}_2y_2\bar{y}_3(0) + \\
& \bar{x}_1\bar{x}_2y_2y_3(\bar{y}_4) + \\
& \bar{x}_1x_2\bar{y}_2\bar{y}_3(0) + \\
& \bar{y}_1x_2\bar{y}_2y_3(0) + \\
& \bar{x}_1x_2y_2\bar{y}_3(0) + \\
& \bar{x}_1x_2y_2y_3(0) + \\
& x_1\bar{x}_2\bar{y}_2\bar{y}_3(y_4) + \\
& x_1\bar{x}_2\bar{y}_2y_3(y_4) + \\
& x_1\bar{x}_2y_2\bar{y}_3(y_4) + \\
& x_1\bar{x}_2y_2y_3(y_4) + \\
& x_1x_2\bar{y}_2\bar{y}_3(0) + \\
& x_1x_2\bar{y}_2y_3(0) + \\
& x_1x_2y_2\bar{y}_3(0) + \\
& x_1x_2y_2y_3(y_4) \quad .
\end{aligned}
$$

$$
\begin{aligned}
Y_4 = \ & \bar{x}_1\bar{x}_2\bar{y}_2\bar{y}_3(0) + \\
& \bar{x}_1\bar{x}_2\bar{y}_2y_3(y_4) + \\
& \bar{x}_1\bar{x}_2y_2\bar{y}_3(y_4) + \\
& \bar{x}_1\bar{x}_2y_2y_3(0) + \\
& \bar{x}_1x_2\bar{y}_2\bar{y}_3(0) + \\
& \bar{x}_1x_2\bar{y}_2y_3(0) + \\
& \bar{x}_1x_2y_2\bar{y}_3(y_4) + \\
& \bar{x}_1x_2y_2y_3(y_4) + \\
& x_1\bar{x}_2\bar{y}_2\bar{y}_3(0) + \\
& x_1\bar{x}_2\bar{y}_2y_3y_4 + \\
& x_1\bar{x}_2y_2\bar{y}_3(y_4) + \\
& x_1\bar{x}_2y_2y_3(0) + \\
& x_1x_2\bar{y}_2\bar{y}_3(y_4) + \\
& x_1x_2\bar{y}_2y_3(0) + \\
& x_1x_2y_2\bar{y}_3(y_4) + \\
& x_1x_2y_2y_3(0) \quad .
\end{aligned}
$$

We observe that Y_1 is a function of three-variables x_1, x_2 and y_1 only, and is implemented with a three-variable ULM.

Y_2, Y_3 and Y_4 are each a function of five variables x_1, x_2, y_2, y_3 and y_4, and are implemented using five-variable modules as shown in Fig. 10.1. The decomposition has made Y_2, Y_3 and Y_4 independent of y_1, thus making it possible to implement these expressions using one module (one level) for each one; otherwise one would have to use multi-levels.

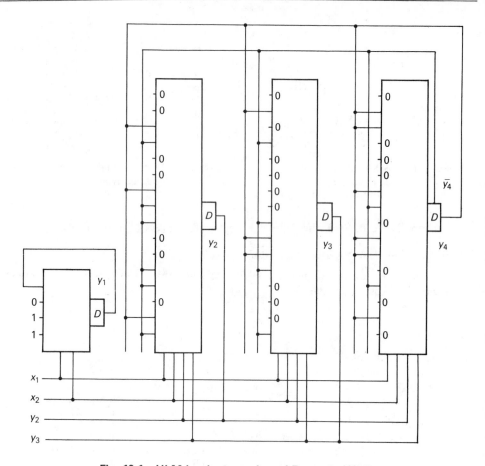

Fig. 10.1   ULM implementation of Example (10.1)

10.3.2 Serial Decomposition

A machine can be decomposed into two components operating in series iff there exists a closed partition π. The predecessor is defined by π and the successor is defined by a partition τ, such that $\pi \cdot \tau = \pi(0)$. This is similar to synchronous machines, only now one should ensure that the proper operation of the cascade connection of the component machines is independent of the order in which the component machines change state for STT assignment. In the case of MTT assignments, the order of the change should be fixed by insertion of suitable delays.

To illustrate the case of serial decomposition for asynchronous systems and how it differs from synchronous systems, we shall consider a machine M with certain properties. STT assignment is employed for efficient operating speed, and decomposition is hoped to produce efficient logic.

Discussion of Machine M   The machine M in Table 10.3 has the following closed partition

$$\pi = \overline{123}, \overline{45}, \overline{67}, \overline{89}$$

Assignment		Present state	Input x	
y_1y_2	y_3y_4		0	1
0 0	0 0	1	①	5
0 0	1 1	2	②	4
0 0	1 0	3	③	5
0 1	0 1	4	7	④
0 1	1 0	5	6	⑤
1 1	1 1	6	⑥	8
1 1	0 0	7	⑦	9
1 0	1 1	8	2	⑧
1 0	1 0	9	3	⑨

Table 10.3   Flow table for machine M with assignment

The machine shall be decomposed into two components operating in series with the predecessor M_1 having as its states the blocks of π. A USTT assignment is shown for M_1, which is expected to be independent of the successor M_2. The state variables y_1 and y_2 of M_1 distinguish the blocks of π. This assignment is used as part of the assignment for the original machine M. The USTT assignment for M is completed by choosing another set of state variables so as to cover all dichotomies not covered by the state variables distinguishing the blocks of π. If the set of state variables of M_1 do not cover any of the dichotomies, the decomposition is trivial, and the second set of state variables realize the original machine.

For machine M, the dichotomies that should be covered for STT assignments are:

(1, 2 8), (1, 3 9), (1, 4 7), (1, 5 6), (2 8, 4 7), (2 8, 5 6)
(3 9, 4 7), (3 9, 5 6), (4 7, 5 6), (2 8, 3 9) ... Column 1

(1 5, 2 4), (1 5, 6 8), (1 5, 7 9), (2 4, 3 5), (2 4, 6 8),
(2 4, 7 9), (3 5, 6 8), (3 5, 7 9), (6 8, 7 9) ... Column 2

The coding given for M_1 which distinguishes the blocks of π, also covers all the dichotomies composed of two blocks of π; these are:

(1, 5 6), (2 8, 4 7), (2 8, 5 6), (3 9, 4 7), (3 9, 5 6),
(1 5, 6 8), (1 5, 7 9), (1, 4 7), (2 4, 6 8), (2 4, 7 9),
(3 5, 6 8), (3 5, 7 9)

The remaining dichotomies must be covered by additional state variables. The coding of the state variables y_3y_4 cover these dichotomies. They correspond to the state variables of the successor component M_2.

y_3y_4 distinguishes the partition $\tau = \overline{17}, \overline{268}, \overline{359}, \overline{4}$. Note that $\pi \cdot \tau = \pi(0)$.
Tables 10.4(*a*) and 10.5(*a*) are the flow tables of M_1 and M_2 respectively.

Table 10.4(a) Flow table for M_1, with assignment

y_1 y_2			$x=0$	$x=1$
0 0	$\overline{123}$	1	①00	2 01
0 1	$\overline{45}$	2	3 11	②01
1 1	$\overline{67}$	3	③11	4 10
1 0	$\overline{89}$	4	1 00	④10

Table 10.4(b) Transition table for machine M_1

Address			Output	
x	y_1	y_2	y_1	y_2
0	0	0	0	0
0	0	1	1	1
0	1	0	0	0
0	1	1	1	1
1	0	0	0	1
1	0	1	0	1
1	1	0	1	0
1	1	1	1	0

Table 10.5(a) Flow table for M_2, with assignment

y_3 y_4			$x\,y_1\,y_2$							
			000	001	011	010	110	111	101	100
0 0	$\overline{17}$	1	①	①	①	–	3	3	3	3
1 1	$\overline{268}$	2	②	②	②	②	②	②	4	4
1 0	$\overline{359}$	3	③	2	2	③	③	③	③	③
0 1	$\overline{4}$	4	–	1	1	–	–	–	④	④

Table 10.5(b) Transition table for M_2.

Address					Output		Address					Output	
x	y_1	y_2	y_3	y_4	y_3	y_4	x	y_1	y_2	y_3	y_4	y_3	y_4
0	0	0	0	0	0	0	1	0	0	0	0	1	0
0	0	0	0	1	–	–	1	0	0	0	1	0	1
0	0	0	1	0	1	0	1	0	0	1	0	1	0
0	0	0	1	1	1	1	1	0	0	1	1	0	1
0	0	1	0	0	0	0	1	0	1	0	0	1	0
0	0	1	0	1	0	0	1	0	1	0	1	0	1
0	0	1	1	0	1	1	1	0	1	1	0	1	0
0	0	1	1	1	1	1	1	0	1	1	1	0	1
0	1	0	0	0	–	–	1	1	0	0	0	1	0
0	1	0	0	1	–	–	1	1	0	0	1	–	–
0	1	0	1	0	1	0	1	1	0	1	0	1	0
0	1	0	1	1	1	1	1	1	0	1	1	1	1
0	1	1	0	0	0	0	1	1	1	0	0	1	0
0	1	1	0	1	0	0	1	1	1	0	1	–	–
0	1	1	1	0	1	1	1	1	1	1	0	1	0
0	1	1	1	1	1	1	1	1	1	1	1	1	1

From these assignments the following relations are obtained:

$$Y_1 = \bar{x}y_2 + xy_1$$
$$Y_2 = \bar{x}y_2 + x\bar{y}_1$$
$$Y_3 = \bar{x}y_3 + xy_1 + x\bar{y}_2 + y_3\bar{y}_4$$
$$Y_4 = xy_4 + \bar{x}y_1y_2 + y_3y_4$$

These expressions can be expanded and implemented using ULMs or, ROMs can be used as shown in Fig. 10.2. The first step is to reorganize the transition tables shown in Tables 10.4(b) and 10.5(b) which belong to submachines M_1 and M_2 respec-

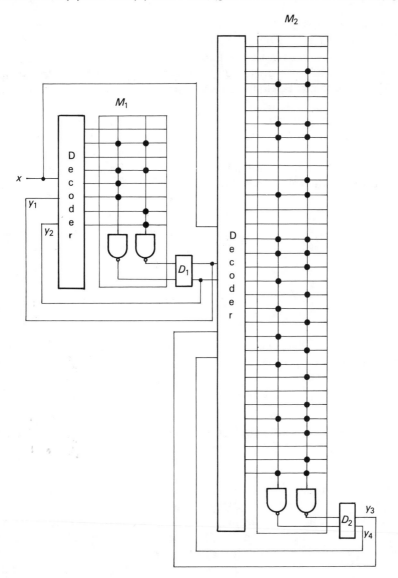

Fig. 10.2   Implementation of machine M using two small ROM modules

tively. The ROMs are programmed as shown in Fig. 10.2 directly from the transition tables. Needless to say, PLA modules could have been used, and are recommended if one or both components have a large number of inputs.

When implementing the previous examples, we assumed that the devices are ideal. As we indicated in Chapter 3, the devices are not ideal and a delay in the decoder could cause a hazard which is not acceptable in an asynchronous system.

Consider Y_2 in relation to machine M; from Table 10.4(b) Y_2 may be written as follows:

$$Y_2 = \bar{x}\,\bar{y}_1 y_2 + \bar{x}\,y_1 y_2 + x\,\bar{y}_1\bar{y}_2 + x\,\bar{y}_1 y_2$$

If the system is in state ②, the term $x\,\bar{y}_1 y_2$ holds Y_2 high. When x changes to 0, the system should move from 101 to 001 and the term $\bar{x}\bar{y}_1 y_2$ keeps Y_2 high. If however the change in x sets the term $x\,\bar{y}_1 y_2$ to 0 before setting the term $\bar{x}\bar{y}_1 y_2$ to 1, for a short duration δt (the delay through the inverter) Y_2 will be 0, which is the wrong output.

This hazard can be eliminated by using an inertial delay of more than δt. This solution is not always desirable since the addition of delays offsets the advantage of using asynchronous circuits by slowing the speed of the system. If a PLA is used instead, the hazard can be overcome by adding an extra term as shown dotted in the K-map below. This term holds the output at 1 while the change in x propagates through the inverter and thus eliminates the hazard. Similarly the hazards in the other three next-state variables are eliminated before the PLAs are programmed.

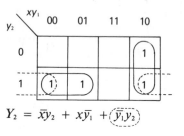

$$Y_2 = \bar{x}y_2 + x\bar{y}_1 + \widehat{y_1 y_2}$$

This suggests that, in cases like this, the PLA has an advantage over the ROM when used for asynchronous systems.

10.4   UNGER'S THREE-CHANGE RULE FOR ELIMINATING NON-USEFUL PARTITIONS

Let a unique y-state y^i be assigned to each row i. Let $T(i, j)$ refer to the variables distinguishing the states assigned to rows i and j.

If in a given flow table the next state of i under input I is j, i.e. $S_iI = S_j$ then for every state S whose y-state is in $T(i, j)$ the next state SI must equal j in order for the resulting circuit to be STT and free of critical races. Hence if $i \to j$, and $k \to m$ are transitions in the same column of a flow table and $m \neq j$, then $T(i, j)$ and $T(k, m)$ must be disjoint. Unger stated the following:

(i)   A row assignment for a SOC flow table has no critical races iff, for every pair of transitions $i{\to}j$ and $k{\to}m$ that appear in the same column such that $j{\neq}m$ and $i{\neq}j$ or $k{\neq}m$, the sets $T(i, j)$ and $T(k, m)$ are disjoint.

(ii) A row assignment for a SOC flow table is a valid USTT assignment iff, for every pair of transitions $i{\to}j$ and $k{\to}m$ that appear in the same column and are such that $j{\neq}m$, the associated dichotomy (ij, km) is covered by at least one y-variable of the assignment.

Consider a USTT assignment for a system with SOC flow table, in which only one input variable at a time may change. Using the three-change rule, we soon detect an essential hazard in the following table:

	I_1	I_2	
1	①	2	
2	3	②	
3	③	f	$f \neq 2$

Assume the system is in state ① to start with. After the first input change, the system moves to state 2 then ②. After the second input change, the system moves to state 3 then ③. After the third input change, the system moves to state f, where $f{\neq}2$.

This is an indication of essential hazard, but from above, we see that 12 and $3f$ must be disjoint, i.e. 12 and $3f$ must not be in a block of a closed partition, because if they are, then in the component based on that partition, 12 and $3f$ will have the same code, which contradicts with the requirement for a valid USTT assignment. This fact can therefore be used to restrict the choice of useful s.p. partitions for this type of asynchronous system, by making a list of all pairs that must not be in a block, and then eliminating all partitions not adhering to this requirement.

Another look at the general pattern that leads to essential hazards as shown in the above table can also reveal other interesting results. In the process of computing s.p. partitions described in Chapter 8, we start by identifying pairs of states, and finding the implied pairs under each input. If we identify 12 then we have 13 as the first implied pair. The pair 13 also implies 13 and $2f$. We can see now, that whatever $2f$ may imply, 12, 13, and $2f$ will result in $123f$ being in a single block, while we already established that 12 and $3f$ must be disjoint. Hence, there is no need to consider 12 when looking for pairs of states to start the search for s.p. partitions.

Let us look again at Table 10.3. We start at state ①, and we find that three input changes result in the following transitions $1{\to}5$, $5{\to}6$ and $6{\to}8$. Since we end up in state 8 instead of state 5, we have essential hazard. The pairs 15 and 68 must be disjoint, and therefore there is no need to test the pair 15 when computing the s.p. partitions since,

15 implies 16 and 55
16 implies 16 and 58.

Regardless of what 58 may imply, we still have 1568 in one block. Similarly, we can discard the following pairs:

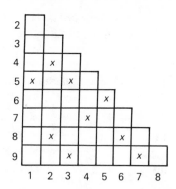

Fig. 10.3 Showing a reduction in the number of computations required to compute the s.p. partition for machine M

24, 35, 47, 56, 68, 79, 28, and 39.

In Fig. 10.3, which illustrates the procedure for finding the basic partitions, we see that a quarter of the possible initial identifications are eliminated already. These are marked x.

We also indicated in Chapter 8 that if a pair of states i, j results in an I-partition, any other pair that has ij in its list of implied pairs will also result in an I-partition. Similarly, any pair that results in one of the forbidden pairs marked x can also be neglected. This will further simplify the searching process for s.p. paritions.

Having detected all possible sequences that lead to essential hazards, one might make a list of the forbidden pairs and instruct the computer not to compute the partitions for these pairs. For hand computations, neglecting all the pairs shown above can obviously save the designer some time and effort.

11

Hybrid Design Techniques
for Sequential Circuits

11.1   INTRODUCTION

So far, MSI devices have been considered independently, although this is not always the most economical way. In real life problems, mixed modules are often used to implement logic systems.

Referring to Example (8.1), assignment (i), we observe that the transitions between the blocks of $\tau_i = \pi_1$, which gave rise to the predecessor in Fig. 8.8, occur in the following sequence.

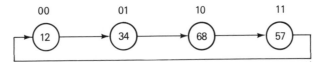

Assigning the blocks of this partition to the binary count sequence, a two-bit binary counter can be used to realize the input-independent component. The state variables y_1 and y_2 are assigned to the counter. This counter plus a 16 × 1 ROM and a latch could be used to implement the system, as will be shown later.

Another situation where a hybrid combination could prove useful, is the case of finite-memory machines. When a finite memory machine is to be implemented using a shift register, a ROM may be used to implement the combinational logic which determines the output.

11.2   LARGE PARTITIONS FOR AUTONOMOUS CLOCKS

In Chapter 8 autonomous clocks were briefly mentioned. The introduction to this chapter makes it necessary to have another look at this topic.

If a machine possesses an input-consistent partition τ_i and several closed partitions $\pi_1, \pi_2, \ldots, \pi_n$ such that: $\pi_j \geqslant \tau_i$ for all $j = 1, 2, \ldots, n$ then the autonomous clock corresponding to the smallest π_j is refered to as the maximal autonomous clock.

If the machine is strongly connected, any component induced by a closed partition on the states of the machine is also strongly connected. In a strongly connected machine, for every pair of states S_i and S_j, there exists an input sequence which takes the machine from S_i to S_j. From this we see that the autonomous clock of a strongly connected machine is also strongly connected, and it is also periodic. The period is given by $k \leqslant \# (\pi_j)$ where $\# (\pi_j)$ is the number of states of the clock.

It is also known that if we have a strongly connected clock with period k, and we let j be an integer that divides k, then there is a j block partition π_j with s.p. on the states of the clock which gives rise to a strongly connected clock with period j. This result can be used for finding a clock of a smaller period and smaller number of states, embedded in the maximal clock — which means effectively that one could choose the clock to match the counter available.

In the process of finding larger partitions for the clock, one must take care to preserve the successor relationship of the blocks. To illustrate this, consider the partition below with four blocks.

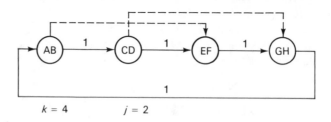

$$k = 4 \qquad\qquad j = 2$$

A larger partition having $j = 2$ blocks can be derived, and as can be seen above the successor relationships are preserved.

Hence, we have the partition $\overline{\text{ABEF}}\ \overline{\text{CDGH}}$

We notice that in the original partition, and the new larger partition, the successors are in a common block.

CD is the successor of AB.
EF is the successor of CD.
GH is the successor of EF.
AB is the successor of GH.

The blocks to be combined are joined by broken lines of length equal to j units. As another example consider the partition below with six blocks.

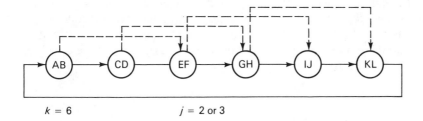

$k = 6$ $j = 2$ or 3

When $j = 2$

The length of the dotted lines joining any two blocks is 2, which results in the following partition $\overline{\text{ABEFIJ}}\ \overline{\text{CDGHKL}}$

Similarly when $j = 3$, we expect a three-block partition.

The length of the broken line joining the blocks to be combined is 3 as shown below.

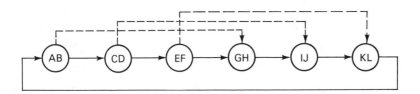

This results in the following partition $\overline{\text{ABGH}}\ \overline{\text{CDIJ}}\ \overline{\text{EFKL}}$

Using this procedure, one can find the partition of one's choice, to suit the available component.

For an illustration, we return to Example (8.1). This example had the following input-consistent partition:

$$\tau_i = \overline{1\,2}\,,\overline{3\,4}\,,\overline{6\,8}\,,\overline{5\,7}$$

Assigning the blocks of this partition to the binary count sequence, we can use a two-bit counter. The state variables y_1 and y_2 are assigned to this code. The third state variable y_3 is then fixed, and is found to depend on the input variable x, and all the state variables. The assignment is shown in Fig. 11.1(a), and the circuit diagram is shown in Fig. 11.1(b).

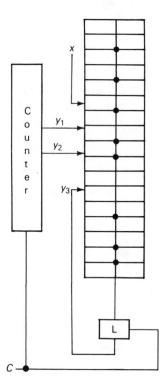

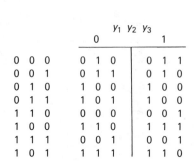

y_1 y_2 y_3

0			1	
0 0 0	0 1 0		0 1 1	
0 0 1	0 1 1		0 1 0	
0 1 0	1 0 0		1 0 0	
0 1 1	1 0 1		1 0 0	
1 1 0	0 0 0		0 0 1	
1 0 0	1 1 0		1 1 1	
1 1 1	0 0 1		0 0 1	
1 0 1	1 1 1		1 1 0	

Fig. 11.1(a)   A state assignment for Example (8.1)

Fig. 11.1(b)   Implementation of Example (8.1) based on the assignment in Fig. 11.1(a) using a two-bit counter and a ROM.

If the blocks are added as described above, we get a larger input-consistent partition while still preserving the substitution property. The partition is given by:

$$\tau_i' = \overline{1\,2\,6\,8}, \overline{3\,4\,5\,7}$$

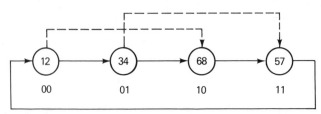

The state variable y_1 distinguishing the blocks of this partition requires a one-bit counter (flip-flop), followed by a ROM.

Using the assignment in Table 8.5, we obtain the circuit diagram of Fig. 11.1(c).

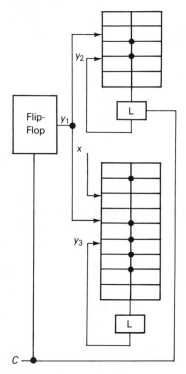

Fig. 11.1(c)   Implementation of Example (8.1) based on the assignment in Table 8.5, using a flip-flop and two ROMs.

11.3   HYBRID TECHNIQUES

11.3.1 Design Example

The following discussion forms an example on design of a system from its state diagram [Example (11.1)].

Consider the state diagram given in Fig. 11.2 representing a synchronous system with five input variables, start S, N, P, Q, and R. Eight outputs are required to show when the system is in one of the states A, B, ..., H. No output is required when the system is in state Z. That is to say, when the system is in state A, $A = 1$, when the system is in state B, $B = 1$, and so on. Another requirement is that the system must remain in certain states for a specified period of time as follows:

$$A \ldots 2\tau \quad ; \quad B \ldots \tau \quad ; \quad C \ldots 3\tau \quad ;$$
$$D \ldots 2\tau \quad ; \quad E \ldots \tau \quad ; \quad F \ldots 4\tau \quad ;$$
$$G \ldots \tau \quad ; \quad H \ldots \tau \quad ; \quad Z \ldots \tau \quad .$$

where τ is the clock period.

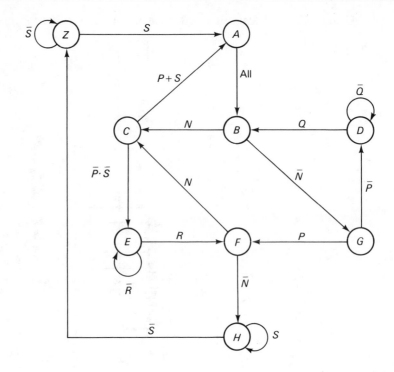

Fig. 11.2   State diagram for Example (11.1)

The first thing we notice is that even though it has a large number of inputs, the transition between states is dependent upon the change of one or at most two primary inputs. Such a system is more conveniently represented by a state diagram than by a state table, since the state table will have a maximum of 32 columns. Transitions between states are labelled with the condition that must be true to cause that transition. For example, a transition from state G to state F would take place if $P = 1$ irrespective of the value of other input variables.

To ensure that the machine remains in a particular state A say for two periods, A is replaced by two equivalent states A' and A''. This will result in the 9 original states increasing to 16 states as shown in Fig. 11.3. Both before and after this process, four state variables are required for the state assignment.

11.3.2 System Implementation

Here we decide to use a ROM to implement the system. A direct implementation would require 12×2^9 bits of storage, which is rather high. The need for such a large ROM is due to the large number of inputs. An obvious conclusion would be to try and decompose the system. Unfortunately, it is not an easy matter to compute the s.p. partitions for large systems, especially if the designer has no access to a computer program. Further, this particular example has no non-trivial s.p. partitions. This

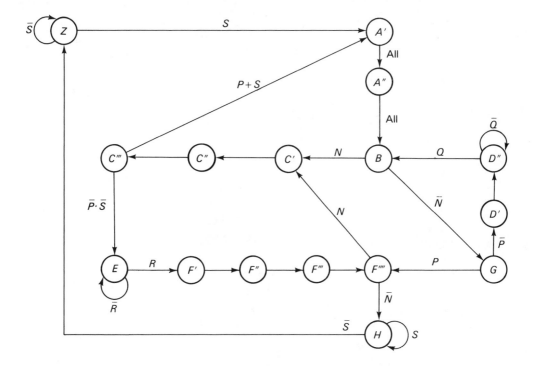

Fig. 11.3 State diagram for Example (11.1) allowing the system to rest in certain states

leads us to the conclusion that other devices have to be included, and some other design techniques have to be considered.

One way of reducing the number of inputs is to use a multiplexer module at the input. The ROM will see only the one output of the multiplexer as shown in the block diagram of Fig. 11.4. An arbitrary code is given to the states as in Table 11.1. The output is derived by using a demultiplexer. We note here that the way the outputs of the demultiplexer are tied together is possible with some modules (open collecter type). When this is not possible, some additional gating may be required. This implementation requires a 4×2^5-bit ROM, a multiplexer, a demultiplexer and a four-bit parallel shift register.

11.3.3 State-Input Pair Addressable ROM (19)

Example (11.1) was implemented by using a multiplexer which helped reduce the number of inputs to the ROM. This was possible, because only one input condition was influential at any time, i.e. N or $\bar{N}$, S or $\bar{S}$, $\bar{P} \cdot \bar{S}$ or $\overline{\bar{P} \cdot \bar{S}}$ etc. It is more appropriate however to look at more general techniques. This can be done by defining a portion of the ROM word to name the input for each state. This part is called the *test* part. Also a *link* part defines two next states chosen on the basis of the input

	y_4	y_3	y_2	y_1
Z	0	0	0	0
A'	0	0	0	1
A''	0	0	1	0
B	0	0	1	1
C'	0	1	0	0
C''	0	1	0	1
C'''	0	1	1	0
D'	0	1	1	1
D''	1	0	0	0
E	1	0	0	1
F'	1	0	1	0
F''	1	0	1	1
F'''	1	1	0	0
F''''	1	1	0	1
G	1	1	1	0
H	1	1	1	1

Table 11.1 Arbitrary assignment for Example (11.1)

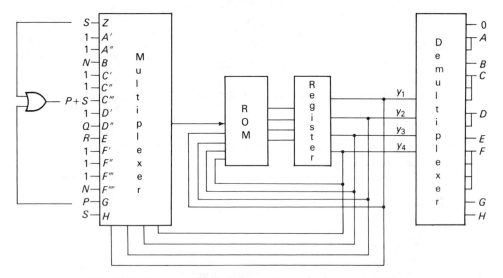

Fig. 11.4 One possible implementation of Example (11.1)

selected by the test part. This way, each address selects a word describing a *state-input* pair. In Fig. 11.5 the multiplexer takes the test part of the ROM word and selects a corresponding input as its output.

The switch, called a *data selector*, uses combinational logic with link F and link T as inputs. Link F (False) is the required next state if the selected input is 0. Link T (True) is the required next state if the selected input is 1. The output of the switch is either link F or link T depending on whether the multiplexer output is 0 or 1 respectively.

Since the instruction (output) is a function only of the states, the ROM word

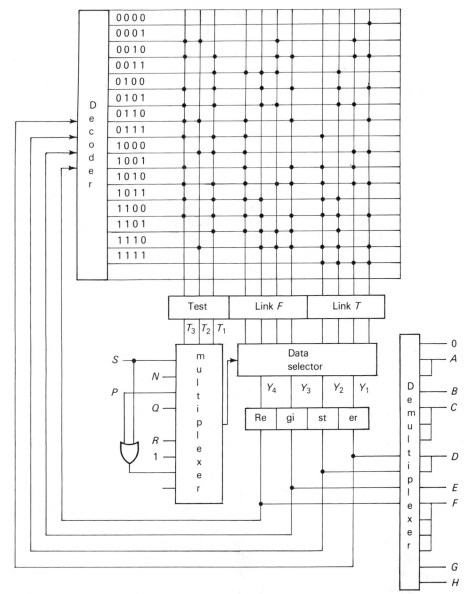

Fig. 11.5 A hybrid implementation of Example (11.1) based on the state-input pair technique

length can be shortened by using combinational logic to obtain the necessary outputs from the states. In the implementation of Example (11.1) shown in Fig. 11.5 the instruction part is discarded, and the output is obtained by using a demultiplexer module as in the last implementation. In general, the output for each state might not be unique, and some additional logic may be required. The state assignment is the same as in the last implementation given in Table 11.1. The test code is given in Table 11.2 and it is arbitrarily based on the binary count sequence.

	T_3	T_2	T_1
S	0	0	0
N	0	0	1
P	0	1	0
Q	0	1	1
R	1	0	0
All	1	0	1
$P+S$	1	1	0

**Table 11.2   Test code
for input conditions**

To explain the program, let us look at word 0000, i.e. the code for state Z. At this state only input S could cause a transition. The test code for S is 000. Therefore the first test word is programmed as 000. The next state is Z if the input is false. The first link F word is programmed as 0000. The next state is A' if the input is true. Thus the link T word is programmed as 0001 (the code for A'), as given in Table 11.1. The rest of the ROM is similarly programmed, and the final circuit diagram is given in Fig. 11.5. This implementation uses one 11×2^4-bit ROM, a three-variable (eight-way) multiplexer, a four-variable (16-way) demultiplexer, a four-bit parallel shift register, and a two-way four-bit data selector.

Needless to say, other structures are possible, e.g. two inputs could be tested at each clock pulse and one of four possible next states is selected (four links). The ROM word can be reduced in size by using only the link-T field. The link F is called an assumed address and is provided for by adding 1 to the present address.

11.4   SYSTEMS WITH MANY VARIABLES

Reading through the last few sections, one can see other possibilities, and other variations, that offer the designer a set of economic trade-offs in meeting his objectives. In Example (11.1) a multiplexer was used in order to reduce the number of inputs to the ROM module. This was possible, since only one input condition was responsible for any one transition. A more general case may arise where few inputs are relevant at a time out of a large number of input conditions. In such a case, a PLA might be used to encode the inputs. We note that, whatever the number of inputs, any state could only go to one of n next states for an n-state machine. Therefore systems having a very large number of inputs can normally be simplified by encoding the inputs. In what follows some helpful techniques are suggested that could prove useful when speed is not of vital importance.

11.4.1 Systems Having Many Inputs not Simultaneously Relevant (1)

Consider a sequential system having a large number of inputs p, but only a small number of the variables are actually inducing a transition at any time.

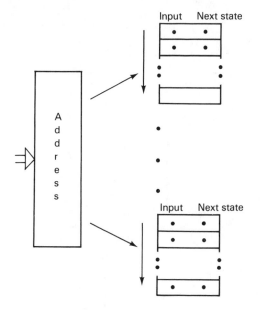

**Fig. 11.6   An illustration of page selection
by the state variables**

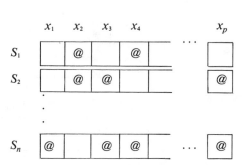

**Table 11.3   A state table in which transitions
are marked @.
Blocks are viewed along the
rows, blanks are DON'T CARE
states**

A state table of n states can be viewed as a stack of n state tables as shown in Table 11.3. This may be thought of as a row decomposition. Learning from microprogramming techniques, we let the present state select what is called a page. This is done by letting the state variables decide the most significant digits (msd) of the addressed decoder. In each page we store only the inputs relevant to that state, and the corresponding next state as shown in Fig. 11.6.

Let m be the maximum number of relevant inputs for any row. This way, we still require $\lceil \log_2(n) \rceil$ state variables, but only $\lceil \log_2(m) \rceil$ input variables are necessary instead of $\lceil \log_2(p) \rceil$. In cases where $p \gg m$, a considerable saving is possible. As can be seen in Fig. 11.6 the input is stored in the memory side by side with the corresponding next state. The obvious penalty is that speed is reduced from the need to search through the stored inputs and compare them with the incoming inputs in order to find a match.

In Fig. 11.7 the present state is assigned to the msd of the address decoder, and selects a portion of the storage (a page). The inputs stored and the incoming inputs are compared. If they are the same, the corresponding next state is loaded into the address register. The counter is reset. If the inputs are not the same, the counter is incremented by one and the above procedure is repeated. The output is obtained using appropriate logic, and the input line shown broken at the foot of the diagram is necessary only in a Mealy machine. In some cases the designer might choose to store the output in the memory instead of using separate logic.

We note that, if a page is partially used, it pays to store the inputs and the

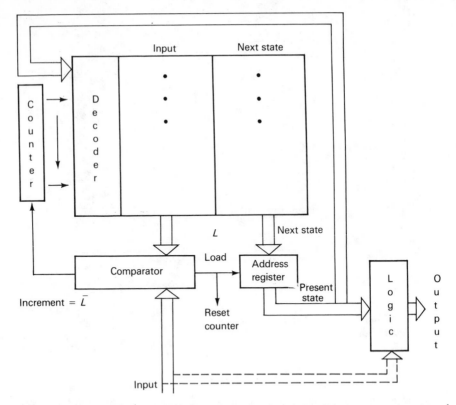

Fig. 11.7   A block diagram showing how the inputs and next states are stored in a ROM

corresponding next states at the top of the page, so as to speed up the scanning process.

Total storage capacity in bits of the system in Fig. 11.7 is given by:

$$M = [\lceil \log_2 p \rceil + \lceil \log_2 n \rceil] \times 2^{[\lceil \log_2 m \rceil + \lceil \log_2 n \rceil]}$$

The procedure can be summarized as follows:

(i)   Let the present state select a page.
(ii)   Compare the current input with the stored inputs.
(iii) Once a match is found, read the next state, and use it to select the next page, and so on.

11.4.2 Systems Having a Large Number of States

As a counterpart to Section 11.4.1, we consider the case where the number of states is high, but the number of input variables is small. For this, we might look at the state table along the input columns, as shown in Table 11.4. The input variables

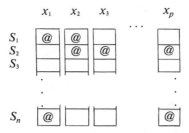

Table 11.4   **A state table in which transitions are marked @.**
Blocks are viewed along the columns, blanks are
DON'T CARE states

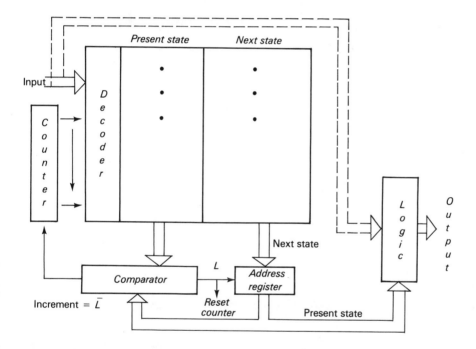

Fig. 11.8   **A block diagram showing how the present and next states are stored**
in a ROM

are used to select the page where the present and next states are stored. The comparator compares the present state at the output of the address register with the contents of the memory at the chosen page. If a match is detected, the next state is loaded into the address register and the counter is reset. Otherwise the counter is incremented and the search is continued until a match is found. As before the output may be obtained by using appropriate logic or may be stored in the memory, and the primary input line shown broken is required only for Mealy machines. A system of this kind is shown in Fig. 11.8.

11.4.3 Implied Input Storage

In the last two sections, a counter was used to find a match for the present state or input. Now we attempt to do without the counter, but we are still dealing with situations where only few input conditions are relevant at any particular state.

To illustrate the procedure, assume that only inputs x_1, x_3 and x_6 can cause a transition when the system is in state A. When in state B, another subset of the inputs, say x_1 and x_{10}, cause the transitions and so on, as shown in Fig. 11.9.

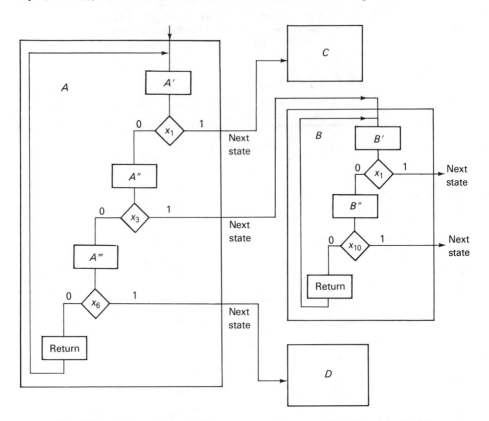

Fig. 11.9   A flow chart to illustrate special procedure (section 11.4.3)

State A is now replaced by A', A'' and A''', while state B is replaced by B' and B'', and so on. The next step is to assign binary codes to these states, such that the code for any state is an increment of the previous state.

Assuming the final total number of states is between 17 and 32, then five state variables y_1, ..., y_5 are required, as shown in Table 11.5.

Once the machine enters a particular state, it will stay in that state comparing the external input to all inputs relevant to the state. If the input does not correspond to any of the relevant inputs, a fault condition can be indicated, or as is done in this case, the machine stays in that state waiting. This is shown by a return state as shown in the flow diagram.

	y_1	y_2	y_3	y_4	y_5
A'	0	0	0	0	0
A''	0	0	0	0	1
A'''	0	0	0	1	0
Return	0	0	0	1	1
B'	0	0	1	0	0
B''	0	0	1	0	1
Return	0	0	1	1	0
C'	0	0	1	1	1
.					
.					
.					
D'	0	1	0	1	1
.					
.					
.					

	T_1	T_2	T_3	T_4
x_1	0	0	0	0
x_2	0	0	0	1
x_3	0	0	1	0
x_4	0	0	1	1
x_5	0	1	0	0
x_6	0	1	0	1
.				
.				
.				

Table 11.5(a)   Next-state code   **Table 11.5(b)   Input test code**

In the block diagram (Fig. 11.10) four test lines are shown, assuming that four variables are enough to code the input conditions. A five-variable multiplexer is used for the inputs. If the external input corresponds to the input under test, the output of the multiplexer will load the next state into the address register (AR); otherwise the register is incremented. For example, when in state A'' and the input is x_3 (see Fig. 11.9), the code for the next state B' will be loaded in the AR or else the AR is incremented so that the contents of it correspond to the code of A'''. Therefore only the next states A', B', C', ... are stored in the memory device.

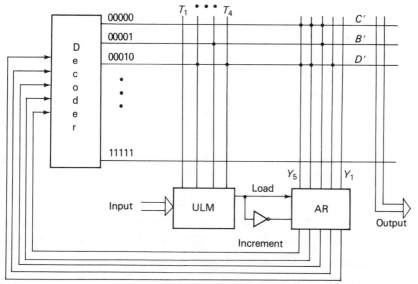

Fig. 11.10   A block diagram illustrating the implementation of the procedure of 11.4.3.   See the state and input codes and the flow chart of Fig. 11.9

The obvious disadvantage is the speed. The system might have to wait in a particular state for more than one clock pulse. The other less serious disadvantage is that the next-state variables are increased. We know that the state variables increase by one every time the states double in number.

It is also clear that the technique, which leans heavily on the assumed address technique already mentioned, is not suitable for all situations. The designer has to notice that test lines are introduced, which increase the word length, but the number of words is reduced because the primary input variables are no longer required at the input of the memory device. This however must be weighted against the increase in state variables. The advantages are summarized as follows:

(i)   In cases where the number of inputs is large, but only few are relevant at any one time, the storage requirements are reduced, and more efficient use of the memory is envisaged. But we must keep in mind that, for such systems, it is not very easy to analyze large tables or test for decomposition.

(ii) The assignment is very simple. Each state is the increment of the state before it.

(iii) The outputs, even if they depend on the primary inputs as in Mealy machines, can be stored in the ROM. This is true because the effect of the inputs is already taken care of in each state location.

(iv) If the system is required to stay in a particular state for a certain number of periods, this can be arranged by letting the input causing the transition be down the chain. If input I is to cause a transition after two periods, then:

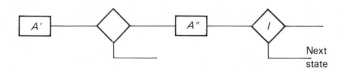

(v)   Since the test code is arbitrary, one might choose the code to correspond to the present state code if the number of present-state variables is sufficient. This way the test lines can be eliminated, and the present state lines are used for both purposes.

Example (11.2)

To illustrate the technique described above, consider the machine given in Fig. 11.11. From the state diagram, a flow chart is drawn as shown in Fig. 11.12. The chart indicates that when the machine is in state S_1 say, only x_1, x_2 and x_3 can cause transitions. The state is in effect split into three states S_1', S_1'', S_1'''. The code for the present state S_1' is 0000. This code addresses location 0000 of the ULM multiplexer. If the input is x_1, the output of the ULM is high. The load signal will load the next state S_2' (which is 0100) into the address register.

If on the other hand, the external input is not x_1, the increment signal will increment the contents of the address register and the code at the ULM control is that of S_1'' (which is 0001), addressing the second pin of the ULM. If the input is x_3, the next state code of S_3' is loaded, or else the address register is incremented to the code

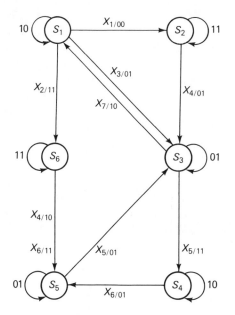

Fig. 11.11 State diagram for Example (11.2)

	y_4	y_3	y_2	y_1
S_1'	0	0	0	0
S_1''	0	0	0	1
S_1'''	0	0	1	0
R_1	0	0	1	1
S_2'	0	1	0	0
$R_{2\cdot}$	0	1	0	1
S_3'	0	1	1	0
S_3''	0	1	1	1
R_3	1	0	0	0
S_4'	1	0	0	1
R_4	1	0	1	0
S_5'	1	0	1	1
R_5	1	1	0	0
S_6'	1	1	0	1
S_6''	1	1	1	0
R_6	1	1	1	1

Table 11.6 State assignment for Example (11.2)

of S_1''', which is 0010. At this stage the system tests for the presence of x_2, which, if present, causes the next state code for S_6' to be loaded; or else the register is incremented to 0011, which is the code for the R_1 state (return to S_1'). The code for R_1 addresses a point in the ULM, where the input is 1. The load signal will load S_1' (0000) again and wait for one of the relevant inputs to occur, and so on.

Fig. 11.13 shows the implementation of this example using an ROM, an address register, a data selector and a multiplexer. The main features of this implementation can be summarized as follows:

(i) Simple state assignment based on the binary count sequence.
(ii) No primary input variables are applied to the ROM. Therefore a smaller ROM can be used. This is very important when the number of external inputs is high.
(iii) No test lines are used in this case. The present state variables are used instead to do the job of the test lines. This means that a shorter word length is required.
(iv) Even though no primary inputs are applied to the ROM address, the output of the machine can be stored in the ROM. Note that a data selector is used to determine the output.

Compared to a direct ROM implementation, this implementation suffers from the following drawbacks:

(i) Slower operation.
(ii) A multiplexer is used to detect the presence of the external primary inputs.
(iii) Four state variables are required as compared with three required by the original system.

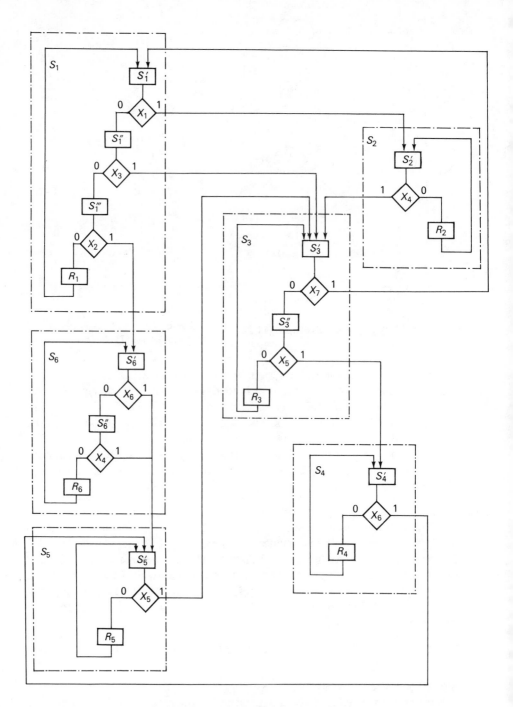

Fig. 11.12   Flow chart for Example (11.2)

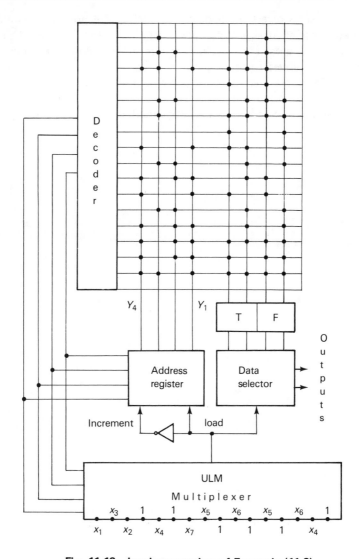

Fig. 11.13   Implementation of Example (11.2)

11.5   CONCLUSION

In the last few chapters, devices were used separately to implement sequential circuits. This helps to give a clearer picture of the effect of decomposition, if a decomposition is possible, on the hardware requirements. In this chapter, situations where it is advantageous to mix various modules are discussed.

Besides direct implementations, ideas are presented to indicate some of the many possible design techniques available to the designer. The objective is not to recommend a particular technique, but rather to point out some of the options which can be employed to suit the situation at hand.

The versatility of present-day logic devices and the availability of computer aid are making the designer's task that much easier. Nevertheless the designer's own intuition and experience will always be the major factor in any successful design.

References and Bibliography

1. Almaini, A.E.A., 'Sequential systems with a large input redundancy', *International Journal of Electronics*, **56** (1984), 533–538.
2. Almaini, A.E.A., 'Sequential machine implementation using ULMs', *IEEE Trans. on Computers*, **C–27** (1978), 951–960.
3. Almaini, A.E.A., Moosa, M.A., and Aziz, N.M., 'CAD of groups of exclusive logic functions', *Digital Processes*, **6** (1980), 227–243.
4. Almaini, A.E.A. and Woodward, M.E., 'An approach to the control variable selection problem for ULMs', *Digital Processes*, **3** (1977), 189–206.
5. Almaini, A.E.A. and Woodward, M.E., 'Computer program for S.P. partitions of sequential machines', *Electronic Letters*, **10** (1974), 445–446.
6. Anderson, J.L., 'Multiplexers double as logic circuits', *Electronics*, 29 Oct. 1969, 100–105.
7. Armstrong, D.B., 'On the efficient assignment of internal codes to sequential machines', *IRE Trans. on Electronic Computers*, **EC11** (1962), 611–622.
8. Armstrong, D.B., 'A programmed algorithm for assigning internal codes to sequential machines', *IRE Trans. on Electronic Computers*, **EC11** (1962), 466–472.
9. Ashenhurst, R.L., 'The decomposition of switching functions' (*Proc. International Symp. on Theory of Switching*), in Annals of Computation Laboratory, Harvard University, **29** (1959), 74–116.
10. Bartee, T.C., *Digital Computer Fundamentals* (McGraw-Hill, 1981).
11. Bartee, T.C., 'Computer design of multiple-output logical networks', *IRE Trans. on Electronic Computers*, **EC10** (1961), 21–30.
12. Bennett, L.A.M., 'The application of map-entered variables to the use of multiplexers in the synthesis of logic functions', *International Journal of Electronics*, **45** (1978), 373–379.
13. Besslich, Ph.W., 'Efficient computer method for EX-OR logic design', *IEE Proc.* **130E** (1983), 203–206.
14. Bimson, A., 'Bipolar gate arrays for system complexities between 500 and 10,000 gates', *Proc. First International Conf. on Semi-custom ICs*, (1981), 137–151.
15. Blakeslee, T.R., *Digital Design with Standard MSI and LSI* (Wiley, 1979).
16. Boole, G., *An Investigation of the Laws of Thought*, (Macmillan, 1854; reissued by Dover, 1958).
17. Brubaker, T.A. and Becker, J.C., 'Multiplication using Logarithms implemented with ROM', *IEEE Trans. on Computers*, *C – 24* (1975), 761–765.
18. Bursky, D. (ed.), *Memory Systems Design and Applications* (Wiley, 1981).
19. Clare, C.R., *Designing Logic Systems using State Machines* (McGraw-Hill, 1973).
20. Curtis, H.A., *A New Approach to the Design of Switching Circuits* (Van Nostrand, 1962).
21. Curtis, H.A., 'Multiple reduction of variable dependency of sequential machines', *J. ACM*, **9** (1962), 324–344.
22. Das. S.R. and Shang, C.L., 'On detecting total or partial symmetry of switching functions', *IEEE Trans. on Computers*. **C–20** (1971), 352–355.
23. Davies, R.D., 'The case for cMOS', *IEEE Spectrum*, **20** Oct. 1983, 26–32.

24. Davio, M., Deschamps, J.P. and Thayse, A., *Digital Systems with Algorithm Implementation* (Wiley, 1983).

25. Davis, W.A., 'Single shift register realization for sequential machines', *IEEE Trans. on Computers*, **C−17** (1968), 421−431.

26. Davis, W.A., 'Logical design using shift registers', *IEEE Trans. on Computers* (correspondence), **C−18** (1969), 958−959.

27. Dietmeyer, D. and Su, Y., 'Logic design automation of fan-in limited NAND networks', *IEEE Trans. on Computers*, **C−18** (1969), 11−22.

28. Dolotta, T.A. and McCluskey, E.J., 'The coding of internal states of sequential circuits', *IEEE Trans. on Electronic Computers*, **EC13** (1964), 549−562.

29. Eicheiberger, E.B., 'Hazard detection in combinational and sequential switching circuits', *Proc. Fifth Annual Symposium on Switching Theory and Logical Design*, Princeton University, Oct. 1964, 111−120.

30. Ektare, A.E. and Mital, D.C., 'Probabilistic approach to multiplexer logic circuit design', *Electronics Letters*, **16** (1980), 686−687.

31. 'Designing with ULAs', *Electronic Engineering/AMJ Joint Course*, March 1982, 53−57; April 1982, 65−70; May 1982, 49−54.

32. *Electronic International*. Special issue on cMOS, April 1984.

33. Epley, D.L. and Wang, P.T., 'On state assignments and sequential machine decomposition from S.P. partitions', *Proc. Fifth Annual Symposium on Switching Theory and Logical Design*, Princeton University, Oct. 1964, 228−233.

34. Fletcher, W.I. and Despain, A.M., 'Simplify combinational logic circuits with programmable ROMs', *Electronic Design*, **13** 24 June 1971, 72−73, and 'Simplify sequential circuit design with PROMs, *Electronic Design*, **14** 8 July 1971, 70−72.

35. Fletcher, W.I., *An Engineering Approach to Digital Design*, (Prentice-Hall, 1980).

36. Floyd, T.L., *Digital Fundamentals* (Merrill, 1982).

37. Friedman, A.D., Graham, R.L. and Ullman, J.D., 'Universal single transition time asynchronous state assignments', *IEEE Trans. on Computers*, **C−18** 1969, 541−547.

38. Gavlan, N., 'Field PLAs simplify logic designs', *Electronic Design*, **18** 1 Sept. 1975, 84−90.

39. Gindraux, L. and Catlin, G., 'CAE stations simulators tackle 1 million gates', *Electronic Design*, **31** Nov. 1983, 127−136.

40. Gioffi, G., Constantini, E. and Dijiulio, S., 'A new approach to the decomposition of sequential systems', *Digital Processes*, **3** Spring 1977, 35−48.

41. Haring, D.R., *Sequential Circuit Synthesis: State Assignment Aspects* (M.I.T. Press, 1965).

42. Hartmanis, J., 'On the state assignment problem for sequential machines I', *IRE Trans. on Electronic Computers*, **EC10** 1961, 157−165.

43. Hartmanis, J. and Stearns, R.E., *Algebraic Structure Theory of Sequential Machines* (Prentice-Hall, 1966).

44. Hartmanis, J. and Stearns, R.E., 'Some dangers in state reduction of sequential machines', *Information and Control*, **5** 1962, 252−260.

45. Henderson, R. and Brown, J., 'How to design with programmable logic arrays', *New Electronics*, **7** 25 June 1974, 14−18.

46. Hill, F.J. and Peterson, C.R., *Introduction to Switching Theory and Logical Design* (Wiley, 1974).

47. Hillman, A., 'Non-volatile memory selection mandate careful trade-offs', *EDN*, **8** April 1983, 135−139.

48. Huffman, D.A., 'The synthesis of sequential switching circuits', *Journal of the Franklin Institute*, **257** (1954), 161−190; and **257** (1954), 275−303.

49. Huffman, D.A., 'A study of memory requirements of sequential switching circuits', *M.I.T. Research Laboratory of Electronics Technical Report*, 1955, 293.

50. Humphrey, W.S., *Switching Circuits with Computer Applications*, (McGraw-Hill, 1958).

51. Jayashri, T. and Basu, D., 'Cellular array for multiplication of signed binary numbers', *The Radio and Electronic Engineer*, **44** (1974), 18−20.

52. Kang, S. and Van Cleemput, W.M., 'Automatic PLA synthesis from a DDL-P description', *Proc. 18th Conference on Design Automation*, 1981, 391–397.
53. Karnaugh, M., 'The map for synthesis of combinational logic circuits', *AIEE Trans.*, **72** (1953), Pt.1,593–599.
54. Kinney, L.L., 'Decomposition of asynchronous sequential switching circuits', 1968, Ph.D. Dissertation, University of Iowa.
55. Kitson, B., Laws, D. and Miller, W., Programmable logic chip rivals gate arrays in flexibility, *Electronic Design*, **31** Dec. 1983, 95–102.
56. Kjelkerud, E., 'A computer program for the synthesis of switching circuits by decomposition', *IEEE Trans. on Computers*, **C–21** (1972), 568–573.
57. Kohavi, Z., 'Secondary state assignment for sequential machines', *IEEE Trans. on Electronic Computers*, **EC13** (1964), 193–203.
58. Kohavi, Z. and Smith, E.J., 'Decomposition of sequential machines', *Proc. Sixth Annual Symposium on Switching Theory and Logical Design*, Ann Arbor, Mich., Oct. 1965.
59. Kohavi, Z., 'Reduction of output dependency on sequential machines', *IEE Trans. on Electronic Computers*, **EC14** 1965. 932–935.
60. Kohavi, Z., *Switching and Finite Automata Theory* (McGraw-Hill, 1970).
61. Kvamme, F., 'Standard read only memories simplify complex logic design', *Electronics*, **5** Jan. 1970, 88–94.
62. Lee, S.C., *Modern Switching Theory and Digital Design* (Prentice-Hall, 1978).
63. Levey, L.S. and Freeman, M., 'Every finite-state machine can be simulated (realized) by a synchronous (asynchronous) binary feedback shift register machine', *IEEE Trans. on Computers*, **C–23** (correspondence), Feb. 1974, 124–128.
64. Lewin, D., *Theory and Design of Digital Computer Systems* (Nelson, 1980).
65. Lewin, D., 'Outstanding Problems in Logic Design', *The Radio and Electronic Engineer*, **44** (1974), 9–17.
66. Lewin, D., '*Logical Design of Switching Circuits*' (Nelson, 1974).
67. Li, H.F., 'Variable selection in logic synthesis using multiplexers', *Int. J. Electronics*, **49** (1980), 185–195.
68. Lind, L.F. and Nelson, J.C.C., *Analysis and Design of Sequential Digital Systems* (The Macmillan Press, 1979).
69. Liu, C.L., '*K*th order finite automation', *IEEE Trans. on Computers*, **EC–12** (1963), 470–475.
70. Liu, C.L., 'Sequential machine realization using feedback shift registers', *Proc. 5th Annual Symposium on Switching Circuit Theory and Logical Design*, Princeton University, (1964), 204–227.
71. Liu, C.N., 'A state variable assignment method for asynchronous sequential machines', *J. ACM*, **10** (1963), 209–216.
72. Lloyd, A.M., 'Design of multiplexer universal logic module network using spectral technique', *IEE Proc.* **127E** (1980), 31–36.
73. Majumder, D.D. and Das, J., *Digital Computers Memory Technology* (Wiley, 1980).
74. Marcus, M.P., 'Derivation of maximal compatibles using Boolean algebra', *IBM Journal*, **8** (1964), 537–538.
75. Martin, R.L., *Studies in Feedback Shift Register Synthesis of Sequential Machines*, (M.I.T. Press, 1969).
76. McCluskey, E.J., 'Minimization of Boolean Functions', *Bell System Technical Journal*, **35** (1956), 1417–1444.
77. McCluskey, E.J. and Unger, S.H., 'A note on the number of internal variable assignments for sequential switching circuits', *IRE Trans. on Electronic Computers*, **EC8** (1959), 439–440.
78. Mead, C., Conway, L., Introduction to VLSI Systems (Addison-Wesley, 1980).
79. Mealy, G.H., 'A method for synthesizing sequential circuits', *Bell System Technical Journal* **34** (1955), 1045–1079.
80. Moore, E.F., Gedanken experiments on sequential machines in *Automata Studies* (Princeton University Press, 1956).

81. Mukhopadhyay, A., 'Detection of total or partial symmetry of a switching function with the use of decomposition charts', *IEEE Trans. on Electronic Computers*, **EC12** 1963, 553–557.

82. Nashelsky, L., *Introduction to Digital Technology* (Wiley, 1983).

83. Nichols, J., 'A logical next step for read only memories', *Electronics*, 12 June 1967, 111–113.

84. Oberman, R.M.M., *Digital Circuits for Binary Arithmetic* (Macmillan, 1979).

85. Page, E.W., 'Matrix method for systematic sequential circuit design', *Electronics Letters*, **12** 22 Jan 1976, 49–50.

86. Posa, J.G., 'Memories', *Electronics*, **54** 20 Oct. 1981, 130–138.

87. Priel, U. and Holland, P., 'Application of high speed programmable logic array', *Computer Design*, **12** Dec. 1973, 94–96.

88. Prioste, J.E., 'Replacing sequential logic with read only memories', *Electronic Engineering*, **48** Jan. 1976, 35–38.

89. Quine, W.V., 'The problem of simplifying truth function', *Amer. Math. Monthly*, **59** (1952), 521–531.

90. Roth, J.P., 'Minimization over Boolean trees', *I.B.M. J. Res. Dev.*, **5** (1960), 543–548.

91. Sawin, D.H., 'Optimization of asynchronous circuit realizations', *IEEE Trans. on Computers* (correspondence), **C–23** (1974), 186–188.

92. Shannon, C.E., 'A symbolic analysis of relay and switching circuits', *AIEE Trans.* **57** (1938), 713–723.

93. Shannon, C.E., 'The synthesis of two-terminal switching circuits', *Bell Syst. Tech. Journal*, **28** (1949), 54–98.

94. Shiva, S.G., 'Automatic hardware synthesis', *Proc. IEEE*, **71** (1983), 76–87.

95. Sholl, H.A., 'Design of asynchronous sequential networks using ROMs', *IEEE Trans. on Computers*, **C–24** (1975), 195–206.

96. Smith, E.J. and Kohavi, Z., 'Synthesis of multiple sequential machines', *Proc. Seventh Annual Symposium Switching and Automata Theory*, Berkeley, California, 1966.

97. Smith, R.J., 'Generation of internal state assignment for large asynchronous sequential machines', *IEEE Trans. on Computers*, **C–23** (1974), 924–932.

98. Smith, R.J., Tracey, J.H., Schoeffel, N.L. and Maki, G.K., 'Automation in the design of asynchronous sequential circuits', *AFIPS Conf. Proc.*, **22** (1968), 55–60.

99. Special issue on Computer Aided Design, *Proc. IEEE*, **69** (1981), No. 10.

100. Special issue on VLSI Design, *Proc. IEEE*, **71** (1983), No. 1.

101. Stearns, R.E. and Hartmanis, J., 'On the state assignment problem for sequential machines II', *IRE Trans. on Electronic Computers*, **EC10** (1961), 593–603.

102. Su, S. and Nam, C., 'Computer aided synthesis of multiple-output multi-level NAND networks with fan-in and fan-out constraints', *IEEE Trans. on Computers*, **C–20** (1971), 1445–1455.

103. Su, Y. and Dietmeyer, D., 'Computer reduction of two level multiple output switching circuits', *IEEE Trans. on Computers*, **C–18** (1969), 58–63.

104. Su, H.T., 'Shift register realization of asynchronous sequential machines', *Information and Control*, **15** (1959), 226–234.

105. Tan, C.J., 'State assignment for asynchronous sequential machines', *IEEE Trans. on Computers*, **C–20** (1971), 382–391.

106. Tan, C.J., Menon, P.R. and Friedman, A.D., 'Structural simplification and decomposition of asynchronous sequential circuits', *IEEE Trans. on Computers*, **C–18** (1969), 830–838.

107. Tracey, J.H., 'Internal state assignment for asynchronous sequential machines', *IEEE Trans. on Electronic Computers*, **EC15** (1966), 551–560.

108. Turmaine, B., 'Gate arrays for the eighties', *Electronic Technology*, **15** (1981), 149–152.

109. Uimari, D., 'PROMs – a practical alternative to random logic', *Electronic Products*, 21 Jan. 1974, 75–91.

110. Unger, S.H., 'Hazards and delay in asynchronous sequential switching circuits', *IEE Trans. on Circuit Theory*, **CT6** (1959), 12–25.

111. Unger, S.H., *Theory of Asynchronous Sequential Switching Circuits* (Wiley, 1969).
112. Unger, S.H., 'Asynchronous sequential switching circuits with unrestricted input changes', *IEEE Trans. on Computers*, **C−20** (1971), 1437−1444.
113. Vingron, P., 'Coherent design of asynchronous circuits', *IEE Proc.*, **130E** (1983), 190−201.
114. Whitehead, D.G., 'Algorithms for logic circuit synthesis by using multiplexers', *Electronic Letters*, **13** (1977), 355−356.
115. Yang. H.C. and Allen, C.R., 'VLSI implementation of an optimized hierarchical multipler', *IEE Proc.* **131G** (1984), 56−60.
116. Yau, S.S. and Tang, Y.S., 'On identification of redundancy and symmetry of switching functions', *IEEE Trans. on Computers*, **C−20** (1971), 1609−1613.
117. Yau, S.S. and Tang, C.K., 'Universal logic circuits and their modular realizations', *AFIPS Proc.*, **32** (1968), 297−305.
118. Yau, S.S. and Tang, C.K., 'Universal logic modules and their applications', *IEEE Trans. on Computers*, **C−19** (1970), 141−149.

Index